Automata, Formal Languages and Algebraic Systems

Automata, Formal Languages and Algebraic Systems

Proceedings of AFLAS 2008

Kyoto, Japan, 20 – 22 September 2008

edited by

Masami Ito
Kyoto Sangyo University, Japan

Yuji Kobayashi
Toho University, Japan

Kunitaka Shoji
Shimane University, Japan

NEW JERSEY • LONDON • SINGAPORE • BEIJING • SHANGHAI • HONG KONG • TAIPEI • CHENNAI

Published by

World Scientific Publishing Co. Pte. Ltd.

5 Toh Tuck Link, Singapore 596224

USA office: 27 Warren Street, Suite 401-402, Hackensack, NJ 07601

UK office: 57 Shelton Street, Covent Garden, London WC2H 9HE

British Library Cataloguing-in-Publication Data
A catalogue record for this book is available from the British Library.

AUTOMATA, FORMAL LANGUAGES AND ALGEBRAIC SYSTEMS
Proceedings of AFLAS 2008

ISBN-13 978-981-4317-60-3
ISBN-10 981-4317-60-8

Printed in Singapore.

PREFACE

The *International Workshop on Automata, Formal Languages and Algebraic Systems* (AFLAS 2008) was held at the Kansai Seminar House, Kyoto, Japan during the period September 20-22, 2008 as a satellite workshop of the Twelfth International Conference on Developments in Language Theory (Kyoto, September 16-19, 2008).

The workshop was organized under the sponsorship of Kyoto Sangyo University and with the financial support of Japan Society for the Promotion of Science.

The organizing committee consisted of the following members: M. Ito (Kyoto, Japan), P. Leupold (Kassel, Germany), Y. Kobayashi (Funabashi, Japan), K. Shoji (Matsue, Japan), F.M. Toyama (Kyoto, Japan), Y. Tsujii (Kyoto, Japan).

The topics of the workshop were: semigroups, codes and cryptography, automata and formal languages, word- and term-rewriting systems, ordered structures and categories, combinatorics on words, complexity and computability, molecular computing and quantum computing.

The number of participants was 37 from 13 different countries. There were 24 lectures during the sessions.

This volume contains mainly the papers based on lectures given at the workshop. All papers have been refereed. The editors express their gratitude to all contributors of this volume including the referees.

The organizers and editors would like to express their thanks to Kyoto Sangyo University, Japan Society for the Promotion of Science, the World Scientific Publishing Company and the Kansai Seminar House for providing the conditions to realize the workshop. We are also grateful to Kayoko Tsuji, Yoshiyuki Kunimochi and Shinnosuke Seki for their assistance during the workshop.

June 30, 2010

Masami Ito
Yuji Kobayashi
Kunitaka Shoji
Editors

CONTENTS

SOLIDIFYABLE MINIMAL CLONES OF PARTIAL OPERATIONS

S. BUSAMAN

Department of Mathematics and Computer Science,
Prince of Songkla University,
94000 Pattani, Thailand
[*] *E-mail: bsaofee@bunga.pn.psu.ac.th*

K. DENECKE

Institute of Mathematics, University of Potsdam,
Potsdam, Germany
[*] *E-mail: kdenecke@rz.uni-potsam.de*

Partial operations occur in the algebraic description of partial recursive functions and Turing machines (cf. A. I. Mal'cev[8]). Similarly to total operations superposition operations can also be defined on sets of partial operations. A clone of partial operations is a set of partial operations defined on the same base set A which is closed under superposition and contains all total projections. The collection of all clones of partial operations defined on a set A forms a complete lattice. For a finite nonempty set A this lattice is atomic and dually atomic. A partial algebra is said to be strongly solid if every strong identity of $\mathcal{A}$ is satisfied as a strong hyperidentity in $\mathcal{A}$, i.e. if it is satisfied after any replacement of operation symbols by derived term operations of $\mathcal{A}$ of the corresponding arity. A clone C of partial operations is called strongly solidifyable if there is a partial algebra $\mathcal{A}$ such that C is equal to the clone of all term operations of $\mathcal{A}$. In this paper we determine all minimal strongly solidifyable clones of partial operations defined on a finite nonempty set A.

Keywords: Partial algebra; Hyperidentity, Unsolid strong variety, Fluid strong variety.

1. Introduction

Let A be a nonempty finite set. For every positive integer n an n-ary partial operation on A is a map $f^A : domf^A \to A$ where $domf^A \subseteq A^n$, i.e. $domf^A$ is an n-ary relation on A, called the domain of f^A. Let $P^n(A)$ be the set of all $n - ary$ partial operations defined on the set A

and let $P(A) := \bigcup_{n=1}^{\infty} P^n(A)$ be the set of all partial operations on A. Let $O(A) \subset P(A)$ be the set of all total operations defined on A, i.e. $O(A) := \bigcup_{n=1}^{\infty} O^n(A)$ with $O^n(A) := \{f^A \in P^n(A) \mid domf^A = A^n\}$. For $n, m \geq 1$, $f^A \in P^n(A)$ and $g_1^A, \ldots, g_n^A \in P^m(A)$, we define the superposition of f^A and $g_1^A, \ldots, g_n^A$, denoted by $S_m^{n,A}(f^A, g_1^A, \ldots, g_n^A) \in P^m(A)$, by setting $domS_m^{n,A}(f^A, g_1^A, \ldots, g_n^A) := \{(a_1, \ldots, a_m) \in A^m \mid (a_1, \ldots, a_m) \in \bigcap_{i=1}^{n} domg_i^A$ and $(g_1^A(a_1, \ldots, a_m), \ldots, g_n^A(a_1, \ldots, a_m)) \in domf^A\}$ and

$$S_m^{n,A}(f^A, g_1^A, \ldots, g_n^A)(a_1, \ldots, a_m) := f^A(g_1^A(a_1, \ldots, a_m), \ldots, g_n^A(a_1, \ldots, a_m))$$

for all $(a_1, \ldots, a_m) \in domS_m^{n,A}(f^A, g_1^A, \ldots, g_n^A)$.
Let $D \subseteq A^n$ be an n-ary relation on A. Then for every positive integer n and each $1 \leq i \leq n$ we denote by $e_{i,D}^{n,A}$ the n-ary i-the partial projection defined by

$$e_{i,D}^{n,A}(x_1, \ldots, x_n) = x_i$$

for all $(x_1, \ldots, x_n) \in D$.
Let $J_A := \{e_{i,D}^{n,A} \mid 1 \leq i \leq n \text{ and } D = A^n\}$ be the set of all total projections defined on A and let J_A^n be the set of all total n-ary projections defined on A.

Definition 1.1. A partial clone C on A is a superposition closed subset of $P(A)$ containing J_A. A proper partial clone is a partial clone C containing an n-ary operation f^A with $domf^A \neq A^n$. If $C \subseteq O(A)$ then C is called a total clone.

Partial clones can be regarded as subalgebras of the heterogeneous algebra

$$((P^n(A))_{n\in\mathbb{N}^+}; (S_m^{n,A})_{m,n\in\mathbb{N}^+}, (J_A^n)_{n\in\mathbb{N}^+})$$

where $\mathbb{N}^+$ is the set of all positive integers.
This remark shows that the set of all partial clones on A, ordered by inclusion, forms an algebraic lattice $\mathcal{L}_{P(A)}$ in which arbitrary infimum is the set-theoretical intersection. For $F \subseteq P(A)$ by $\langle F \rangle$ we denote the least partial clone containing F.

A *partial algebra* $\mathcal{A} = (A; (f_i^A)_{i\in I})$ of type $\tau = (n_i)_{i\in I}$ is a pair consisting of a set A and an indexed set $(f_i^A)_{i\in I}$ of partial operations where f_i^A is $n_i - ary$. Let $PAlg(\tau)$ be the class of all partial algebras of type τ.

Definition 1.2. Let $\mathcal{A} = (A; (f_i^A)_{i\in I})$ be a partial algebra of a given type τ. To every partial algebra $\mathcal{A}$ we assign the partial clone generated by $\{f_i^A \mid i \in I\}$, denoted by $T(\mathcal{A})$. The set $T(\mathcal{A})$ is called clone of all term operations of the algebra $\mathcal{A}$.

We notice that we want to define terms over partial algebras in such a way that the set of all partial operations induced by these terms is precisely the clone of all term operations of $\mathcal{A}$. Such terms are defined in the following way:
Let $X_n = \{x_1, \ldots, x_n\}$ be an n-element alphabet and let X be an arbitrary countable alphabet. Let $\{f_i \mid i \in I\}$ be a set of operation symbols of type τ, where each f_i has arity n_i and where $X \cap \{f_i \mid i \in I\} = \emptyset$ and $X_n \cap \{f_i \mid i \in I\} = \emptyset$. We need additional symbols $\varepsilon_j^k \notin X$, for every $k \in \mathbb{N}^+ := \mathbb{N} \setminus \{0\}$ and $1 \le j \le k$. The set of all n-ary terms of type τ over X_n is defined inductively as follows (see[2]):

(i) Every $x_i \in X_n$ is an n-ary term of type τ.
(ii) If $w_1, \ldots, w_k$ are n-ary terms of type τ, then $\varepsilon_j^k(w_1, \ldots, w_k)$ is an n-ary term of type τ for all $1 \le j \le k$ and all $k \in \mathbb{N}^+$.
(iii) If $w_1, \ldots, w_{n_i}$ are n-ary terms of type τ and if f_i is an n_i-ary operation symbol, then $f_i(w_1, \ldots, w_{n_i})$ is an n-ary term of type τ.

Let $W_\tau^C(X_n)$ be the set of all n-ary terms of type τ defined in this way. Then $W_\tau^C(X) := \bigcup_{n=1}^{\infty} W_\tau^C(X_n)$ denotes the set of all terms of this type.
We notice that for convenience we will denote the variables from X or from X_n also by x, y, z, etc. Every n-ary term $w \in W_\tau^C(X_n)$ induces an n-ary term operation $w^{\mathcal{A}}$ of any partial algebra $\mathcal{A} = (A; (f_i^A)_{i\in I})$ of type τ. For $a_1, \ldots, a_n \in A$, the value $w^{\mathcal{A}}(a_1, \ldots, a_n)$ is defined in the following inductive way (see[2]):

(i) If $w = x_i$ then $w^{\mathcal{A}} = x_i^{\mathcal{A}} = e_i^{n,A}$, where $e_i^{n,A}$ is the n-ary total projection on the i-th component.
(ii) If $w = \varepsilon_j^k(w_1, \ldots, w_k)$ and we assume that $w_1^{\mathcal{A}}, \ldots, w_k^{\mathcal{A}}$ are the term operations induced by the terms $w_1, \ldots, w_k$ and that $w_i^{\mathcal{A}}(a_1, \ldots, a_n)$ are defined for $1 \le i \le k$, then $w^{\mathcal{A}}(a_1, \ldots, a_n)$ is defined and $w^{\mathcal{A}}(a_1, \ldots, a_n) = w_j^{\mathcal{A}}(a_1, \ldots, a_n)$.
(iii) Now assume that $w = f_i(w_1, \ldots, w_{n_i})$ where f_i is an n_i-ary operation symbol, and assume that $w_j^{\mathcal{A}}(a_1, \ldots, a_n)$ are defined, with values $w_j^{\mathcal{A}}(a_1, \ldots, a_n) = b_j$ for $1 \le j \le n_i$. If $f_i^A(b_1, \ldots, b_{n_i})$

is defined, then $w^{\mathcal{A}}(a_1,\ldots,a_n)$ is defined and $w^{\mathcal{A}}(a_1,\ldots,a_n) = f_i^A(w_1^{\mathcal{A}}(a_1,\ldots,a_n),\ldots,w_{n_i}^{\mathcal{A}}(a_1,\ldots,a_n))$.

Let $T^n(\mathcal{A})$ be the set of all term operations induced by the terms from $W_\tau^C(X_n)$ on the partial algebra $\mathcal{A}$ and let $T(\mathcal{A}) := \bigcup_{n=1}^{\infty} T^n(\mathcal{A})$.

We denote by $\mathrm{ar} f$ the arity of the partial operation f. Any mapping $\varphi = (\varphi^{(n)})_{n\in\mathbb{N}^+} : C \to C'$ from a clone $C \subseteq P(A)$ into $C' \subseteq P(B)$ is a clone homomorphism if

(i) $\mathrm{ar} f = \mathrm{ar}\varphi(f)$ for $f \in C$,
(ii) $\varphi(e_i^{n,A}) = e_i^{n,B}$ $(1 \le i \le n \in \mathbb{N}^+)$,
(iii) $\varphi(S_m^{n,A}(f^A, g_1^A,\ldots,g_n^A)) = S_m^{n,B}(\varphi(f^A),\varphi(g_1^A),\ldots,\varphi(g_n^A))$ for $f^A \in C^{(n)}$ and $g_1^A,\ldots,g_n^A \in C^{(m)}$.

(Here $\varphi(f^A)$ means $\varphi^{(n)}(f^A)$ where f^A is n-ary). We recall that term operations on $\mathcal{A}$ satisfy the same compatibility condition with respect to clone homomorphisms as fundamental operations of $\mathcal{A}$.

Lemma 1.3. *Let $\varphi : T(\mathcal{A}) \to T(\mathcal{B})$ be a clone homomorphism defined by $\varphi(f_i^A) = f_i^B$ for all $i \in I$. Then $\varphi(t^{\mathcal{A}}) = t^{\mathcal{B}}$ for all $t \in W_\tau^C(X)$.*

Proof. We will give a proof by induction on the complexity of the term t.
(i) If $t = x_i$, then

$$\varphi(t^{\mathcal{A}}) = \varphi(x_i^{\mathcal{A}}) = \varphi(e_i^{n,A}) = e_i^{n,B} = x_i^{\mathcal{B}}.$$

(ii) If $t = \varepsilon_j^k(t_1,\ldots,t_k)$ and if we assume that $\varphi(t_i^{\mathcal{A}}) \mid_D = t_i^{\mathcal{B}} \mid_D$ where D is the intersection of all domains of $\varphi(t_i^{\mathcal{A}})$ and $t_i^{\mathcal{B}}$, $1 \le i \le k$, then

$$\begin{aligned} \varphi(t^{\mathcal{A}}) \mid_D &= \varphi(\varepsilon_j^k(t_1,\ldots,t_k)^{\mathcal{A}}) \mid_D \\ &= \varphi(t_j^{\mathcal{A}}) \mid_D \\ &= t_j^{\mathcal{B}} \mid_D \\ &= \varepsilon_j^k(t_1,\ldots,t_k)^{\mathcal{B}} \mid_D \\ &= t^{\mathcal{B}} \mid_D. \end{aligned}$$

(iii) If $t = f_i(t_1,\ldots,t_{n_i})$ and if we assume that $\varphi(t_j^{\mathcal{A}}) \mid_D = t_j^{\mathcal{B}} \mid_D$ where D is the intersection of all domains of $\varphi(t_j^{\mathcal{A}})$ and $t_j^{\mathcal{B}}$, $1 \le j \le n_i$, then

$$
\begin{aligned}
\varphi(t^{\mathcal{A}}) \mid_D &= \varphi(f_i(t_1,\ldots,t_{n_i})^{\mathcal{A}}) \mid_D \\
&= \varphi(S_n^{n_i,A}(f_i^A, t_1^{\mathcal{A}},\ldots,t_{n_i}^{\mathcal{A}})) \mid_D \\
&= S_n^{n_i,B}(\varphi(f_i^A), \varphi(t_1^{\mathcal{A}}),\ldots,\varphi(t_{n_i}^{\mathcal{A}})) \mid_D \\
&= S_n^{n_i,B}(\varphi(f_i^A), \varphi(t_1^{\mathcal{A}}) \mid_D,\ldots,\varphi(t_{n_i}^{\mathcal{A}}) \mid_D) \\
&= S_n^{n_i,B}(f_i^{\mathcal{B}}, t_1^{\mathcal{B}} \mid_D,\ldots,t_{n_i}^{\mathcal{B}} \mid_D) \\
&= f_i(t_1,\ldots,t_{n_i})^{\mathcal{B}} \mid_D \\
&= t^{\mathcal{B}} \mid_D .
\end{aligned}
$$

■

For terms we need to define a superposition operation $\overline{S}_n^m$, as follows. Let $w_1,\ldots,w_m$ be n-ary terms and let t be an m-ary term. Then we define an n-ary term $\overline{S}_n^m(t, w_1,\ldots,w_m)$ inductively by the following steps:

(i) For $t = x_j$, $1 \leq j \leq m$ (m-ary variable), we define $\overline{S}_n^m(x_j, w_1,\ldots,w_m) = w_j$.
(ii) For $t = \varepsilon_j^k(s_1,\ldots,s_k)$ we set $\overline{S}_n^m(t, w_1,\ldots,w_m) = \varepsilon_j^k(\overline{S}_n^m(s_1, w_1,\ldots,w_m),\ldots,\overline{S}_n^m(s_k, w_1,\ldots,w_m))$, where $s_1,\ldots,s_k$ are m-ary, for all $k \in \mathbb{N}^+$ and $1 \leq j \leq k$.
(iii) For $t = f_i(s_1,\ldots,s_{n_i})$ we set $\overline{S}_n^m(t, w_1,\ldots,w_m) = f_i(\overline{S}_n^m(s_1, w_1,\ldots, w_m),\ldots,\overline{S}_n^m(s_{n_i}, w_1,\ldots,w_m))$, where $s_1,\ldots,s_{n_i}$ are again m-ary.

This defines an operation

$$\overline{S}_n^m : W_\tau^C(X_m) \times (W_\tau^C(X_n))^m \to W_\tau^C(X_n),$$

which describes the superposition of terms.

The *term clone* of type τ is the heterogeneous algebra

$$Clone\mathcal{T}^c := ((W_\tau^C(X_n))_{n\in\mathbb{N}^+};\ (\overline{S}_n^m)_{n,m\in\mathbb{N}^+},\ \ (e_j^k)_{k\in\mathbb{N}^+,\ 1\leq j\leq k}).$$

Let $\mathcal{A}$ be a partial algebra of type τ and let $T(\mathcal{A})$ be the clone of term operations of $\mathcal{A}$. We define a family $\varphi = (\varphi^{(n)})_{n\in\mathbb{N}^+}$ of mappings, $\varphi^{(n)} : W_\tau^C(X_n) \to T^n(\mathcal{A})$, by setting $\varphi^{(n)}(t) = t^{\mathcal{A}}$, the n-ary term operation induced by t. It is easy to see that φ has the following properties ([13]):

(i) $\varphi^{(n)}(x_i) = e_i^{n,A}$, $1 \leq i \leq n$, $n \in \mathbb{N}^+$,
(ii) $\varphi^{(n)}(\overline{S}_n^m(s, t_1,\ldots,t_m)) \mid_D = S_n^m(\varphi^{(m)}(s), \varphi^{(n)}(t_1),\ldots,\varphi^{(n)}(t_m)) \mid_D$, for $n \in \mathbb{N}^+$, where D is the intersection of the domains of all $t_i^{\mathcal{A}}$, $1 \leq i \leq m$, where s is m-ary, and $t_1,\ldots,t_m$ are n-ary.

Definition 1.4. ([13]) Let $\{f_i \mid i \in I\}$ be a set of operation symbols of type τ and $W_\tau^C(X)$ be the set of all terms of this type. A mapping $\sigma : \{f_i \mid i \in I\} \longrightarrow W_\tau^C(X)$ which maps each n_i-ary operation symbol f_i to a term of arity n_i is called a *hypersubstitution* of type τ.

Any hypersubstitution σ of type τ can be extended to a map $\widehat{\sigma} : W_\tau^C(X) \longrightarrow W_\tau^C(X)$ defined for all terms, in the following way ([13]):

(i) $\widehat{\sigma}[x_i] = x_i$ for every $x_i \in X_n$,
(ii) $\widehat{\sigma}[\varepsilon_j^k(s_1, \ldots, s_k)] = \overline{S}_n^k(\varepsilon_j^k(x_1, \ldots, x_k), \widehat{\sigma}[s_1], \ldots, \widehat{\sigma}[s_k])$, where $s_1, \ldots, s_k \in W_\tau^C(X_n)$,
(iii) $\widehat{\sigma}[f_i(t_1, \ldots, t_{n_i})] = \overline{S}_n^{n_i}(\sigma(f_i), \widehat{\sigma}[t_1], \ldots, \widehat{\sigma}[t_{n_i}])$, where $t_1, \ldots, t_{n_i} \in W_\tau^C(X_n)$.

Let $Var(t)$ be the set of all variables occurring in the term t.

Definition 1.5. ([9]) The hypersubstitution σ is called *regular* if $Var(\sigma(f_i)) = \{x_1, \ldots, x_{n_i}\}$, for all $i \in I$.

Let $Hyp_R^C(\tau)$ be the set of all regular hypersubstitutions of type τ.

Definition 1.6. ([12]) A pair $t_1 \approx t_2 \in W_\tau^C(X)^2$ is called a *strong identity* in a partial algebra $\mathcal{A}$ (in symbols $\mathcal{A} \underset{s}{\models} t_1 \approx t_2$) if and only if the right hand side is defined whenever the left hand side is defined and both are equal, i.e. when both sides are defined, then the induced partial term operations $t_1^{\mathcal{A}}$ and $t_2^{\mathcal{A}}$ are equal.

Let $K \subseteq PAlg(\tau)$ be a class of partial algebras of type τ and $\Sigma \subseteq W_\tau^C(X)^2$. Consider the connection between $PAlg(\tau)$ and $W_\tau^C(X)^2$ given by the following two operators $Id^s : \mathcal{P}(PAlg(\tau)) \rightarrow \mathcal{P}(W_\tau^C(X)^2)$ and $Mod^s : \mathcal{P}(W_\tau^C(X)^2) \rightarrow \mathcal{P}(PAlg(\tau))$ with
$Id^s K := \{s \approx t \in W_\tau^C(X)^2 \mid \forall \mathcal{A} \in K \ (\mathcal{A} \underset{s}{\models} s \approx t)\}$ and

$Mod^s\Sigma := \{\mathcal{A} \in \mathcal{PA}\updownarrow\}(\tau) \mid \forall \int \approx \sqcup \in \pm(\mathcal{A} \underset{s}{\models} s \approx t)\}$.

Clearly, the pair (Mod^s, Id^s) is a Galois connection between $PAlg(\tau)$ and $W_\tau^C(X)^2$. We have two closure operators $Mod^s Id^s$ and $Id^s Mod^s$ and their sets of fix points.

Definition 1.7. Let $V \subseteq PAlg(\tau)$ be a class of partial algebras of type τ. The class V is called a *strong variety* of partial algebras if $V = Mod^s Id^s V$.

For $\mathcal{A} \in PAlg(\tau)$, $V(\mathcal{A})$ is called the strong variety generated by the single algebra $\mathcal{A}$ (i.e. $V(\mathcal{A}) = Mod^s Id^s \mathcal{A}$).

Definition 1.8. A *strong identity* $s \approx t$ in a partial algebra $\mathcal{A}$ is called a *strong hyperidentity* in $\mathcal{A}$ (in symbols $\mathcal{A} \underset{shyp}{\models} t_1 \approx t_2$) if and only if $\hat{\sigma}[s] \approx \hat{\sigma}[t]$ are strong identities in $\mathcal{A}$ for every $\sigma \in Hyp_R^C(\tau)$.

The next concept which we have to introduce is the concept of a totally symmetric and totally reflexive relation:

Definition 1.9. A relation $R \subseteq A^n$ on the set A is called *totally symmetric* if for all permutations s on $\{1, \ldots, n\}$

$$(a_1, \ldots, a_n) \in R \Leftrightarrow (a_{s(1)}, \ldots, a_{s(n)}) \in R$$

and *totally reflexive* if $R \supseteq \iota_n$ where ι_n is defined by

$$\iota_n := \{(a_1, \ldots, a_n) \in A^n \mid a_i = a_j \text{ and } 1 \leq i < j \leq n\}.$$

R is called trivial if $R = A^n$.
A binary totally reflexive and totally symmetric relation is reflexive and symmetric in the usual sense.

2. Equivalent Strong Varieties of Partial Algebras

The concept of a hypersubstitution can be generalized to a mapping which assigns operation symbols of one type to terms of a different type.

Definition 2.1. ([13]) Let $\tau = (f_i)_{i \in I}$, $\tau' = (g_j)_{j \in J}$ be arbitrary types. A mapping

$${}_{\tau}^{\tau'}\sigma : \{f_i \mid i \in I\} \to W_{\tau'}^C(X),$$

(with arf_i=ar $\sigma(f_i)$), which assigns to every n_i-ary operation symbol f_i of type τ an n_i-ary term $\sigma(f_i) \in W_{\tau'}^C(X)$, is called a *$(\tau, \tau')$-hypersubstitution.*

Definition 2.2. ([13]) The (τ, τ')-hypersubstitution ${}_{\tau}^{\tau'}\sigma$ is called *regular* if $Var({}_{\tau}^{\tau'}\sigma(f_i)) = \{x_1, \ldots, x_{n_i}\}$ for all operation symbols f_i of type τ.

Let $Hyp_R^C(\tau, \tau')$ denote the set of all regular (τ, τ')-hypersubstitutions and let ${}_{\tau}^{\tau'}\sigma_R$ be some member of $Hyp_R^C(\tau, \tau')$.
Any regular (τ, τ')-hypersubstitution ${}_{\tau}^{\tau'}\sigma_R$ can be extended to a map

$${}_{\tau}^{\tau'}\hat{\sigma}_R : W_{\tau}^C(X) \to W_{\tau'}^C(X)$$

defined for all terms, in the following way:

(i) ${}^{\tau'}_{\tau}\widehat{\sigma}_R[x_i] = x_i$ whenever $x_i \in X$;
(ii) ${}^{\tau'}_{\tau}\widehat{\sigma}_R[\varepsilon_j^k(t_1,\ldots,t_k)] = \varepsilon_j^k({}^{\tau'}_{\tau}\widehat{\sigma}_R[t_1],\ldots,{}^{\tau'}_{\tau}\widehat{\sigma}_R[t_k])$;
(iii) ${}^{\tau'}_{\tau}\widehat{\sigma}_R[f_i(t_1,\ldots,t_{n_i})] = \overline{S'}^{n_i}_n({}^{\tau'}_{\tau}\sigma_R(f_i),{}^{\tau'}_{\tau}\widehat{\sigma}_R[t_1],\ldots,{}^{\tau'}_{\tau}\widehat{\sigma}_R[t_{n_i}])$.

Lemma 2.3. *([13]) Let ${}^{\tau'}_{\tau}\sigma_R \in Hyp_R^C(\tau,\tau')$. Then*

$$ {}^{\tau'}_{\tau}\widehat{\sigma}_R[\overline{S}^m_n(t,t_1,\ldots,t_m)] = \overline{S'}^m_n({}^{\tau'}_{\tau}\widehat{\sigma}_R[t],{}^{\tau'}_{\tau}\widehat{\sigma}_R[t_1],\ldots,{}^{\tau'}_{\tau}\widehat{\sigma}_R[t_m]). $$

Since the extension ${}^{\tau'}_{\tau}\widehat{\sigma}_R$ of the regular (τ,τ')-hypersubstitution ${}^{\tau'}_{\tau}\sigma_R$ preserves arities, every extension ${}^{\tau'}_{\tau}\widehat{\sigma}_R$ defines a family of mappings

$$ {}^{\tau'}_{\tau}\widehat{\sigma}_R = (\eta^{(n)} : W_\tau^C(X_n) \to W_{\tau'}^C(X_n))_{n\in\mathbb{N}^+}. $$

Theorem 2.4. *([13]) The extension ${}^{\tau'}_{\tau}\widehat{\sigma}_R$ of a regular (τ,τ')-hypersubstitution ${}^{\tau'}_{\tau}\sigma_R$ defines a homomorphism*
$(\eta^{(n)})_{n\in\mathbb{N}^+} : Clone\tau^c \to Clone\tau'^c$ *where*
$Clone\tau^c := ((W_\tau^C(X_n))_{n\in\mathbb{N}^+};(\overline{S}^m_n)_{m,n\in\mathbb{N}^+},(e_j^k)_{k\in\mathbb{N}^+,1\le j\le k})$ *and*
$Clone\tau'^c := ((W_{\tau'}^C(X_n))_{n\in\mathbb{N}^+};(\overline{S'}^m_n)_{m,n\in\mathbb{N}^+},(e'^k_j)_{k\in\mathbb{N}^+,1\le j\le k})$.

Using our new concept of a hypersubstitution we can define a relation between strong varieties of partial algebras of different types.

Definition 2.5. ([13]) Let $V \subseteq PAlg(\tau)$ and $V' \subseteq PAlg(\tau')$ be strong varieties of type τ and τ', respectively. Then V and V' are called *equivalent*, in symbols $V \sim V'$, if there exist a regular (τ,τ')-hypersubstitution ${}^{\tau'}_{\tau}\sigma_R$ and a regular (τ',τ)-hypersubstitution ${}^{\tau}_{\tau'}\sigma_R$ such that for all $t,t_1,t_2 \in W_\tau^C(X)$ and $t',t'_1,t'_2 \in W_{\tau'}^C(X)$:
(a) $V \underset{s}{\models} t_1 \approx t_2 \Rightarrow V' \underset{s}{\models} {}^{\tau'}_{\tau}\widehat{\sigma}_R[t_1] \approx {}^{\tau'}_{\tau}\widehat{\sigma}_R[t_2]$;

(a′) $V' \underset{s}{\models} t'_1 \approx t'_2 \Rightarrow V \underset{s}{\models} {}^{\tau}_{\tau'}\widehat{\sigma}_R[t'_1] \approx {}^{\tau}_{\tau'}\widehat{\sigma}_R[t'_2]$;

(b) $V \underset{s}{\models} {}^{\tau}_{\tau'}\widehat{\sigma}_R[{}^{\tau'}_{\tau}\widehat{\sigma}_R[t]] \approx t$;

(b′) $V' \underset{s}{\models} {}^{\tau'}_{\tau}\widehat{\sigma}_R[{}^{\tau}_{\tau'}\widehat{\sigma}_R[t']] \approx t'$.

Lemma 2.6. *Let ${}^{\tau'}_{\tau}\sigma_{R_1}$ and ${}^{\tau'}_{\tau}\sigma_{R_2}$ be regular (τ,τ')-hypersubstitutions and $\mathcal{A} \in PAlg(\tau')$. If ${}^{\tau'}_{\tau}\sigma_{R_1}(f_i)^{\mathcal{A}} = {}^{\tau'}_{\tau}\sigma_{R_2}(f_i)^{\mathcal{A}}$ for all $i \in I$, then ${}^{\tau'}_{\tau}\widehat{\sigma}_{R_1}[t]^{\mathcal{A}} = {}^{\tau'}_{\tau}\widehat{\sigma}_{R_2}[t]^{\mathcal{A}}$ for $t \in W_\tau^C(X)$.*

Proof. We will give a proof by induction on the complexity of the term t.
(i) If $t = x_i \in X$, then ${}^{\tau'}_{\tau}\widehat{\sigma}_{R_1}[t]^{\mathcal{A}} = x_i^{\mathcal{A}} = {}^{\tau'}_{\tau}\widehat{\sigma}_{R_2}[t]^{\mathcal{A}}$.
(ii) If $t = \varepsilon_j^k(t_1,\ldots,t_k)$ and if we assume that ${}^{\tau'}_{\tau}\widehat{\sigma}_{R_1}[t_i]^{\mathcal{A}}|_D = {}^{\tau'}_{\tau}\widehat{\sigma}_{R_2}[t_i]^{\mathcal{A}}|_D$

where D is the intersection of all domains of ${}_{\tau}^{\tau'}\widehat{\sigma}_{R_1}[t_i]^{\mathcal{A}}$ and ${}_{\tau}^{\tau'}\widehat{\sigma}_{R_2}[t_i]^{\mathcal{A}}$ for $1 \leq i \leq k$, then

$$\begin{aligned}
{}_{\tau}^{\tau'}\widehat{\sigma}_{R_1}[t]^{\mathcal{A}} \mid_D &= \varepsilon_j^k({}_{\tau}^{\tau'}\widehat{\sigma}_{R_1}[t_1], \ldots, {}_{\tau}^{\tau'}\widehat{\sigma}_{R_1}[t_k])^{\mathcal{A}} \mid_D \\
&= {}_{\tau}^{\tau'}\widehat{\sigma}_{R_1}[t_j]^{\mathcal{A}} \mid_D \\
&= {}_{\tau}^{\tau'}\widehat{\sigma}_{R_2}[t_j]^{\mathcal{A}} \mid_D \\
&= \varepsilon_j^k({}_{\tau}^{\tau'}\widehat{\sigma}_{R_2}[t_1], \ldots, {}_{\tau}^{\tau'}\widehat{\sigma}_{R_2}[t_k])^{\mathcal{A}} \mid_D \\
&= {}_{\tau}^{\tau'}\widehat{\sigma}_{R_2}[t]^{\mathcal{A}} \mid_D.
\end{aligned}$$

(iii) If $t = f_i(t_1, \ldots, t_{n_i})$ and if we assume that ${}_{\tau}^{\tau'}\widehat{\sigma}_{R_1}[t_j]^{\mathcal{A}} \mid_D = {}_{\tau}^{\tau'}\widehat{\sigma}_{R_2}[t_j]^{\mathcal{A}} \mid_D$ where D is the intersection of all domains of ${}_{\tau}^{\tau'}\widehat{\sigma}_{R_1}[t_j]^{\mathcal{A}}$ and ${}_{\tau}^{\tau'}\widehat{\sigma}_{R_2}[t_j]^{\mathcal{A}}$, for $1 \leq j \leq n_i$, then

$$\begin{aligned}
{}_{\tau}^{\tau'}\widehat{\sigma}_{R_1}[t]^{\mathcal{A}} \mid_D &= \overline{S'}_n^{n_i,\mathcal{A}}({}_{\tau}^{\tau'}\sigma_{R_1}(f_i)^{\mathcal{A}}, {}_{\tau}^{\tau'}\widehat{\sigma}_{R_1}[t_1]^{\mathcal{A}}, \ldots, {}_{\tau}^{\tau'}\widehat{\sigma}_{R_1}[t_{n_i}]^{\mathcal{A}}) \mid_D \\
&= \overline{S'}_n^{n_i,\mathcal{A}}({}_{\tau}^{\tau'}\sigma_{R_1}(f_i)^{\mathcal{A}}, {}_{\tau}^{\tau'}\widehat{\sigma}_{R_1}[t_1]^{\mathcal{A}} \mid_D, \ldots, {}_{\tau}^{\tau'}\widehat{\sigma}_{R_1}[t_{n_i}]^{\mathcal{A}} \mid_D) \\
&= \overline{S'}_n^{n_i,\mathcal{A}}({}_{\tau}^{\tau'}\sigma_{R_2}(f_i)^{\mathcal{A}}, {}_{\tau}^{\tau'}\widehat{\sigma}_{R_2}[t_1]^{\mathcal{A}} \mid_D, \ldots, {}_{\tau}^{\tau'}\widehat{\sigma}_{R_2}[t_{n_i}]^{\mathcal{A}} \mid_D) \\
&= \overline{S'}_n^{n_i,\mathcal{A}}({}_{\tau}^{\tau'}\sigma_{R_2}(f_i)^{\mathcal{A}}, {}_{\tau}^{\tau'}\widehat{\sigma}_{R_2}[t_1]^{\mathcal{A}}, \ldots, {}_{\tau}^{\tau'}\widehat{\sigma}_{R_2}[t_{n_i}]^{\mathcal{A}}) \mid_D \\
&= {}_{\tau}^{\tau'}\widehat{\sigma}_{R_1}[t]^{\mathcal{A}} \mid_D.
\end{aligned}$$

■

Lemma 2.7. *For every mapping $h : \{f_i \mid i \in I\} \to T(\mathcal{A})$, $\mathcal{A} \in PAlg(\tau')$, which maps the n_i-ary operation symbol f_i of type τ to an n_i-ary term operation from $T(\mathcal{A})$, there exists a regular (τ, τ')-hypersubstitution ${}_{\tau}^{\tau'}\sigma_R$ such that $h(f_i) = {}_{\tau}^{\tau'}\sigma_R(f_i)^{\mathcal{A}}$ for all $i \in I$.*

Proof. Let a mapping $h : \{f_i \mid i \in I\} \to T(\mathcal{A})$ with $h(f_i) = t_i^{\mathcal{A}}$ where $t_i \in W_{\tau'}^C(X_{n_i})$ be given. Then we can consider a regular (τ, τ')-hypersubstitution ${}_{\tau}^{\tau'}\sigma_R : \{f_i \mid i \in I\} \to W_{\tau'}^C(X)$ defined by ${}_{\tau}^{\tau'}\sigma_R(f_i) = t_i$, for $i \in I$ and we get that $h(f_i) = t_i^{\mathcal{A}} = {}_{\tau}^{\tau'}\sigma_R(f_i)^{\mathcal{A}}$ for $i \in I$. ■

Lemma 2.8. *If $\mathcal{A} \in PAlg(\tau)$, $\mathcal{B} \in PAlg(\tau')$, then for every clone homomorphism $\gamma : T(\mathcal{A}) \to T(\mathcal{B})$ there exists a regular (τ, τ')-hypersubstitution ${}_{\tau}^{\tau'}\sigma_R$ such that $\gamma(t^{\mathcal{A}}) = {}_{\tau}^{\tau'}\widehat{\sigma}_R[t]^{\mathcal{B}}$ for every $t \in W_{\tau}^C(X)$.*

Proof. Let $\mathcal{A} \in PAlg(\tau)$, $\mathcal{B} \in PAlg(\tau')$ and $\gamma : T(\mathcal{A}) \to T(\mathcal{B})$ be a clone homomorphism. Since γ preserves the arity, we can consider a mapping $h : \{f_i \mid i \in I\} \to T(\mathcal{B})$ with $h(f_i) = \gamma(f_i^{\mathcal{A}})$, for $i \in I$ which preserves the arity and by Lemma 2.7, we have a regular (τ, τ')-hypersubstitution ${}_{\tau}^{\tau'}\sigma_R$ such that $h(f_i) = {}_{\tau}^{\tau'}\sigma_R(f_i)^{\mathcal{B}}$, for $i \in I$. Then we get that $\gamma(f_i^{\mathcal{A}}) = {}_{\tau}^{\tau'}\sigma_R(f_i)^{\mathcal{B}}$, for $i \in I$. We want to show that $\gamma(t^{\mathcal{A}}) = {}_{\tau}^{\tau'}\widehat{\sigma}_R[t]^{\mathcal{B}}$ for $t \in W_{\tau}^C(X)$. We will give a proof by induction on the complexity of the term t.

(i) If $t = x_i$, then

$$\gamma(t^{\mathcal{A}}) = \gamma(x_i^{\mathcal{A}}) = \gamma(e_i^{n,\mathcal{A}}) = e_i^{n,\mathcal{B}} = {}_{\tau}^{\tau'}\widehat{\sigma}_R[x_i]^{\mathcal{B}} = {}_{\tau}^{\tau'}\widehat{\sigma}_R[t]^{\mathcal{B}}.$$

(ii) If $t = \varepsilon_j^k(t_1, \ldots, t_k)$ and if we assume that $\gamma(t_i^{\mathcal{A}}) \mid_D = {}^{\tau'}_{\tau}\widehat{\sigma}_R[t_i]^{\mathcal{B}} \mid_D$ where D is the intersection of all domains of $\gamma(t_i^{\mathcal{A}})$ and ${}^{\tau'}_{\tau}\widehat{\sigma}_R[t_i]^{\mathcal{B}}$, $1 \leq i \leq k$, then

$$\begin{aligned}
\gamma(t^{\mathcal{A}}) \mid_D &= \gamma(\varepsilon_j^k(t_1, \ldots, t_k)^{\mathcal{A}}) \mid_D \\
&= \gamma(t_j^{\mathcal{A}}) \mid_D \\
&= {}^{\tau'}_{\tau}\widehat{\sigma}_R[t_j]^{\mathcal{B}} \mid_D \\
&= \varepsilon_j^k({}^{\tau'}_{\tau}\widehat{\sigma}_R[t_1]^{\mathcal{B}} \mid_D, \ldots, {}^{\tau'}_{\tau}\widehat{\sigma}_R[t_k]^{\mathcal{B}} \mid_D) \\
&= \varepsilon_j^k({}^{\tau'}_{\tau}\widehat{\sigma}_R[t_1]^{\mathcal{B}}, \ldots, {}^{\tau'}_{\tau}\widehat{\sigma}_R[t_k]^{\mathcal{B}}) \mid_D \\
&= {}^{\tau'}_{\tau}\widehat{\sigma}_R[t]^{\mathcal{B}} \mid_D.
\end{aligned}$$

(iii) If $t = f_i(t_1, \ldots, t_{n_i})$ and if we assume that $\gamma(t_j^{\mathcal{A}}) \mid_D = {}^{\tau'}_{\tau}\widehat{\sigma}_R[t_j]^{\mathcal{B}} \mid_D$ where D is the intersection of all domains of $\gamma(t_j^{\mathcal{A}})$ and ${}^{\tau'}_{\tau}\widehat{\sigma}_R[t_j]^{\mathcal{B}}$, $1 \leq j \leq n_i$, then

$$\begin{aligned}
\gamma(t^{\mathcal{A}}) \mid_D &= \gamma(f_i(t_1, \ldots, t_{n_i})^{\mathcal{A}}) \mid_D \\
&= \gamma(S_n^{n_i, A}(f_i^{\mathcal{A}}, t_1^{\mathcal{A}}, \ldots, t_{n_i}^{\mathcal{A}})) \mid_D \\
&= S_n^{n_i, B}(\gamma(f_i^{\mathcal{A}}), \gamma(t_1^{\mathcal{A}}), \ldots, \gamma(t_{n_i}^{\mathcal{A}})) \mid_D \\
&= S_n^{n_i, B}(\gamma(f_i^{\mathcal{A}}), \gamma(t_1^{\mathcal{A}}) \mid_D, \ldots, \gamma(t_{n_i}^{\mathcal{A}}) \mid_D) \\
&= S_n^{n_i, B}({}^{\tau'}_{\tau}\sigma_R(f_i)^{\mathcal{B}}, {}^{\tau'}_{\tau}\widehat{\sigma}_R[t_1]^{\mathcal{B}} \mid_D, \ldots, {}^{\tau'}_{\tau}\widehat{\sigma}_R[t_{n_i}]^{\mathcal{B}} \mid_D) \\
&= {}^{\tau'}_{\tau}\widehat{\sigma}_R[f_i(t_1, \ldots, t_{n_i})]^{\mathcal{B}} \mid_D \\
&= {}^{\tau'}_{\tau}\widehat{\sigma}_R[t]^{\mathcal{B}} \mid_D.
\end{aligned}$$

■

Proposition 2.9. *Let $\mathcal{A} \in PAlg(\tau)$, $\mathcal{B} \in PAlg(\tau')$ be partial algebras and let $V := V(\mathcal{A})$ and $V' := V(\mathcal{B})$ be the strong varieties generated by $\mathcal{A}$ and by $\mathcal{B}$, respectively. Then we have $V \sim V'$ if and only if $T(\mathcal{A}) \cong T(\mathcal{B})$, i.e. if the clones $T(\mathcal{A})$ and $T(\mathcal{B})$ are isomorphic.*

Proof. Let $\tau = (f_i)_{i \in I}$, $\tau' = (g_j)_{j \in J}$. Let $V \sim V'$. Then there are regular hypersubstitutions ${}^{\tau'}_{\tau}\sigma_R$, ${}^{\tau}_{\tau'}\sigma_R$ satisfying Definition 2.5 (a) – (b'). Then $\gamma : T(\mathcal{A}) \to T(\mathcal{B})$ with $t^{\mathcal{A}} \mapsto {}^{\tau'}_{\tau}\widehat{\sigma}_R[t]^{\mathcal{B}}$ is well-defined (because of $s^{\mathcal{A}} = t^{\mathcal{A}} \Rightarrow {}^{\tau'}_{\tau}\widehat{\sigma}_R[s]^{\mathcal{B}} = {}^{\tau'}_{\tau}\widehat{\sigma}_R[t]^{\mathcal{B}}$) and by Lemma 2.3 we get that γ is a clone homomorphism. Moreover, γ is injective by Definition 2.5 (a') and (b) since

$${}^{\tau'}_{\tau}\widehat{\sigma}_R[s]^{\mathcal{B}} = {}^{\tau'}_{\tau}\widehat{\sigma}_R[t]^{\mathcal{B}} \Rightarrow {}^{\tau}_{\tau'}\widehat{\sigma}_R[{}^{\tau'}_{\tau}\widehat{\sigma}_R[s]]^{\mathcal{A}} = {}^{\tau}_{\tau'}\widehat{\sigma}_R[{}^{\tau'}_{\tau}\widehat{\sigma}_R[t]]^{\mathcal{A}} \Rightarrow s^{\mathcal{A}} = t^{\mathcal{A}},$$

and γ is surjective by Definition 2.5 (b') since

$$t'^{\mathcal{B}} = {}^{\tau'}_{\tau}\widehat{\sigma}_R[{}^{\tau}_{\tau'}\widehat{\sigma}_R[t']]^{\mathcal{B}} = \gamma({}^{\tau}_{\tau'}\widehat{\sigma}_R[t']^{\mathcal{A}}).$$

Conversely, let $T(\mathcal{A}) \cong T(\mathcal{B})$ and let $\gamma : T(\mathcal{A}) \to T(\mathcal{B})$ be a clone isomorphism. Then there exist $t_i \in W_{\tau'}^C(X_{n_i})$, $s_j \in W_{\tau}^C(X_{n_j})$ such that $\gamma(f_i^{\mathcal{A}}) = t_i^{\mathcal{B}}$, $\gamma^{-1}(g_j^{\mathcal{B}}) = s_j^{\mathcal{A}}$. We define the regular hypersubstitutions ${}^{\tau'}_{\tau}\sigma_R : f_i \mapsto t_i$, ${}^{\tau}_{\tau'}\sigma_R : g_j \mapsto s_j$. By Lemma 2.8 we have $\gamma(t^{\mathcal{A}}) = {}^{\tau'}_{\tau}\widehat{\sigma}_R[t]^{\mathcal{B}}$,

$\gamma^{-1}(t'^{\mathcal{B}}) =^{\tau}_{\tau'} \widehat{\sigma}_R[t']^{\mathcal{A}}$ for $t \in W^C_{\tau}(X)$ and $t' \in W^C_{\tau'}(X)$. We are going to show that ${}^{\tau'}_{\tau}\sigma_R$, ${}^{\tau}_{\tau'}\widehat{\sigma}_R$ fulfil Definition 2.5 (a) – (b'), which implies $V \sim V'$.

(a) $V \underset{s}{\models} s \approx t \Rightarrow s^{\mathcal{A}} = t^{\mathcal{A}} \Rightarrow^{\tau'}_{\tau} \widehat{\sigma}_R[s]^{\mathcal{B}} = \gamma(s^{\mathcal{A}}) = \gamma(t^{\mathcal{A}}) =^{\tau'}_{\tau} \widehat{\sigma}_R[t]^{\mathcal{B}} \Rightarrow$
$V \underset{s}{\models} {}^{\tau'}_{\tau}\widehat{\sigma}_R[s] \approx^{\tau'}_{\tau} \widehat{\sigma}_R[t]$.

Analogously we obtain for (a') (using γ^{-1} instead of γ):

$$(b){}^{\tau}_{\tau'}\widehat{\sigma}_R[{}^{\tau'}_{\tau}\widehat{\sigma}_R[t]]^{\mathcal{A}} = \gamma^{-1}({}^{\tau'}_{\tau}\widehat{\sigma}_R[s]^{\mathcal{B}}) = \gamma^{-1}(\gamma(t^{\mathcal{A}})) = t^{\mathcal{A}},$$

i.e. $V \underset{s}{\models} {}^{\tau}_{\tau'}\widehat{\sigma}_R[{}^{\tau'}_{\tau}\widehat{\sigma}_R[t]] \approx t$.

In a similar way we conclude for (b'). ■

3. Minimal Partial Clones

Let A be a finite set. The lattice $\mathcal{L}_{P(A)}$ of all partial clones is atomic ([1]). There are only finitely many minimal partial clones (atoms). In[1] all of them are determined up to the knowledge of the minimal clones in the lattice $\mathcal{L}_{O(A)}$ of all total clones. Unfortunately, in general the total minimal clones are unknown. Lots of work has been done to determine all minimal clones of total operations defined on a finite set ([4,11]). We will use the following theorem (see[1]):

Theorem 3.1. *The lattice $\mathcal{L}_{P(A)}$ of all partial clones on a finite set A is atomic and contains a finite number of atoms. $C \in \mathcal{L}_{P(A)}$ is a minimal partial clone if and only if C is a minimal total clone or C is generated by a proper partial projection with a nontrivial totally reflexive and totally symmetric domain.*

Example 3.2. For a set F of operations defined on the same set let $\langle F\rangle$ be the clone generated by F. For the two-element set $A = \{0,1\}$ the total minimal clones are the following ones ([10]): $\langle\wedge\rangle$, $\langle\vee\rangle$, $\langle x+y+z\rangle$, $\langle m\rangle$, $\langle c^1_0\rangle$, $\langle c^1_1\rangle$, $\langle N\rangle$, where $\wedge, \vee, N$ denote the conjunction, disjunction and negation. The symbol $+$ denotes the addition modulo 2 and c^1_0, c^1_1 are the unary constant operations with the value 0 and 1, respectively. We denote by m a ternary operation defined by $m(x,y,z) = (x \wedge y) \vee (y \wedge z) \vee (x \wedge z)$. Note that we write $\langle\wedge\rangle$ instead of $\langle\{\wedge\}\rangle$. Since for $n > 2$ every totally symmetric and totally reflexive relation on $\{0,1\}$ is trivial, we have exactly the following proper partial minimal clones on $\{0,1\}$: $\langle e^2_{1,\{(00),(11)\}}\rangle$, $\langle e^1_{1,\{0\}}\rangle$, $\langle e^1_{1,\{1\}}\rangle$,

$\langle e^1_{1,\emptyset}\rangle$. Altogether we have 11 minimal partial clones of operation defined on the set $\{0,1\}$.
In ([4]) all total minimal clones on a three-element set are determined. There are 84 total minimal clones on $\{0,1,2\}$. Further we have exactly the proper partial minimal clones generated by unary partial projections with the domains $\{0\},\{1\},\{2\},\{0,1\},\{0,2\},\{1,2\},\emptyset$, and the proper partial minimal clones generated by binary projections with the domains $\{(0,0),(1,1),(2,2)\},\{(0,0),(1,1),(2,2),(0,1),(1,0)\},\{(0,0),(1,1),(2,2),(0,2),(2,0)\},\{(0,0),(1,1),(2,2),(1,2),(2,1)\},\{(0,0),(1,1),(2,2),(0,1),(1,0),(0,2),(2,0)\},\{(0,0),(1,1),(2,2),(0,1),(1,0),(1,2),(2,1)\},\{(0,1),(1,0),(0,2),(2,0)\}$. Since for $n > 3$ every totally symmetric and totally reflexive relation on $\{0,1,2\}$ is trivial, we have to consider totally symmetric and totally reflexive at most ternary relations. Since the relations have to be totally symmetric by identification of variables one obtains binary proper partial projections except in the case that the domain is $\{(0,0,0),(1,1,1),(2,2,2)\}$. In this case by identification of variables one obtains the proper partial binary projection with domain $\{(0,0),(1,1),(2,2)\}$. Altogether we have 98 partial minimal clones on $\{0,1,2\}$.
For $|A| > 4$ not all total minimal clones are known. By ([11]) each total minimal clone can be generated by an operation f of one of the following types:
(1) f is unary and $f^2 = f$ or $f^p = id$ for some prime number p,
(2) f is binary and idempotent,
(3) f is a ternary majority operation ($f(x,x,y) = f(x,y,x) = f(y,x,x) = x$),
(4) f is the ternary operation $x + y + z$ in a Boolean group,
(5) f is a semiprojection (i.e. ar $f = n \geq 3$ and there exists an element $i \in \{1,\ldots,n\}$ such that $f(a_1,\ldots,a_n) = a_i$ whenever $a_1,\ldots,a_n$ are not pairwise different).

4. Strongly Solidifyable Partial Clones

Definition 4.1. The partial algebra $\mathcal{A}$ is called *strongly solid* if every strong identity is a strong hyperidentity of $\mathcal{A}$.

Example 4.2. Consider the three-element partial algebra $\mathcal{A} = (\{0,1,2\}; f^A)$ of type (1) with $dom f^A = \{1,2\}$ and $f^A(1) = 1$, $f^A(2) = 0$. Every strong identity of $\mathcal{A}$ can be derived from the strong identity $f^2(x) = f^3(x)$ $(f^n(x) = f(\ldots(f(x))\ldots))$. The unary terms over $\mathcal{A}$ are $\varepsilon^1_1(x)$, $f(x)$ and $f^2(x)$. Each of them fulfils $f^2(x) = f^3(x)$. That means, $f^2(x) = f^3(x)$

is a strong hyperidentity and since all strong identities of $\mathcal{A}$ can be derived from $f^2(x) = f^3(x)$ every strong identity is a strong hyperidentity and $\mathcal{A}$ is strongly solid.

Now we give some conditions under which $\mathcal{A}$ is not strongly solid.

Proposition 4.3. *Let $\mathcal{A} = (A; (f_i^A)_{i \in I})$ be a partial algebra with $|A| \geq 2$. Then $\mathcal{A}$ is not strongly solid if it satisfies one of the following conditions:*

(i) *There is a binary commutative operation under the fundamental operations,*
(ii) *there is a total constant operation under the fundamental operations,*
(iii) *there is a nowhere defined (discrete) operation under the fundamental operations,*
(iv) *$\mathcal{A}$ satisfies a strong identity $s \approx t$ with $Left(s) \neq Left(t)$ or $Right(s) \neq Right(t)$, where $Left(s)$ and $Right(s)$ denote the first and the last vaiable, respectively occurring in the term s.*
(v) *$\mathcal{A}$ satisfies a strong identity of the form $f(x_{s_1(1)}, \ldots, x_{s_1(n)}) \approx f(x_{s_2(1)}, \ldots, x_{s_2(n)})$ with mappings $s_1, s_2 : \{1, \ldots, n\} \to \{1, \ldots, n\}$, $n \geq 2$, such that $s_1(i) \neq s_2(i)$ for all $i = 1, \ldots, n$.*

Proof. We show that $\mathcal{A}$ is not strongly solid indicating a strong identity which is not a strong hyperidentity.
(i) Let f^A be a binary commutative fundamental operation of $\mathcal{A}$. Commutativity of f^A means: $f(x, y) \approx f(y, x)$ is a strong identity. The strong identity $f(x, y) \approx f(y, x)$ is not a strong hyperidentity. This becomes clear if we substitute for the binary operation symbol f in $f(x, y)$, $f(y, x)$ the term $\varepsilon_1^2(x, y)$.
(ii),(iii) A total, constant or nowhere defined unary operation f^A satisfies the strong identity $f(x) \approx f(y)$. The strong identity $f(x) \approx f(y)$ is not a strong hyperidentity. This is evident if we substitute for f in $f(x) \approx f(y)$ the term $\varepsilon_1^1(x)$. If f^A is an n-ary total, constant or nowhere defined operation and $n > 1$, then $f(x_1, x_2 \ldots, x_n) \approx f(x_2, x_1, \ldots, x_n)$ is a strong identity but not a strong hyperidentity. We see this if we substitute for the n-ary operation symbol f in $f(x_1, x_2 \ldots, x_n) \approx f(x_2, x_1, \ldots, x_n)$ the term $\varepsilon_1^n(x_1, \ldots, x_n)$.
(iv) This becomes clear if we substitute for all n-ary operation symbols occurring in terms s, t the term $\varepsilon_1^n(x_1, \ldots, x_n)$ (or the term $\varepsilon_n^n(x_1, \ldots, x_n)$ in the second case in which $Right(s) \neq Right(t)$).

(v) In this case we get the proof substituting for all n-ary operation symbols ($n > 1$) in $f(x_{s_1(1)}, \ldots, x_{s_1(n)}) \approx f(x_{s_2(1)}, \ldots, x_{s_2(n)})$ the term $\varepsilon_j^n(x_1, \ldots, x_n)$ for $j = 1, \ldots, n$. ■

Definition 4.4. A partial clone $C \subseteq P(A)$ is called *strongly solidifyable* if there exists a strongly solid algebra $\mathcal{A}$ with $C = T(\mathcal{A})$.

From Proposition 4.3, we get some criterions for partial clones to be not strongly solidifyable.

Proposition 4.5. *Let $C \subseteq P(A)$ be a partial clone, $|A| \geq 2$. If C satisfies one of the following conditions (1)-(4), then C is not strongly solidifyable.*

(1) *C contains a binary commutative operation,*
(2) *C contains a total constant operation,*
(3) *C contains a nowhere defined operation,*
(4) *there exists an $f^A \in C^{(n)}, n \geq 2$, and mappings $s_1, s_2 : \{1, \ldots, n\} \to \{1, \ldots, n\}, n \geq 2$, such that $s_1(i) \neq s_2(i)$ for all $i = 1, \ldots, n$ and $f(x_{s_1(1)}, \ldots, x_{s_1(n)}) \approx f(x_{s_2(1)}, \ldots, x_{s_2(n)})$ is a strong identity in $\mathcal{A}$.*

Proof. If $\mathcal{A}$ is a partial algebra such that $T(\mathcal{A}) = C$, and if C has one of the properties (1) - (4), then $T(\mathcal{A})$ has the same property. We can assume that $\mathcal{A}$ has one of the operations requested in conditions (1) - (4) under its fundamental operations. By Proposition 4.3 the partial algebra $\mathcal{A}$ cannot be strongly solid. ■

Since clones of partial operations are total algebras, we can characterize solidifyable clones in the same way as it was done in[6] for clones of total algebras.

Theorem 4.6. *C is strongly solidifyable if and only if C is a free algebra, freely generated by $\{f_i^A \mid i \in I\}$.*

Proof. Assume that C is strongly solidifyable. Then there exists a strongly solid partial algebra $\mathcal{A} = (A; (f^A)_{i \in I})$ such that $C = T(\mathcal{A})$. Let $F^{n,A} := \{f_j^A \mid j \in I$ and f_j^A is n-ary $\}$. Consider an arbitrary sequence $\varphi := (\varphi^{(n)})_{n \in \mathbb{N}^+}$ of mappings with $\varphi^{(n)} : F^{n,A} \to T^n(\mathcal{A})$. For every $n \in \mathbb{N}^+$ and every n-ary f_j^A, there are n-ary term operations $t_j^{\mathcal{A}} \in T(\mathcal{A})$ with $\varphi^{(n)}(f_j^A) = t_j^{\mathcal{A}}$. This allows us to define a regular hypersubstitution σ_R with $\sigma_R(f_j) = t_j$, $j \in I$. Then we have $\varphi^{(n)}(f_j^A) = \sigma_R(f_j)^{\mathcal{A}}$, $j \in I$.

Let $\overline{\varphi^{(n)}}(t^{\mathcal{A}}) = \widehat{\sigma}_R[t]^{\mathcal{A}}$ for any $t \in W_\tau^C(X_n)$. Then $(\overline{\varphi^{(n)}})_{n\in\mathbb{N}^+}$ is the extension of $(\varphi^{(n)})_{n\in\mathbb{N}^+}$ since $\overline{\varphi^{(n)}}(f^A) = \widehat{\sigma}_R[f_i(x_1,\ldots,x_{n_i})]^{\mathcal{A}} = \sigma_R(f_i)^{\mathcal{A}}$ and $\overline{\varphi} = (\overline{\varphi^{(n)}})_{n\in\mathbb{N}^+}$ is an endomorphism because of

$$\begin{aligned}\overline{\varphi^{(n)}}(S_m^{n,A}(t^{\mathcal{A}},t_1^{\mathcal{A}},\ldots,t_n^{\mathcal{A}})) &= \overline{\varphi^{(n)}}(S_m^n(t,t_1,\ldots,t_n)^{\mathcal{A}})\\ &= \widehat{\sigma}_R[S_m^n(t,t_1,\ldots,t_n)]^{\mathcal{A}}\\ &= S_m^n(\widehat{\sigma}_R[t],\widehat{\sigma}[t_1],\ldots,\widehat{\sigma}[t_n])^{\mathcal{A}}\\ &\quad \text{by Lemma 2.3}\\ &= S_m^n(\widehat{\sigma}_R[t]^{\mathcal{A}},\widehat{\sigma}[t_1]^{\mathcal{A}},\ldots,\widehat{\sigma}[t_n]^{\mathcal{A}})\\ &= S_m^{n,A}(\overline{\varphi^{(n)}}(t^{\mathcal{A}}),\overline{\varphi^{(n)}}(t_1^{\mathcal{A}}),\ldots,\overline{\varphi^{(n)}}(t_n^{\mathcal{A}}))\\ &\quad \text{for every } n \geq 1.\end{aligned}$$

Therefore any mapping $(\varphi^{(n)})_{n\in\mathbb{N}^+}$ can be extended to an endomorphism of C and C is a free algebra, freely generated by $\{f_i^A \mid i \in I\}$.
Conversely, let C be a free algebra, freely generated by $\{f_i^A \mid i \in I\}$ (i.e. for every map $\varphi : \{f_i^A \mid i \in I\} \to C$ there is a homomorphism (clone homomorphism $\overline{\varphi} : \langle\{f_i^A \mid i \in I\}\rangle \to C$). Then we have that $C = \langle\{f_i^A \mid i \in I\}\rangle = T(\mathcal{A})$, where $\mathcal{A} = (A; (f_i^A)_{i\in I})$ is a partial algebra. The next step is to show that $\mathcal{A}$ is strongly solid. Let $\sigma_R : \{f_i \mid i \in I\} \to W_\tau^C(X)$ be a regular hypersubstitution. Consider a mapping $\gamma : \{f_i^A \mid i \in I\} \to C = T(\mathcal{A})$ with $\gamma(f_i^A) = \sigma_R(f_i)^{\mathcal{A}}$. Then γ can be extended to a clone endomorphism $\overline{\gamma} : \langle\{f_i^A \mid i \in I\}\rangle \to C$ and by Lemma 2.8 for every term $t \in W_\tau^C(X)$ we have

$$\begin{aligned}s \approx t \in Id^s\mathcal{A} &\Rightarrow s^{\mathcal{A}} = t^{\mathcal{A}}\\ &\Rightarrow \overline{\gamma}(s^{\mathcal{A}}) = \overline{\gamma}(t^{\mathcal{A}})\\ &\Rightarrow \widehat{\sigma}_R[s]^{\mathcal{A}} = \widehat{\sigma}_R[t]^{\mathcal{A}}\\ &\Rightarrow \widehat{\sigma}_R[s] \approx \widehat{\sigma}_R[t] \in Id^s\mathcal{A}.\end{aligned}$$

Therefore $\mathcal{A}$ is strongly solid. ∎

Proposition 4.7. *Let $C, C' \subseteq P(A)$ be clones of partial algebras. If $C \cong C'$ and C is strongly solidifyable then C' is also strongly solidifyable.*

Proof. Since C is strongly solidifyable, there is a partial algebra $\mathcal{A} = (A; (f_i^A)_{i\in I})$ such that $C = T(\mathcal{A}) = \langle\{f_i^A \mid i \in I\}\rangle$. Since $C \cong C'$, there is an isomorphism $\varphi : T(\mathcal{A}) \to C'$ which maps the generating system of $T(\mathcal{A})$ to a generating system of C'. Therefore $C' = \langle\{\varphi(f_i^A) \mid i \in I\}\rangle$ and we get that C' is a free algebra, freely generated by $\{\varphi(f_i^A) \mid i \in I\}$. By Theorem 4.6, we have that C' is strongly solidifyable. ∎

From the definition of strongly solidifyable clones, from Proposition 2.9 and Proposition 4.7, we have that

Corollary 4.8. *If $\mathcal{A}$ is strongly solid and $V(\mathcal{A}) \sim V(\mathcal{B})$, then $\mathcal{B}$ is strongly solid.*

Now we want to determine all strongly solidifyable partial clones generated by a single unary operation f^A. A partial algebra $\mathcal{A} = (A; f^A)$, $(|A| \geq 2)$, where f^A is a unary operation on A is called *mono-unary*. Every strong identity of a mono-unary partial algebra has the form

$$f^k(x) \approx f^l(x) \quad (k, l \in \{0, 1, \ldots\})$$

or

$$f^k(x) \approx f^k(y) \quad (k \in \{1, 2, \ldots\}).$$

Obviously, identities of the second form cannot be strong hyperidentities because when substituting for the unary operation symbol the term $\varepsilon_1^1(x)$ we would get $\varepsilon_1^1(x) \approx \varepsilon_1^1(y)$ (i.e. $x \approx y$) in contradiction to $|A| > 1$.
For a partial unary operation $f^A : A \multimap A$ let $Imf^A := \{f^A(a) \mid a \in A\}$ be the image of f^A and let $\lambda(f^A)$ denote the least non-negative m such that $Im(f^A)^m = Im(f^A)^{m+1}$.

Example 4.9.

1. Consider the three-element partial algebra $\mathcal{A} = (\{0, 1, 2\}; f^A)$ of type (1) with $domf^A = \{1, 2\}$ and $f^A(1) = 0$, $f^A(2) = 1$. Then we have

	f^A	$(f^A)^2$	$(f^A)^3$
0	−	−	−
1	0	−	−
2	1	0	−

and $\lambda(f^A) = 3$.

2. Consider the three-element partial algebra $\mathcal{A} = (\{0, 1, 2\}; f^A)$ of type (1) with $domf^A = \{0, 2\}$ and $f^A(0) = 0$, $f^A(2) = 0$. Then we have

	f^A	$(f^A)^2$
0	0	0
1	−	−
2	0	0

and $\lambda(f^A) = 1$. Then $|Im(f^A)^{\lambda(f^A)}| = |Im(f^A)^1| = 1$.

Corollary 4.10. *The partial clone generated by the mono-unary partial operation f^A contains a constant if and only if* $|Im(f^A)^{\lambda(f^A)}| = 1$.

Then we have:

Proposition 4.11. *A mono-unary partial algebra $\mathcal{A} = (A; f^A)$, ($|A| \geq 2$), is strongly solid if and only if $|Im(f^A)^{\lambda(f^A)}| > 1$ (i.e. $T(\mathcal{A})$ contains no constant and no nowhere defined partial operation).*

Proof. Assume $|Im(f^A)^{\lambda(f^A)}| > 1$. Then the powers $(f^A)^m$ are not constant and not nowhere defined operations. Every strong identity of $\mathcal{A}$ is of the form $f^k(x) \approx f^l(x)$. The powers $(f^A)^m$ and the identity operation are the only unary operations of $T(\mathcal{A})$ and satisfy this identity since

$$((f^A)^m)^k(x) = ((f^A)^k)^m(x) = ((f^A)^l)^m(x) = ((f^A)^m)^l(x).$$

Thus every strong identity is a strong hyperidentity, i.e. $\mathcal{A}$ is strongly solid. If $|Im(f^A)^{\lambda(f^A)}| \leq 1$ then $(f^A)^{\lambda(f^A)}$ is a nowhere defined operation or $(f^A)^{\lambda(f^A)}$ is constant. In this case $f^k(x) \approx f^k(y)$ is a strong identity in $\mathcal{A}$ but not a strong hyperidentity in $\mathcal{A}$. This becomes clear when substituting for the unary operation symbols the term $\varepsilon_1^1(x)$. Then we get $\varepsilon_1^1(x) \approx \varepsilon_1^1(y)$ (i.e. $x \approx y$), a contradiction to $|A| > 1$. ■

If we want to determine all solidifyable minimal partial clones following Theorem 3.1 we have to investigate the proper partial minimal clones, i.e. the clones generated by a proper partial projection with a nontrivial totally reflexive and totally symmetric domain. We can restrict our investigation to one projection $e^n_{i,D}$ for every totally reflexive and totally symmetric domain D and every n since $e^n_{j,D} \in \langle e^n_{i,D} \rangle$ and $e^n_{i,D} \in \langle e^n_{j,D} \rangle$ for each $1 \leq i, j \leq n$ and thus $\langle e^n_{i,D} \rangle = \langle e^n_{j,D} \rangle$.
We consider the following cases:
(i) $2 < n \leq |A|$.
Choose $i = 1$. Then $\tilde{e}^n_{1,D}(x_1, x_2, x_3, x_4, \ldots, x_n) \approx \tilde{e}^n_{1,D}(x_1, x_3, x_2, x_4, \ldots, x_n)$ where $\tilde{e}^n_{1,D}$ is an operation symbol corresponding to the operation $e^n_{1,D}$, is a strong identity of the algebra $\mathcal{A} = (A; e^n_{1,D})$. Indeed, if $(x_1, x_2, x_3, x_4, \ldots, x_n) \in dom\ e^n_{1,D} (= D)$, then $(x_1, x_3, x_2, x_4, \ldots, x_n) \in D$ since D is totally symmetric and conversely. Further, in the case that both sides are defined, the values agree. The equation $f(x_1, x_2, x_3, x_4, \ldots, x_n) \approx f(x_1, x_3, x_2, x_4, \ldots, x_n)$ is not a strong hyperidentity of $\mathcal{A} = (A; e^n_{1,D})$ since when substituting for the operation symbol f the term $\varepsilon^n_2(x_1, \ldots, x_n)$ we would get $e_2^{n,A}(a_1, \ldots, a_n) \neq e_3^{n,A}(a_1, \ldots, a_n)$ because of $|A| > 2$. This means that $\mathcal{A}$ is not strongly solid. In a similar way for any other $1 < i \leq n$

and any totally symmetric and totally reflexive $D \subseteq A^n$ we get that $(A; e^n_{i,D})$ is not strongly solid. Therefore, the clones $\langle e^n_{i,D}\rangle$ with $n < 2$ and $1 \leq i \leq n$ are not strongly solidifyable.
(ii) $2 = n \leq |A|$.
Let $D \neq \iota_2$, i.e. D is different from the diagonal $\iota_2 = \{(a,a) \mid a \in A\}$. Now we consider the equation

$$\tilde{e}^2_{1,D}(x_1, \tilde{e}^2_{1,D}(x_1, x_2)) \approx \tilde{e}^2_{1,D}(x_1, \tilde{e}^2_{1,D}(x_2, x_1)).$$

Assume that the left hand side is defined, i.e. $(x_1, x_2) \in D$. Then $\tilde{e}^2_{1,D}(x_1, x_2) \approx x_1$ and $(x_1, x_1) \in D$ because of the reflexivity of D. Since D is symmetric we get $(x_2, x_1) \in D$ and therefore $\tilde{e}^2_{1,D}(x_2, x_1) \approx x_2$. From $(x_1, x_2) \in D$ we get that the right hand side is defined. In the same way we get that the left hand side is defined whenever the right hand side is defined and both sides agree. On the other hand, $f(x_1, f(x_1, x_2)) \approx f(x_1, f(x_2, x_1))$ is not a strong hyperidentity of $\mathcal{A} = (A; e^2_{1,D})$ since when we substitute for the operation symbol f the term $\varepsilon^2_2(x_1, x_2)$ we would get $e^{2,A}_2(x_1, x_2) = e^{2,A}_1(x_1, x_2)$ i.e. A would be a one-element set. If D is the diagonal ι_2 we have no contradiction. In this case $e^2_{1,D}$ is commutative and by Proposition 4.3(i) we conclude that $\mathcal{A}$ is not strongly solid. In a similar way we get also that $\langle e^2_{2,D}\rangle$ is not strongly solidifyable and therefore clones of the form $\langle e^2_{i,D}\rangle$ when $i \in \{1,2\}, D = \iota_2$, are not strongly solidifyable.
(iii) $n = 1$.
At first we consider the case that $D \neq \emptyset$. Then all strong identities of the ([13]) algebra $(A; e^1_D)$ can be derived from the strong identity $\tilde{e}^1_D(x_1) \approx [\tilde{e}^1_D]^2(x_1)$. Clearly, the equation $f(x_1) \approx f^2(x_1)$ is a strong hyperidentity of $\mathcal{A} = (A; e^1_D)$. If $D = \emptyset$, then e^1_D is the discrete unary operation satisfying the strong identity $\tilde{e}^1_D(x_1) \approx \tilde{e}^1_D(x_2)$ for all $x_1, x_2 \in A$. The equation $f(x_1) \approx f(x_2)$ is not a strong hyperidentity. This is evident if we substitute for f in $f(x_1) \approx f(x_2)$ the term $\varepsilon^1_1(x)$.

Together with Theorem 3.1 we get our result:

Theorem 4.12. *A minimal partial clone C of partial operations on A (A finite, $|A| \geq 2$) is strongly solidifyable if and only if C has one of the following forms*

(1) *C is generated by a unary operation f^A different from the unary empty operation and satisfying $(f^A)^2 = f^A$ or $(f^A)^p = id$ where p is a prime number, id the identity operation on A and C contains no constant operation.*

(2) *C is generated by a binary operation g^A which fulfils the identities*

$$g(x_1, x_1) \approx x_1,\ g(g(x_1, x_2), x_3) \approx g(x_1, f(x_2, x_3)) \approx g(x_1, x_3).$$

Proof. We consider two cases:
case 1. C is generated by a proper partial projection with a nontrivial totally reflexive and totally symmetric domain. Then by the remarks before Theorem 4.12 C cannot be strongly solidifyable;
case 2. C is a total minimal clone. Then C is generated by an operation f of one of the types (1) - (4):
(1) f is unary and $f^2 = f$ or $f^p = id$ for some prime number p. Similar to Proposition 4.11, we get that $\mathcal{A}$ is a solid algebra and C is strongly solidifyable.
(2) The operation f is binary and idempotent. If the binary operation f satisfies $f(x_1, x_1) \approx x_1$ and $f(x_1, f(x_2, x_3)) \approx f(x_1, x_3)$, then $\langle f \rangle$ is the clone of a rectangular band and since rectangular bands are solid, $\langle f \rangle$ is strongly solidifyable. Conversely, assume that C is minimal, strongly solidifyable and of type (2). Then there exists a solid algebra $\mathcal{A}$ with $C = T(\mathcal{A})$. We may assume that the type of $\mathcal{A} = (A; f^A)$ is (n) since C is minimal and is generated by only one operation which is not a projection. By identification of variables, we get a binary operation $g(x_1, x_2) := f(x_1, x_2, \ldots, x_2)$ which belongs to C. Clearly, g cannot be a projection, otherwise $\mathcal{A}$ satisfies the identity $g(x_1, x_2) \approx x_1$ or the identity $g(x_1, x_2) \approx x_2$. This contradicts the solidity of $\mathcal{A}$. Therefore $\langle g \rangle = C$ and then $(A; g^A)$ is also solid. Let t be an arbitrary binary term over $(A; g^A)$ such that $leftmost(t) = rightmost(t) = x_1$. Assume that $t^{\mathcal{A}}$ is not a projection, then $t^{\mathcal{A}}$ generates C. This means, we can obtain g^A from $t^{\mathcal{A}}$ by superposition and then the term t can be produced by g and variables x_1, x_2 and this gives an equation of the form $g(x_1, x_2) \approx f(x_1, x_2, \ldots, x_2, x_1)$. Since $\mathcal{A}$ is a solid algebra, this cannot be an identity in $\mathcal{A}$ and thus $t^{\mathcal{A}}$ is a projection and the term t satisfies $t(x_1, x_2, \ldots, x_2, x_1) \approx x_1$. Therefore g satisfies the identities $g(x_1, x_1) \approx x_1$ and $g(x_1, g(x_2, x_1)) \approx x_1$.
(3) f is a ternary majority operation ($f(x_1, x_1, x_2) \approx f(x_1, x_2, x_1) \approx f(x_2, x_1, x_1) \approx x_1$). Then the identity $f(x_2, x_1, x_1) \approx x_1$ is not a hyperidentity of $\mathcal{A} = (A; f^A)$ since when we substitute for the operation symbol the term $\varepsilon_1^3(x_1, x_2, x_3)$, we get a contradiction.
(4) f is the ternary operation $x_1 + x_2 + x_3$ in a Boolean group. Then we have that $x_1 + x_1 + x_2 \approx x_2 \approx x_2 + x_1 + x_1$ is an identity. The identity $x_1 + x_1 + x_2 \approx x_2$ is not a hyperidentity. This becomes clear if we substitute for the operation symbol the term $\varepsilon_1^3(x_1, x_2, x_3)$.

(5) f is a semiprojection (i.e. ar $f = n \geq 3$ and there exists an element $i \in \{1, \ldots, n\}$ such that $f(x_1, \ldots, x_n) = x_i$ whenever $x_1, \ldots, x_n$ are not pairwise different). Then we have that $f(x_1, x_2, \ldots, x_n) = x_i = f(x_2, x_1, \ldots, x_n)$ where $i \in \{1, \ldots, n\}$. So, the identity $f(x_1, x_2, \ldots, x_n) \approx f(x_2, x_1, \ldots, x_n)$ is not a hyperidentity since when we substitute for the operation symbol the term $\varepsilon_1^n(x_1, \ldots, x_n)$, we get $x_1 \approx x_2$. ■

In[6] the concept of the degree of representability $degr(C)$ for a clone of total operations is introduced. We generalize this concept to clones of partial operations.

Definition 4.13. Let $C \subseteq P(A)$ be a clone of partial operations. Then the degree of representability $degr(C)$ is the smallest cardinality $|A'|$ such that there is a clone $C' \subseteq P(A')$ with $C \cong C'$.

Proposition 4.14. *Let C be a strongly solidifyable minimal partial clone.*
(i) If $C = \langle f \rangle$, $f^2 = f$ and dom $f \subset A$ then $degr(C) = 2$.
(ii) If $C = \langle f \rangle$, $f^2 = f$ and dom $f = A$ then $degr(C) = 3$.
(iii) If $C = \langle f \rangle$, $f^p = id$ then $degr(C) = p$, where p is a prime number.
(iv) If $C = \langle f \rangle$ and f is binary then $degr(C) = 4$.

Proof. (i) If $f^2 = f$ and *dom* $f \subset A$ then $C \cong T(\mathcal{A})$ where $\mathcal{A} = (\{0,1\}; f_0)$ with $f_0(0) = 0$ and *dom* $f_0 = \{0\}$ since in each case the Cayley table of the clone has the form

	id	f
id	id	f
f	f	f

and thus $C^{(1)} \cong T^{(1)}(\mathcal{A})$. Since C and $T(\mathcal{A})$ are generated by its unary operations we get

$$\langle C^{(1)} \rangle = C \cong T(\mathcal{A}) = \langle T^{(1)}(\mathcal{A}) \rangle.$$

(ii), (iii) and (iv) were proved in.[6] ■

References

1. F. Börner, L. Haddad, R. Pöschel, *Minimal partial clones*, Preprint, 1990.
2. F. Börner, *Varieties of Partial Algebras*, Beiträge zur Algebra und Geometrie, Vol. 37 (1996), No. 2, 259-287.
3. P. Burmeister, *A Model Theoretic Oriented Approach to Partial Algebras*, Akademie-Verlag, Berlin 1986.

4. B. Csákány, *All minimal clones on the three-element set*, Acta Cybernetica (Szeged), 6 (1983), 227-238.
5. K. Denecke, *On the characterization of primal partial algebras by strong regular hyperidentities*, Acta Math. Univ. Comenianae, Vol.LXIII, 1 (1994), 141-153.
6. K. Denecke, D. Lau, R. Pöschel, D. Schweigert, *Solidifyable clones*, General Algebra and Applications, Heldermann-Verlag, Berlin 1992.
7. E. Graczynska, D. Schweigert, *Hyperidentities of given type Algebra Universalis*, 27 (1990), 305-318.
8. A.I. Mal'cev, *Algorithms and Recursive Functions*, Wolters Nordhoff Publishing, 1970.
9. J. Płonka, *On Hyperidentities of some Varieties*, General Algebra and Discrete Mathematics, Heldermann-Verlag, Berlin 1995, 199-214.
10. E.L. Post, *The two-valued iterative systems of mathematica logic*, Ann. Math. Studies 5, Princeton Univ. Press (1941).
11. I.G. Rosenberg, *Minimal clones I: The five types*, Lectures in Universal Algebra, Colloqu. Math. Soc. J. Bolyai 43, 1983, 405-427.
12. B. Staruch, B. Staruch, *Strong Regular Varieties of Partial Algebras*, Algebra Universalis, 31 (1994), 157-176.
13. D. Welke, *Hyperidentitäten Partieller Algebren*, Ph.D.Thesis, Universität Potsdam, 1996.

Received: August 17, 2009

Revised: October 15, 2009

A NOVEL CRYPTOSYSTEM BASED ON FINITE AUTOMATA WITHOUT OUTPUTS

P. DÖMÖSI*

Institute of Mathematics and Informatics, College of Nyíregyháza,
Sóstói út 31./B, Nyíregyháza, H-4400, Hungary
** E-mail: domosi@nyf.hu*
www.nyf.hu

In this paper we introduce a novel cryptosystem based on finite automata without outputs. For encryption and decryption the apparatus uses the same secret keys, which have the transition matrix of a key-automaton without outputs and with an initial state and final states. To each character in the character set of the plaintext there is one or more final states of the key automaton assigned. During encryption the plaintext is read in sequentially character by character and the key automaton assigns to each plaintext character a character string, whose length is adjustable within a given length range. The apparatus creates the ciphertext by linking these character strings together. During decryption the key automaton starting from the initial state reads in the ciphertext character by character and decryption is accomplished by linking together the plaintext characters associated with certain final states, which provides the plaintext in its original form.

Keywords: Cryptosystem; Finite Automata without Outputs.

1. Introduction

Automata theory provides a natural basis for designing cryptosystems and several such systems have been designed. Some of them are based on Mealy automata or their generalization, while others are based on cellular automata.

Almost all cryptosystems can be modeled with Mealy machines (as sequential machines) or generalized sequential machines.[1,4,13,19–22,25] A further generation of the cryptosystems based on Mealy machines is the family of public key FAPKC and FAPKC-3 systems.[23,24]

Almost from the very beginning of research into cellular automata, there have been serious attempts at cryptographic applications.[3,5,6,11,12,14,17,18,26]

The subject matter of the present work is a cryptographic apparatus with a Rabin-Scott automaton as key for encoding and decoding of information.

2. Preliminaries

We start with some standard concepts and notations. All concepts not defined here can be found in.[15,16] By an *alphabet* we mean a finite nonempty set. The elements of an *alphabet* are called *letters.* A *word* over an alphabet Σ is a finite string consisting of letters of Σ. A word over a binary alphabet is called a *bit string.* The string consisting of zero letters is called the *empty word,* written by λ. The *length* of a word w, in symbols $|w|$, means the number of letters in w when each letter is counted as many times it occurs. By definition, $|\lambda| = 0$. At the same time, for any set H, $|H|$ denotes the cardinality of H. In addition, for every nonempty word w, denote by $\overrightarrow{w}$ the last letter of w. ($\overrightarrow{\lambda}$ is not defined.) If $u = x_1 \cdots x_k$ and $v = x_{k+1} \cdots x_\ell$ are words over an alphabet Σ (with $x_1, \ldots, x_k, x_{k+1}, \ldots, x_\ell \in \Sigma$), then their *catenation* $uv = x_1 \cdots x_k x_{k+1} \cdots x_\ell$ is also a word over Σ. In this case we also say that u is a *prefix* of uv and v is a *suffix* of uv. Catenation is an associative operation and, by definition, the empty word λ is the identity with respect to catenation: $w\lambda = \lambda w = w$ for any word w. For every word w, put $w^0 = \lambda$, moreover, $w^n = ww^{n-1}, n \geq 1$. Let Σ^* be the set of all words over Σ, moreover, let $\Sigma^+ = \Sigma^* \setminus \{\lambda\}$. Σ^* and Σ^+ are the *free monoid* and the *free semigroup,* respectively, generated by Σ under catenation. Subsets of Σ^* are called *(formal) languages.* In particular, we put $\Sigma^0 = \{\lambda\}, \Sigma^n = \{w : |w| = n\}, n \geq 1$, and $\Sigma^{(0)} = \Sigma^0, \Sigma^{(n)} = \{w : |w| \leq n\}, n \geq 1$.

By an *automaton* we mean a finite Rabin-Scott automaton, i.e. a deterministic finite initial automaton without outputs supplied by a set of final states which is a subset of the state set. In more details, an automaton is an algebraic structure $\mathcal{A} = (A, a_0, A_F, \Sigma, \delta)$ consisting of the nonempty and finite *state set* A, the nonempty and finite *input set* Σ, a *transition function* $\delta : A \times \Sigma \to A$, the initial state $a_0 \in A$ and the (not necessarily nonempty) set $A_F \subseteq A$ of final states. The elements of the state set are the *states,* the elements of A_F are *the final states*, and the elements of the input set are the *input signals.* It may happen that the initial state is a final state as well (this is not excluded). An element of A^+ is called a *state word* [a] and an element of Σ^* is called an input word. State and input words are also called *state strings* and *input strings,* respectively. If a state string $a_1 a_2 \cdots a_s$ $(a_1, \ldots, a_s \in A)$ has at least three elements, the states $a_2, a_3, \ldots, a_{s-1}$ are also called inter-

[a] The empty word is not considered as a state word.

mediate states. It is understood that δ is extended to $\delta^* : A \times \Sigma^* \to A^+$ with $\delta^*(a, \lambda) = a$, $\delta^*(a, xq) = \delta(a, x)\delta^*(\delta(a, x), q), a \in A, x \in \Sigma, q \in \Sigma^*$. In other words, $\delta^*(a, \lambda) = a$ and for every nonempty input word $x_1x_2 \cdots x_s \in \Sigma^+$ (where $x_1, x_2, \ldots, x_s \in \Sigma$) there are $a_1, \ldots, a_s \in A$ with $\delta(a, x_1) = a_1, \delta(a_1, x_2) = a_2, \ldots, \delta(a_{s-1}, x_s) = a_s$ such that $\delta^*(a, x_1 \cdots x_s) = a_1 \cdots a_s$.

In the sequel, we will consider the transition of an automaton in this extended form and thus we will denote it by the same Greek letter δ.

If $\overrightarrow{\delta(a, w)} = b$ holds [b] for some $a, b \in A, w \in \Sigma^*$ then we say that w *takes* the automaton from its state a into the state b, and we also say that the automaton *goes* from the state a into the state b under the effect of w. We say that $z \in \Sigma^+$ is a *dummy* string with respect to the input word $u \in \Sigma^*$ if for every nonempty prefix w of z, $\overrightarrow{\delta(a_0, uw)} \notin A_F$ (including $\overrightarrow{\delta(a_0, uz)} \notin A_F$).

Finally, for every pair $a, b \in A$ of states define the language $L_{a,b} \subseteq \Sigma^*$ of input words which take the automaton from the state a into the state b without intermediate final states. In formula, let $L_{a,b} =$

$$\{w \in \Sigma^* \mid \overrightarrow{\delta(a, w)} = b, \forall u, v \in \Sigma^* : (w = uv \;\&\; u, v \neq \lambda) \Rightarrow \overrightarrow{\delta(a, u)} \notin A_F\}.$$

In addition, for every pair i, j of positive integers with $i \leq j$, put $L_{a,b}^{i,j} = \{pq | p \in \Sigma^{i-1}, q \in L_{c,b} \cap \Sigma^{(j-i+1)}, c = \overrightarrow{\delta(a, p)}\}$.

3. A Novel Cipher

The working of the considered system mainly differs from the most of the stream ciphers : it does not generate the ciphertexts in such a way that the plaintext bit stream is combined with a cipher bit stream by an exclusive-or operation (XOR). On the other hand, it has the main property of the stream ciphers : the plaintext digits are encrypted one at a time, and the transformation of successive digits varies during the encryption.

The key is an automaton having the property that for every state pair, whose first element is the initial state or a final state, its second element is any final state, there are several distinct input strings such that the last element of the state string, assigned to the first element of the state pair and the given input string by the generalized transition function, is the same as the second element of the state pair and none of the intermediate elements of the state string is a final state.

---

[b] Using the above notation $\overrightarrow{z}$ for a given nonempty word z, $\overrightarrow{\delta(a, w)} = b$ means that the last letter of the state word $\delta(a, w)$ is equal to the state b.

3.1. *Key Automaton and Random Ciphertext Blocks*

Let us given a pair of alphabets Π, Σ called, in order, *a plaintext alphabet* and *a ciphertext alphabet.* Consider an automaton $\mathcal{A} = (A, a_0, A_F, \Sigma, \delta)$ with $|A_F| \geq |\Pi|$, a surjective mapping $\varphi : A_F \to \Pi$, and a triplet d, s_{min}, s_{max} of positive integers having $s_{min} \leq s_{max}$. We say that $\mathcal{A}$ is a *key automaton* (with respect to $\Pi, \Sigma, \varphi, d, s_{min}, s_{max}$) if for every $a \in \{a_0\} \cup A_F, b \in A_F$, there are not fewer than d input words with length at least s_{min} and at most s_{max} taking the automaton from its state a into the state b without intermediate final states. In formula, for every pair $a, b \in A$, it is assumed that $|L_{a,b}^{\mathbf{s_{min}},\mathbf{s_{max}}}| \geq d$.

Put for every $y \in \Pi$, $\varphi^{-1}(y) = \{a \in A_F \mid \varphi(a) = y\}$ as usual and let $i_1 \cdots i_k$ be a *plaintext* with $i_1, \ldots, i_k \in \Pi$. Consider a list $w_{i_1}, \ldots, w_{i_k}$ of words with $w_{i_1} \in L_{a_0,a_1}^{\mathbf{s_{min}},\mathbf{s_{max}}}, \ldots, w_{i_k} \in L_{a_{k-1},a_k}^{\mathbf{s_{min}},\mathbf{s_{max}}}$ such that, in order, $a_1 \in \varphi^{-1}(i_1), \ldots, a_k \in \varphi^{-1}(i_k)$. Then $w_{i_1} \cdots w_{i_k}$ is a *ciphertext*[c] of $i_1 \cdots i_k$, where $w_{i_1}, \ldots, w_{i_k}$ are called *ciphertext blocks.*

3.2. *Encryption*

Several types of encryption processes can be constructed. One of them may be the following general (but not really effective) one.

- Let $i_1 \cdots i_k$ $(i_1, \ldots, i_k \in \Pi)$ be a plaintext.
- 1. Put $a = a_0$ and $j = 1$.
- 2. Do while end of the plaintext file.
  - 2.1. Read the character i_j in.
  - 2.2. Let $w_{i_j} = \lambda$.
  - 2.3. do while $\neg(s_{min} \leq |w_{i_j}| \leq s_{max}$ and $\overrightarrow{\delta(a, w_{i_j})} \in \varphi^{-1}(i_j))$.
    - 2.3.1. Let x be a random input signal and exchange the word w_{i_j} with $w_{i_j}x$.
    - 2.3.2. If $(|w_{i_j}| = s_{max}$ and $\overrightarrow{\delta(a, w_{i_j})} \notin \varphi^{-1}(i_j)))$ then exchange w_{i_j} with λ.
  - 2.4. Output w_{i_j}.
  - 2.5. Exchange a with $\overrightarrow{\delta(a, w_{i_j})}$ and j with $j + 1$.

Theoretically, the cycle 2.3 of this process may be arbitrarily long.[7,8] Therefore, this process suffers from practical difficulties. By an appropriate type of key automata and a slight modification of the above process, these difficulties can be overcome. (See Section 4.)

---

[c]Every plaintext and every ciphertext is assumed to be nonempty.

3.3. *Decryption*

The decryption process is also quite simple.

- Let $w_{i_1} \dots w_{i_k} (w_{i_1}, \dots, w_{i_k} \in \Sigma^+)$ be a ciphertext.

1. Put $a = a_0$ and $j = 0$.

2. Do while end of the ciphertext file.

2.1. Read the next ciphertext character x in.

2.2. Exchange a with $\delta(a, x)$ and j with $j + 1$.

2.3. If ($a \in A_F$ and $j \geq s_{min}$) then put j=0 and output $\varphi(a)$.

4. Encryption Without Backtracks

The speed of the encryption (and decryption) has a central importance in the field. For this reason, we propose to consider random transition matrices having the property that each of the final states is contained in each of the columns of the transition matrix assigned to the non-final states, moreover, each of the columns of the transition matrix has some (at least one) of the non-final states (and thus the number of input signals should be greater than that of final states). Then the steps 2.3.1 and 2.3.2 of the process in Section 3.2 is worth modifying as follows.

- 2.3.1. Let t be a random positive integer with $s_{min} \leq t \leq s_{max}$ and put $i = 0$.
- 2.3.2.1. Do while i=t-1.
- 2.3.2.1.1. Let x be a random input signal with $\overrightarrow{\delta(a, w_{i_j} x)} \notin A_F$.
- 2.3.2.1.2. Exchange w_{i_j} with $w_{i_j} x$ and i with $i + 1$.
- 2.3.2.2. Let x be a random input signal with $\overrightarrow{\delta(a, w_{i_j} x)} = \varphi^{-1}(i_j)$.
- 2.3.2.3. Exchange the word w_{i_j} with $w_{i_j} x$.

Obviously, by these properties, there is no backtrack search in the generation of random ciphertext blocks. Therefore the encoding algorithm becomes faster. On the other hand, we can prescribe the random length t of the generated ciphertext block in advance, and apart from the last one, we can choose the random input signals of the ciphertext block in several ways.

5. Cryptanalysis

5.1. *Automatic Learning Algorithms*

It is a famous result[2] that there exists a time polynomial and space linear algorithm to identify the canonical automata of k-reversible languages by using characteristic sample sets. Therefore, a really serious attack could be successful against the proposed stream cipher if some of the automata $\mathcal{A}_{b_0,F} = (A, b_0, F, \Sigma, \delta), A \setminus \{b_0\} \neq \emptyset$, $b_0 \in A, F \subseteq A$ based on the key automaton $\mathcal{A} = (A, a_0, A_F, \Sigma, \delta)$ are k-reversible for a nonnegative integer k. The following statement can help in handling this problem.

Theorem 5.1.[10] *Let $\mathcal{A} = (A, a_0, A_F, \Sigma, \delta)$ be an arbitrary automaton. There is no nonnegative integer k for which $\mathcal{A}$ is k-reversible if and only if there are distinct states $a, b \in A$, a nonempty input word $u \in \Sigma^+$, an input word $v \in \Sigma^*$, such that $\overrightarrow{\delta(a, u)} = a, \overrightarrow{\delta(b, u)} = b, \overrightarrow{\delta(a, v)} \neq \overrightarrow{\delta(b, v)}$, and either $\overrightarrow{\delta(a, v)}, \overrightarrow{\delta(b, v)} \in A_F$ or $\overrightarrow{\delta(a, vx)} = \overrightarrow{\delta(b, vx)}$ for some $x \in \Sigma$.* □

By the above statement, given an automaton $\mathcal{A} = (A, a_0, A_F, \Sigma, \delta)$, none of the automata $\mathcal{A}_{b_0,F} = (A, b_0, F, \Sigma, \delta), A \setminus \{b_0\} \neq \emptyset$, $b_0 \in A, F \subseteq A$ are k-reversible for some nonnegative integer k, if for every distinct $a, b \in A$ there are a nonempty input word $u \in \Sigma^+$, an input word $v \in \Sigma^*$, an input signal $x \in \Sigma$ such that $\overrightarrow{\delta(a, u)} = a, \overrightarrow{\delta(b, u)} = b, \overrightarrow{\delta(a, uvx)} = \overrightarrow{\delta(b, uvx)}$. For example, this property automatically holds if there is a row of the transition matrix having permutation of the state set, moreover, there is a reset signal.

5.2. *Adaptive Chosen-Ciphertext Attack, Adaptive Chosen--Plaintext Attack, Adaptive Chosen-Plaintext -Chosen--Ciphertext Attack*

Assume that the ciphertext $w_1 \cdots w_s$ consisting of the unknown ciphertext blocks $w_1, \ldots, w_s \in \Sigma^*$ is given and the cryptanalyst can make an unbounded number of interactive queries, choosing subsequent ciphertexts based on the information from the previous encryptions. Moreover assume, that an upper bound k for the length of the ciphertext blocks is known for the attacker. Then it can be possible to send a series of random strings of length at most k to the cipher system. Sooner or later the attacker will send the string w_1 and then he/she will get an answer consisting of the plaintext character i_1 to which the first block w_1 of the ciphertext was generated. Recall that, either no answer or an answer consisting of more than one plaintext character will arrive whenever the sent message is differs from w_1. If the plaintext consists of one character then we are ready and the attack was successful. Otherwise the attacker can continue the attack for the suffices $w_2 \cdots w_s, w_3 \cdots w_s, \ldots, w_s$ of the ciphertext receiving, in order, the second, third, ..., last character of the plaintext.

In this case, the only possibility of defense is to apply a relatively large automaton, moreover, relatively large numbers for the minimal and maximal block lengths. Obviously, if the length of the ciphertext blocks is on average k, and m is the minimum of the number of non-final states in all column of the transition matrix, then for every plaintext character one can consider at least m^{k-1} ciphertext blocks (even if all ciphertext blocks have the same length). If the number of the states (and also the number of the input signals) in the key automaton is, say, 256, moreover, there are 16 final states, then $m = 238$ can be assumed.[d] Using the above method, then breaking for $k \geq 18$ is really infeasible.

Similarly to the above method, adaptive chosen-plaintext attack and adaptive chosen-plaintext-chosen-ciphertext attack can be constructed to the proposed stream cipher. Similar defenses can be applied as above.

6. Performance

The speed of encryption and decryption does not essentially depend on the size of the key automaton. We applied key automata from 16 up to 256 states and also from 16 up to 256 inputs having the properties discussed in Section 4. The plaintext alphabet and also the set of the final states of the key automaton was the same consisting of 2, 4, or 16 elements.

Testing software simulations of the proposed stream cipher were implemented using a computer program written in C^{++}. The implementation was tested on a conventional laptop Toshiba Tecra A8-104 clocked at 2 GHz with 2 Mbyte L2 of cache and 1 Gbyte RAM under operation system Windows XP. If the minimal length of the ciphertext block is 5, its maximal length is 10, then the implemented system reaches the speed of 600 Kbyte/s as encryption and 800 Kbyte/s as decryption (in relation to the length of the plaintext). Comparing some stream ciphers (see, for example,[9]), the proposed cryptosystem is rather slow at least for the implemented software case.

7. Conclusion

In this paper we introduced a novel cryptosystem based on finite automata without outputs.

[d] We may assume that all columns of the transition matrix assigned to the non-final states contain not more than 17 final states and that one of the input signals is the reset one.

There are a few major issues with the discussed stream cipher.

- There is no serious security analysis.

- The discussed stream cipher is not really efficient, at least for the software case. In comparison with other promising designs and even with the state of the art ciphers (see, e.g., the homepage of the ESTREAM[9] project) the performance of the discussed cipher is rather slow, at least for the software case. A rigorous machine-independent investigation should be necessary to explore the reasons of this drawback.

- The ciphertext may be much longer than the plaintext. An intrinsic question is, how to deal with the ciphertext blowup. In the further research, a concrete measure should be necessary to describe the tradeoff between security and ciphertext blowup.

On the other hand, the discussed cryptosystem has the following advantages :

- Although the work uses a random number generator, it can take random number generators which are proved to be random indeed, or it can use any radioactive or other physical random number sources.

- To each plaintext message there are several corresponding encoded messages such that several encryptions of the same plaintext yield to several distinct ciphertexts.

- Since there are no initial or end markers in the encoded message, the ciphertext blocks cannot be identified without the key-automaton. So, without the key, even the length of the plaintext is difficult to estimate, since block lengths and the number of blocks are not public.

- Because of its inner structure, the proposed cipher is resistant to reused key attack and substitution attack.

Acknowledgments

The author would like to thank Gábor Balázsfalvi, Tibor Csáki, Tamás Gaál, Géza Horváth, Zoltán Mecsei, Benedek Nagy, Andor Pénzes, Heiko Stamer, Tamás Virág for their helpful comments. Special thanks to Gábor Balázsfalvi for developing the software discussed in Section 6 and Heiko Stamer for his important observations and criticism.

References

1. A. Atanasiu: A class of coders based on gsm. Acta Informatica, **29** (1992), 779–791.
2. D. Angluin : Inference of reversible languages. J Assoc. Comput. Mach., **29** (1982), 741-765.

3. F. Bao: Cryptoanalysis of partially known cellular automata. IEEE Trans. on Computers, **53** (2004), 1493–1497.
4. F. Bao and Y. Igarashi: Break finite automata public key cryptosystems. In: Zoltán Fülöp, Ferenc Gécseg, eds., Proc. 22nd Int. Coll. On Automata Languages and Programming - ICALP'95, Szeged, Hungary, July 10-14, 1995, LNCS **944**, Springer-Verlag, Berlin (1995), 147–158.
5. M. E. Bianco and D. A. Reed : Encryption system based on chaos theory. **US P 5,048,086**, 1991.
6. E. Biham: Cryptoanalysis of the chaotic map cryptosystem suggested at EUROCRYPT'91. In: D. W. Davies, ed., Proc. Conf. Advances in Cryptology - EUROCRYPT'91, Workshop on the Theory and Application of Cryptographic Techniques, Brighton, UK, April 8-11, 1991, LNCS 547 Springer-Verlag, Berlin, 1991, 532-534.
7. P. Erdős, A. Rényi: On a new law of large numbers. J. Analyse Math. **23** (1970), 103–111.
8. P. Erdős, P. Révész: On the length of the longest head-run. Topics in information theory (Second Colloq., Keszthely, 1975), 219–228. Colloq. Math. Soc. Janos Bolyai, Vol. **16**, North-Holland, Amsterdam, 1977.
9. ESTREAM PHASE 3, http://www.ecrypt.eu.org/stream/
10. J. Falucskai: On the k-reversibility of finite automata. Annales Mathematicae et Informaticae **36** (2009), 71–75, http://ami.ektf.hu.
11. P. Guan: Cellular automaton public key cryptosystem. Complex Systems, **1** (1987), 51–56.
12. H. A. Gutowitz, Method and Apparatus for Encryption, Decryption, and Authentication Using Dynamical Systems. **US P 5,365,589**, 1994.
13. M. Gysin: One-key cryptosystem based on a finite non-linear automaton. In: E. Dawson and J- Golic, eds., Proc. Int. Conf. Proceedings of the Cryptography: Policy and Algorithms, CPAC'95, Brisbane, Queensland, Australia, July 3-5, 1995. Lecture Notes in Computer Science **1029**, Springer-Verlag, Berlin (1995), 165–163.
14. T. Habutsu, Y. Nishio, I. Sasase, S. Mori: A Secret Key Cryptosystem by Iterating a Chaotic Map. In: D. W. Davies, ed., Proc. Conf. Advances in Cryptology - EUROCRYPT'91, Workshop on the Theory and Application of Cryptographic Techniques, Brighton, UK, April 8-11, 1991, LNCS **547** Springer-Verlag, Berlin (1991), 127–140 .
15. J. E. Hopcroft, R. Motwani, and J. D. Ullman: Introduction to Automata Theory, Languages, and Computation. **3**rd edition, Pearson Addison-Wesley Publishing Company, Inc., Reading, MA, 2006.
16. G. Horváth: The φ factoring algorithm (in Hungarian) Alkalmazott Mat. Lapok **21**, No. 2, 355–364 (2004).
17. J. Kari: Cryptosystems based on reversible cellular automata. Publ: University of Turku, Finland, April, 1992, preprint.
18. H. B. Lin: Elementary Symbolic Dynamics and Chaos in Dissipative Systems. Publ.: World Scientific, Singapore, 1989.

19. T. Meskaten: On finite automaton public key cryptosystems. Publ.: TUCS Technical Report No. **408**, Turku Centre for Computer Science, Turku (2001), 1–42.
20. V. J. Rayward-Smith: Mealy machines as coding devices. In: H. J. Beker and F. C. Piper, eds., Cryptography and Coding, Claredon Press, Oxford, 1989.
21. G. Sullivan, F. Weierud: Breaking German Army Ciphers. In: Cryptologia **24**(3) (2005), 193-?232.
22. R. Tao: On finite automaton one-key cryptosystems. In: R. Anderson, ed., Proc. 1st Fast Software Encryption Workshop - FSE'93. Proceedings of the Security Workshop held in Cambridge, Cambridge, UK, December 9-11, 1993, LNCS **809**, Springer-Verlag, Berlin (1994), 135-148.
23. R. Tao and S. Chen: Finite automata public key cryptosystem and digital signature. Computer Acta **8** (1985), 401-409 (in Chinese).
24. R. Tao, S. Chen and X. Chen: FAPKC3: a new finite automata public key cryptosystem. Publ.: Technical report No. **ISCAS-LCS-95-05**, Laboratory for Computer Science, Institute of Chinese Academy of Sciences, Beijing, 1995.
25. P. Wichmann: Cryptoanalysis of a modified rotor machine. In: J.-J. Quisquarter, J. Vandewalle, eds., Proc. Conf. Advances in Cryptology - EUROCRYPT'89, Workshop on the Theory and Applications of Cryptographic Techniques, Houthalen, Belgium, April 10-13, 1989, LNCS **434**, Springer-Verlag, Berlin (1990), 395–402.
26. S. Wolfram: Cryptography with Cellular Automata. In: C. W. Hugh, ed., Proc. Conf. Advances in Cryptology - CRYPTO'85, Santa Barbara, California, USA, August 18-22, 1985, LNCS **218**, Springer-Verlag, Berlin (1986), 429–432.

Received: January 26, 2009

Revised: June 9, 2010

LINEAR LANGUAGES OF FINITE AND INFINITE WORDS*

Z.ÉSIK[1], M. ITO[2] and W. KUICH[3]

[1]*Dept. of Computer Science, University of Szeged, Hungary*
[2]*Dept. of Mathematics, Kyoto Sangyo University, Kyoto, Japan*
[3]*Inst. for Discrete Mathematics and Geometry, TU Vienna, Austria*

A linear grammar with Büchi acceptance condition is a system

$$G = (N, \Sigma, P, X_0, R)$$

where (N, Σ, P, X_0) is an ordinary linear grammar with nonterminal alphabet N, terminal alphabet Σ, productions P and start symbol X_0, and $R \subseteq N$ is a set of repeated nonterminals. Consider the set of all finite and *infinite* derivation trees rooted X_0 whose leaves are labeled with letters of the terminal alphabet and possibly the empty word. When the tree is infinite, we require that at least one nonterminal letter in R appears infinitely often as the label of a vertex along the unique infinite branch of the tree. The frontier of such a tree determines a finite or infinite word over Σ. The set of all such words is called the linear language of finite and infinite words generated by G. Using results from [1–3] we provide an algebraic characterization of linear languages by rational operations. More specifically, we associate a Conway semiring-semimodule pair (S, V) with each alphabet Σ, where S is a semiring associated with Σ and V is the set of all subsets of infinite words over Σ of appropriate order type, and show that a set in V is linear if and only if it can be generated from certain simple elements of the semiring S by the rational operations.

Keywords: Linear grammar, Conway semiring-semimodule pair.
AMS Classification: 68Q42, 68Q45, 68Q70, 16Y60

1. Linear languages

In this section, we will consider languages of finite and infinite words over an alphabet Σ generated by linear grammars. Let Σ^ω and $\Sigma^{\omega^{op}}$ respectively

*Research supported by grant no. 77öu9 from the Austrian-Hungarian Action Foundation, the HAS-JSPS cooperative grant no. 101, and by grant no. K 75249 from the National Foundation of Hungary for Scientific Research.

denote the set of all ω-words and the set of all ω^{op}-words over Σ, i.e.,

$$\Sigma^{\omega} = \{a_0 a_1 \ldots : a_i \in \Sigma\}$$
$$\Sigma^{\omega^{\mathrm{op}}} = \{\ldots a_1 a_0 : a_i \in \Sigma\}$$

Now let

$$\Sigma^{\omega}\Sigma^{*} = \{uv : u \in \Sigma^{\omega},\ v \in \Sigma^{*}\}$$
$$\Sigma^{*}\Sigma^{\omega^{\mathrm{op}}} = \{vu : u \in \Sigma^{\omega^{\mathrm{op}}},\ v \in \Sigma^{*}\}$$
$$\Sigma^{\omega}\Sigma^{\omega^{\mathrm{op}}} = \{uv : u \in \Sigma^{\omega},\ v \in \Sigma^{\omega^{\mathrm{op}}}\}$$

Finally, let

$$\Sigma^{\infty} = \Sigma^{*} \cup \Sigma^{\omega}\Sigma^{*} \cup \Sigma^{*}\Sigma^{\omega^{\mathrm{op}}} \cup \Sigma^{\omega}\Sigma^{\omega^{\mathrm{op}}}.$$

We will use linear grammars to generate languages which are subsets of Σ^{∞}.

A *linear grammar with Büchi acceptance condition* is a system

$$G = (N, \Sigma, P, X_0, R)$$

where (N, Σ, P, X_0) is an ordinary linear grammar[6] with nonterminal alphabet N, terminal alphabet Σ, productions P and start symbol X_0, and $R \subseteq N$ is a set of *repeated nonterminals*. Consider the set of all *finite and infinite derivation trees* rooted X_0, whose leaves are labeled with letters of the terminal alphabet and possibly the empty word ϵ. Such a tree has a root labeled X_0 and is such that whenever a vertex is labeled X, for some $X \in N$, then there is some production $X \to X_1 \ldots X_k$ in P with $X_i \in N \cup \Sigma$ for all i such that the vertex has k successors, labeled $X_1, \ldots, X_k$, respectively. In particular, when $k = 0$, there is a single successor labeled ϵ. Clearly, each infinite derivation tree has a unique infinite branch. We say that a derivation tree is *successful* if it is either finite or infinite such that at least one nonterminal in R occurs infinitely often as the label of a vertex along the infinite branch of the tree.

The frontier (or yield) of a derivation tree can naturally be seen as a word in Σ^{∞}. The language $L^{\infty}(G)$ generated by G consists of the frontiers of successful derivation trees rooted X_0. We call a language $L \subseteq \Sigma^{\infty}$ *linear* if there is a linear grammar G with Büchi acceptance condition such that $L = L^{\infty}(G)$.

Example 1.1. Suppose that the only productions of G_1 are $X \to aXb$ and $X \to \epsilon$, where a, b are letters in Σ. If $R = \emptyset$ then $L^{\infty}(G_1) = \{a^n b^n : n \geq 0\}$

is a set of finite words. If $R = \{X\}$, then $L^\infty(G_1) = \{a^n b^n : n \geq 0\} \cup \{a^\omega b^{\omega^{\mathrm{op}}}\}$. (Of course, for any finite word u, $u^\omega = uu\ldots$ and $u^{\omega^{\mathrm{op}}} = \{\ldots uu\}$. When u is the empty word, then these words are also empty.)

Example 1.2. Consider the grammar G_2 with productions $X \to aX, X \to Y, Y \to Yb$, where a, b are terminal letters and X is the start symbol. Let $R = \{Y\}$. Then $L^\infty(G_2) = \{a^n b^{\omega^{\mathrm{op}}} : n \geq 0\}$.

Remark 1.1. For each linear grammar G with Büchi acceptance condition there is an equivalent grammar G' generating the same language with no production whose right side is a terminal word. Indeed, let Z be a new nonterminal and replace each production $X \to u$ where u is a terminal word by the productions $X \to uZ$ and $Z \to Z$. Finally, add Z to the set of repeated nonterminals.

In Section 6 we will give an operational characterization of linear languages, similar to the Kleene theorem for ω-regular languages and Büchi automata, cf.[5] This characterization result can be proved in a way which is similar to the aforementioned Kleene theorem. However, our point is that both of them are instances of a more general algebraic result, formulated in Theorem 5.1. In Sections 2, 3 and 4, we will develop the necessary algebraic machinery needed in order to present this general result and the operational characterization of linear languages.

2. ω-monoids

Definition 2.1. An *ω-semigroup*[5] is an ordered pair (S, V) consisting of a semigroup S, a set V, a left action $S \times V \to V$ of S on V, subject to the axiom

$$s(s'v) = (ss')v \tag{1}$$

for all $s, s' \in S$ and $v \in V$, and an *infinite product* operation $S^\omega \to V$, $(s_0, s_1, \ldots) \mapsto s_0 s_1 \cdots \in V$ such that

$$s(s_0 s_1 \cdots) = s s_0 s_1 \cdots \tag{2}$$

$$(s_0 \cdots s_{i_1 - 1})(s_{i_1} \cdots s_{i_2 - 1}) \cdots = s_0 s_1 \cdots \tag{3}$$

for all $s, s_0, s_1, \ldots$ in S and any sequence $0 < i_1 < i_2 < \ldots$. An *ω-monoid* is an ω-semigroup (S, V) such that S is a monoid, moreover, the action is unitary:

$$1v = v \tag{4}$$

for all $v \in V$. A morphism of ω-semigroups $(S, V) \to (S', V')$ is a pair of functions (h_S, h_V) such that h_S is a semigroup morphism $S \to S'$, h_V is a mapping $V \to V'$, and h_S and h_V jointly preserve the action: $h_S(s)h_V(v) = h_V(sv)$ for all $s \in S$ and $v \in V$, moreover,

$$h_S(s_0)h_S(s_1)\cdots = h_V(s_0 s_1 \cdots)$$

for all $s_0, s_1, \ldots \in S$, i.e., morphisms preserve the infinite product. Morphisms of ω-monoids necessarily preserve the multiplicative identity of the monoid component.

An important example of an ω-semigroup is $(\Sigma^+, \Sigma^\omega)$, where Σ is a set, called an alphabet, Σ^+ is the free semigroup of all finite nonempty words over Σ, Σ^ω is the set of all ω-words over Σ, and the action of Σ^+ on Σ^ω is defined by concatenation, so that for any $u \in \Sigma^+$ and $x \in \Sigma^\omega$, ux is the ω-word with a prefix u and corresponding tail x. More generally, consider now two alphabets Σ and Δ. Then in a similar way, we can define the ω-semigroup $(\Sigma^+, \Sigma^*\Delta \cup \Sigma^\omega)$, where $\Sigma^*\Delta$ is the collection of all finite nonempty words over the (disjoint) union $\Sigma \cup \Delta$ starting with a possibly empty word in Σ^* (the free monoid of all finite words over Σ) and ending in a letter in Δ. The action of Σ^+ on $\Sigma^*\Delta \cup \Sigma^\omega$ is defined as above.

Proposition 2.1.[5,7] *For each pair of sets (Σ, Δ), the ω-semigroup $(\Sigma^+, \Sigma^*\Delta \cup \Sigma^\omega)$ is freely generated by (Σ, Δ). In more detail, given any ω-semigroup (S, V) and any pair of functions (h_Σ, h_Δ) with $h_\Sigma : \Sigma \to S$ and $h_\Delta : \Delta \to V$, there is a unique morphism of ω-semigroups $(h_\Sigma^\sharp, h_\Delta^\sharp) : (\Sigma^+, \Sigma^*\Delta \cup \Sigma^\omega) \to (S, V)$ extending h_Σ and h_Δ.*

Indeed, $h_\Sigma^\sharp$ is the unique semigroup morphism $\Sigma^+ \to S$ extending h_Σ, and $h_\Delta^\sharp$ is defined by

$$h_\Delta^\sharp(a_1 \ldots a_n b) = h_\Sigma(a_1) \ldots h_\Sigma(a_n) h_\Delta(b)$$
$$h_\Delta^\sharp(a_1 a_2 \ldots) = h_\Sigma(a_1) h_\Sigma(a_2) \cdots$$

where each a_i is a letter in Σ and b is a letter in Δ.

We now turn to ω-monoids.

Lemma 2.1. *Suppose that (S, V) is an ω-monoid and $s_0, s_1, \ldots$ is an infinite sequence of elements of S. If the infinite sequence $i_0 < i_1 < \ldots$ contains all those indices n for which $s_n \neq 1$ then $s_0 s_1 \cdots = s_{i_0} s_{i_1} \cdots$.*

Proof. There are two cases. If there exist an infinite number of indices n with $s_n \neq 1$ then our claim follows from (3). In the opposite case, let

$j_0, \ldots, j_k$ be the sequence of all those indices n with $s_n \neq 1$. It follows from (2) and (4) that both $s_0 s_1 \cdots$ and $s_{i_0} s_{i_1} \cdots$ are equal to $(s_{j_0} \cdots s_{j_k})1^\omega$ where 1^ω denotes the infinite product $1 \cdot 1 \cdots$. □

We now describe the structure of the free ω-monoids. Given sets Σ and Δ, let $\perp$ be a letter which is not in $\Sigma \cup \Delta$. Let $\Delta_\perp = \Delta \cup \{\perp\}$, and consider the pair $(\Sigma^*, \Sigma^*\Delta_\perp \cup \Sigma^\omega)$, where Σ^* is the free monoid of all words over Σ including the empty word and $\Sigma^*\Delta_\perp$ is given above. The action of Σ^* on $\Sigma^*\Delta_\perp \cup \Sigma^\omega$ is similar to the action defined above, and the infinite product operation is given by

$$u_0 u_1 \cdots = \begin{cases} u_0 \cdots u_n \perp \text{ if } u_{n+1} = u_{n+2} = \ldots = \epsilon \text{ for some } n \geq 0, \\ u_0 u_1 \ldots \quad \text{otherwise.} \end{cases}$$

In particular, $\epsilon\epsilon \cdots = \perp$.

Proposition 2.2. *For any pair of alphabets* (Σ, Δ), *the* ω*-monoid*

$$(\Sigma^*, \Sigma^*\Delta_\perp \cup \Sigma^\omega)$$

is freely generated by (Σ, Δ).

Proof. Let σ_0 be a new letter and let $\Sigma_0 = \Sigma \cup \{\sigma_0\}$. Consider the free ω-semigroup $(\Sigma_0^+, \Sigma_0^*\Delta \cup \Sigma_0^\omega)$ constructed above. Let $h_\Sigma : \Sigma_0 \to \Sigma^*$ be the function which is the identity on Σ and maps σ_0 to ϵ, and let h_Δ be the inclusion of Δ in $\Delta_\perp$. We know that (h_Σ, h_Δ) extends to a unique morphism of ω-semigroups $(h_\Sigma^\sharp, h_\Delta^\sharp) : (\Sigma_0^+, \Sigma_0^*\Delta \cup \Sigma_0^\omega) \to (\Sigma^*, \Sigma^*\Delta_\perp \cup \Sigma^\omega)$. Note that $h_\Sigma^\sharp$ and $h_\Delta^\sharp$ are surjective.

Suppose now that (S, V) is an ω-monoid and $h_S : \Sigma \to S$ and $h_V : \Delta \to V$. First we extend h_S to a function $\Sigma_0 \to S$ by defining $h_S(\sigma_0) = 1$. Then we extend (h_S, h_V) to a morphism of ω-semigroups

$$(h_S^\sharp, h_V^\sharp) : (\Sigma_0^+, \Sigma_0^*\Delta \cup \Sigma_0^\omega) \to (S, V).$$

It is clear that the kernel of $h_\Sigma^\sharp$ is included in the kernel of $h_S^\sharp$. Also, by Lemma 2.1, the kernel of $h_\Delta^\sharp$ is included in the kernel of $h_V^\sharp$. Thus there is a unique ω-semigroup morphism

$$(\overline{h}_S, \overline{h}_V) : (\Sigma^*, \Sigma^*\Delta_\perp \cup \Sigma^\omega) \to (S, V)$$

such that $\overline{h}_S \circ h_\Sigma^\sharp = h_S^\sharp$ and $\overline{h}_V \circ h_\Delta^\sharp = h_V^\sharp$. It is clear that $\overline{h}_S(\epsilon) = \overline{h}_S(h_\Sigma^\sharp(\sigma_0)) = h_S^\sharp(\sigma_0) = 1$. Thus, $(\overline{h}_S, \overline{h}_V)$ is the unique extension of (h_S, h_V) to an ω-monoid morphism. □

3. Completely idempotent semiring-semimodule pairs

Recall from[4] that a *semiring* $(S, +, \cdot, 0, 1)$ consists of a monoid $(S, \cdot, 1)$ and a commutative monoid $(S, +, 0)$ such that multiplication distributes over all finite sums, so that

$$s(s_1 + s_2) = ss_1 + ss_2$$
$$(s_1 + s_2)s = s_1s + s_2s$$
$$0s = 0$$
$$s0 = 0$$

for all $s, s_1, s_2 \in S$. If in addition it holds that

$$s + s = s,$$

for all $s \in S$, then we call S an *idempotent* semiring. When S is a semiring, an S*-semimodule* is a commutative monoid $(V, +, 0)$ together with an action $S \times V \to V$ subject to the conditions

$$(ss')v = s(s'v)$$
$$1v = v$$
$$(s + s')v = sv + s'v$$
$$s(v + v') = sv + sv'$$
$$0v = 0$$
$$s0 = 0$$

for all $s, s' \in S$ and $v, v' \in V$. We call (S, V) a *semiring-semimodule pair.* Morphisms of semirings preserve all operations and constants. A morphism $(S, V) \to (S', V')$ between semiring-semimodule pairs consists of a semiring morphism $h_S : S \to S'$ and a monoid morphism $h_V : V \to V'$ which preserve the action. Note that when (S, V) is a semiring-semimodule pair such that S is idempotent, then V is also idempotent: $v + v = 1v + 1v = (1 + 1)v = 1v = v$ for all $v \in V$.

A commutative idempotent monoid $(S, +, 0)$ is called a *semilattice with* 0. When S is a semilattice with 0, S is (positively) ordered by $s \leq s'$ if and only if $s + s' = s'$, for $s, s' \in S$. We call S a *complete semilattice* if each nonempty subset A of S has a supremum with respect to the semilattice order denoted $\bigvee A$. In particular, S has a greatest element. Since 0 is clearly the least element of S, it follows that each subset of S has a supremum. Morphisms of complete semilattices preserve all suprema.

A *completely idempotent semiring* is a semiring S that is a complete semilattice, moreover, multiplication distributes over all suprema:

$$s(\bigvee A) = \bigvee\{sa : a \in A\}$$
$$(\bigvee A)s = \bigvee\{as : a \in A\}$$

for all $A \subseteq S$ and $s \in S$. A morphism of completely idempotent semirings is semiring morphism which is a complete semilattice morphism.

A *completely idempotent semiring-semimodule pair* is a semiring semimodule pair (S, V) which is an ω-monoid such that S is a completely idempotent semiring, V is a complete semilattice, the action is completely ditributive so that

$$(\bigvee A)v = \bigvee\{av : a \in A\}$$
$$s(\bigvee X) = \bigvee\{sx : x \in X\}$$

for all $s \in S$, $v \in V$, $A \subseteq S$ and $X \subseteq V$, moreover, the infinite product operation $S^\omega \to V$ is completely distributive:

$$A_0 A_1 \cdots = \bigvee\{a_0 a_1 \cdots : a_i \in A_i\}$$

for all $A_0, A_1, \ldots \subseteq S$. A morphism $(S, V) \to (S', V')$ of completely idempotent semiring-semimodule pairs is a morphism (h_S, h_V) of semiring-semimodule pairs which is a morphism of ω-monoids such that h_S and h_V preserve arbitrary suprema.

We now describe a construction of completely idempotent semiring-semimodule pair from an ω-monoid. Suppose that (M, V) is an ω-monoid. Let $P(M, V) = (P(M), P(V))$ where $P(M)$ and $P(V)$ respectively denote the sets of all subsets of M and V. Now $P(M)$, equipped with set union and the complex product operation $(A, B) \mapsto \{ab : a \in A,\ b \in B\}$, is a completely idempotent semiring with $\emptyset$ and $\{1\}$ acting as 0 and 1, respectively. Also, $P(V)$ equipped with set union and the empty set as 0 is a complete semilattice, and the complex action of $P(M)$ on $P(V)$, defined by

$$AX = \{ax : a \in A,\ x \in X\}$$

for $A \subseteq M$ and $X \subseteq V$ is unitary and completely distributive. Define an infinite product operation $P(M)^\omega \to P(V)$ by

$$A_0 A_1 \cdots = \{a_0 a_1 \cdots : a_i \in A_i\}$$

for all $A_0, A_1, \cdots \subseteq M$.

Theorem 3.1. *Suppose that (S, V) is an ω-monoid. Then $P(S, V)$ is a completely idempotent semiring-semimodule pair. Moreover, $P(S, V)$ is* *freely generated by (S,V): Given any completely idempotent semiring-semimodule pair* (S',V') *together with a morphism* $(h_S,h_V): P(S,V) \to (S',V')$ *of* ω*-monoids, there is a unique morphism of completely idempotent semiring-semimodule pairs* $(h_S^\sharp, h_V^\sharp): P(S,V) \to (S',V')$ *extending* (h_S,h_V).

Proof. It is clear that $P(S,V)$ is a semiring-semimodule pair, moreover, $P(S)$ is a completely idempotent semiring, $P(V)$ is a complete semilattice, and the action is completely distributive. It is straightforward to check that the infinite product satisfies the required properties.

Suppose now that we are given a morphism $(h_S,h_V): (S,V) \to (S',V')$ of ω-monoids, where (S',V') is a completely idempotent semiring-semimodule pair. For each $A \subseteq S$ and $X \subseteq V$, define $h_S^\sharp(A) = \bigvee\{h_S(a) : a \in A\}$ and $h_V^\sharp(X) = \bigvee\{h_V(x) : x \in X\}$. It is a routine matter to verify that $h_S^\sharp$ and $h_V^\sharp$ determine a completely idempotent semiring-semimodule pair morphism extending (h_S,h_V). We only prove that $h_S^\sharp$ and $h_V^\sharp$ preserve the infinite product. To this end, let $A_0, A_1, \dots$ be a sequence of subsets of S. Then

$$\begin{aligned}
h_V^\sharp(A_0A_1\cdots) &= h_V^\sharp(\{s_0s_1\cdots : s_i \in A_i\}) \\
&= \bigvee\{h_V(s_0s_1\cdots) : s_i \in A_i\} \\
&= \bigvee\{h_V(s_0)h_V(s_1)\cdots : s_i \in A_i\} \\
&= (\bigvee\{h(s_0) : s_0 \in A_0\})(\bigvee\{h(s_1) : s_1 \in A_1\})\cdots \\
&= h_S^\sharp(A_0)h_S^\sharp(A_1)\cdots
\end{aligned}$$

Since the definition of the extension is forced, the proof is complete. □

Thus in fact the operator P is the object part of a functor from the category of ω-monoids to the category of completely idempotent semiring-semimodule pairs which is the left adjoint of the obvious forgetful functor from the category of completely idempotent semiring-semimodule pairs to the category of ω-monoids.

Corollary 3.1. *For any pair of sets* (Σ,Δ)*, the completely idempotent semiring-semimodule pair* $P(\Sigma^*, \Sigma^*\Delta_\perp \cup \Sigma^\omega)$ *is freely generated by* (Σ,Δ).

Proof. Suppose that (S,V) is a completely idempotent semiring-semimodule pair and (h_S,h_V) is a pair of functions $h_S : \Sigma \to S$ and $h_V : \Delta \to V$, respectively. First extend (h_S,h_V) to an ω-monoid morphism $(\Sigma^*, \Sigma^*\Delta_\perp \cup \Sigma^\omega) \to (S,V)$ by Proposition 2.2, and then extend this morphism by Theorem 3.1 to a morphism of completely idempotent semiring-semimodule pairs $P(\Sigma^*, \Sigma^*\Delta_\perp \cup \Sigma^\omega) \to (S, V)$. Since the extensions were forced, the proof is complete. □

Remark 3.1. A *Wilke algebra*[5,7] (S, V) consists of a semigroup S, a set V, a left action $S \times V \to V$ subject to the equation (1), and a unary omega operation ${}^\omega : S \to V$ which satisfies the following identities:

$$(ss')^\omega = s(s's)^\omega$$
$$(s^n)^\omega = s^\omega$$

for all $s \in S$ and $n \geq 2$. It then follows that $ss^\omega = s^\omega$, for all $s \in S$, since $ss^\omega = s(ss)^\omega = (ss)^\omega = s^\omega$. Each ω-semigroup (S, V) determines a Wilke algebra: Define s^ω as the infinite product $ss\cdots$, for each $s \in S$. The nontrivial fact proved in [7] is that each finite Wilke algebra (S, V) in turn determines an ω-semigroup. For suppose that (S, V) is a finite Wilke algebra and $s_0, s_1, \ldots$ is an ω-sequence of elements of S. Then, by a Ramsey-type argument it follows that there is a sequence $i_1 < i_2 < \ldots$ such that each $s_{i_j} \cdots s_{i_{j+1}-1}$, $j = 1, 2, \ldots$ is a fixed idempotent e of S. Now generalized associativity (2) forces

$$s_0 s_1 \cdots = s_0 \cdots s_{i_1 - 1} e^\omega.$$

In fact, the category of finite ω-semigroups is isomorphic to the category of finite Wilke algebras.

Call a Wilke algebra (S, V) a *Wilke monoid* if S is a monoid and the action is unitary, i.e. when (4) holds. When (S, V) is a finite Wilke monoid it is also a finite ω-monoid, thus $(P(S), P(V))$ is a complete semiring-semimodule pair. Morphisms of Wilke monoids also preserve the multiplicative identity. It follows that the category of finite Wilke monoids is isomorphic to the category of finite ω-monoids.

4. Conway semiring-semimodule pairs

We start by recalling from[1] the definition of a Conway semiring-semimodule pair.

Suppose that S is a semiring. We say that S is a *Conway semiring* if S is equipped with a star operation ${}^* : S \to S$ subject to the *sum-star* and *product-star* identities defined below:

$$(a + b)^* = (a^*b)^*a^*$$
$$(ab)^* = 1 + a(ba)^*b$$

for all $a, b \in S$. It follows that $aa^* + 1 = a^* = a^*a + 1$ holds in all Conway semirings. Morphisms of Conway semirings also preserve the star operation. Suppose that (S, V) is a semiring-semimodule pair. We say that (S, V) is a Conway semiring-semimodule pair if S is a Conway semiring equipped with an omega operation $^\omega : S \to V$, subject to the following *sum-omega* and *product-omega* identities.

$$(a+b)^\omega = (a^*b)^\omega + (a^*b)^*a^\omega$$
$$(ab)^\omega = a(ba)^\omega,$$

for all $a, b \in S$. In particular, $aa^\omega = a^\omega$. Morphisms of Conway semiring-semimodule pairs preserve star and omega. We say that a Conway semiring-semimodule pair (S, V) is *idempotent* if S is idempotent, and we say that a Conway semiring-semimodule pair (S, V) is *ω-idempotent* if $1^* = 1$ holds. Note that any ω-idempotent Conway semiring-semimodule pair is idempotent since $1 + 1 = 11^* + 1 = 1^* = 1$. We call the operations of a Conway semiring-semimodule pair (including the constants 0 and 1) the *rational operations.*

Each completely idempotent semiring-semimodule pair (S, V) gives rise to an ω-idempotent Conway semiring-semimodule pair. Given (S, V), define

$$s^* = \bigvee_{n \geq 0} \sum_{i=0}^{n} s^i$$
$$s^\omega = ss\cdots$$

for all $s \in S$.

Proposition 4.1.[3] *Each completely idempotent semiring-semimodule pair is an ω-idempotent Conway semiring-semimodule pair. Any morphism of completely idempotent semiring-semimodule pairs is a morphism of ω-idempotent Conway semiring-semimodule pairs.*

Thus in particular, for any alphabets Σ, Δ, $(P(\Sigma^*), P(\Sigma^*\Delta_\perp \cup \Sigma^\omega))$ is a Conway semiring-semimodule pair.

Given a Conway semiring-semimodule pair (S, V) and an integer $n \geq 0$, consider the semiring $S^{n\times n}$ of all $n \times n$ matrices over S as well as the monoid V^n. Using the action of S on V, there is a natural action of $S^{n\times n}$ on V^n. Let $A = (A_{i,j}) \in S^{n\times n}$ and $v = (v_i) \in V^n$. We define

$$(Av)_i = \sum_{j=1}^{n} A_{ij} v_j$$

for all i, resulting in a semiring-semimodule pair $(S^{n\times n}, V^n)$. We turn $(S^{n\times n}, V^n)$ into a Conway semiring-semimodule pair. The definition uses induction on n. Consider a matrix $M \in S^{n\times n}$. When $n = 0$, M is the empty matrix and we define M^* to be the empty matrix and M^ω to be the empty vector. When $n = 1$, we have $M = (s)$ for some $s \in S$. We define $M^* = (s^*)$ and $M^\omega = (s^\omega)$. Suppose that $n > 1$ and let $m = n - 1$. Write

$$M = \begin{pmatrix} A & B \\ C & D \end{pmatrix}$$

where $A \in S^{m\times m}$, $B \in S^{m\times 1}$, $C \in S^{1\times m}$ and $D \in S^{1\times 1}$. We define

$$M^* = \begin{pmatrix} (A + BD^*C)^* & (A + BD^*C)^*BD^* \\ (D + CA^*B)^*CA^* & (D + CA^*B)^* \end{pmatrix} \tag{5}$$

Moreover, we define

$$M^\omega = \begin{pmatrix} (A + BD^*C)^\omega + (A + BD^*C)^*BD^\omega \\ (D + CA^*B)^\omega + (D + CA^*B)^*CA^\omega \end{pmatrix} \tag{6}$$

Theorem 4.1.[1] *When (S, V) is a Conway semiring-semimodule pair, so is $(S^{n\times n}, V^n)$, for each n. Moreover, (5) and (6) hold for all decompositions of the matrix M into four blocks such that A and D are square matrices of any dimension.*

When (S, V) is a Conway semiring-semimodule pair, we define

$$M^{\omega_k} = \begin{pmatrix} (A + BD^*C)^\omega \\ D^*C(A + BD^*C)^\omega \end{pmatrix}$$

for all $M = \begin{pmatrix} A & B \\ C & D \end{pmatrix}$ in $S^{n\times n}$, where $A \in S^{k\times k}$, $D \in S^{m\times m}$, etc, so that $k \le n$, $k + m = n$.

Remark 4.1. Suppose that (S, V) is a completely idempotent semiring-semimodule pair and consider a matrix M decomposed into four parts as above. Then for every $1 \le i \le n$ we have that

$$M_i^{\omega_k} = \bigvee\{M_{ii_1}M_{i_1i_2}M_{i_2i_3}\cdots : \exists^\infty j\ i_j \le k\}.$$

5. Automata

Suppose that (S, V) is an ω-idempotent Conway semiring-semimodule pair, $\Sigma \subseteq S$ and $V_0 \subseteq V$. We define a *(Büchi-)automaton in* (S, V) *over* (Σ, V_0) to be a system $\mathbb{A} = (\alpha, A, k)$ where $\alpha \in \{0, 1\}^{1\times n}$, $A \in S^{n\times n}$ whose entries are finite sums of elements in Σ, and k is an integer $\leq n$. Here, α is the *initial vector* and A is the *transition matrix.* Integer k specifies the "repeated nonterminals", see below. The *behavior* of $\mathbb{A}$ is

$$|\mathbb{A}| = \alpha A^{\omega_k}$$

We say that automata $\mathbb{A}$ and $\mathbb{A}'$ are *equivalent* if $|\mathbb{A}| = |\mathbb{A}'|$. It is not hard to see that for every automaton there is an equivalent automaton whose initial vector is a unit vector. We say that some $x \in V$ is *recognizable over* (Σ, V_0) if there is an automaton over (Σ, V_0) whose behavior is x. We let $\mathrm{Rec}_{(S,V)}(\Sigma, V_0)$ denote the set of all recognizable elements over (Σ, V_0). When $V_0 = \{0\}$, we simply write $\mathrm{Rec}_{(S,V)}(\Sigma)$.

We also define rational elements. Suppose that (S, V), Σ and V_0 are as before. We say that some $x \in V$ is *rational over* (Σ, V_0) in (S, V) if x can be generated from (S_0, V_0) by the rational operations, i.e., when x is included in the least Conway subsemiring-subsemimodule pair of (S, V) containing Σ and V_0. Notation: $\mathrm{Rat}_{(S,V)}(S_0, V_0)$, or just $\mathrm{Rat}_{(S,V)}(\Sigma)$ when $V_0 = \{0\}$.

As an variant of a result in [2] we can show:

Theorem 5.1. *For any ω-idempotent Conway semiring-semimodule pair (S, V) and Σ and V_0 as above,* $\mathrm{Rec}_{(S,V)}(\Sigma, V_0) = \mathrm{Rat}_{(S,V)}(\Sigma, V_0)$.

Thus, in particular, $\mathrm{Rec}_{(S,V)}(\Sigma) = \mathrm{Rat}_{(S,V)}(\Sigma)$.

6. Operational characterization of linear languages

Suppose that Σ is a finite or infinite alphabet. The monoid $(\Sigma^*)^{\mathrm{op}}$, equipped with reverse concatenation $(u, v) \mapsto vu$, for all $u, v \in \Sigma^*$, is isomorphic to the free monoid Σ^* of finite words over Σ (so that it is also free). Now consider the monoid $\Sigma^* \times (\Sigma^*)^{\mathrm{op}}$. We let $\Sigma^* \times (\Sigma^*)^{\mathrm{op}}$ act on Σ^∞ by

$$(u, v)x = uxv$$

for all $u, v \in \Sigma^*$ and $x \in \Sigma^\infty$. Since u and v are finite, it holds that $uxv \in \Sigma^\infty$. Then we define an infinite product operation $(\Sigma^* \times (\Sigma^*)^{\mathrm{op}})^\omega \to \Sigma^\infty$ by

$$(u_0, v_0)(u_1, v_1) \cdots = u_0 u_1 \ldots \ldots v_1 v_0$$

The following fact is clear.

Proposition 6.1. *Equipped with the above operations and action,* $(\Sigma^* \times (\Sigma^*)^{\mathrm{op}}, \Sigma^\infty)$ *is an ω-monoid.*

Corollary 6.1. $(S,V) = P((\Sigma^* \times (\Sigma^*)^{\mathrm{op}}, \Sigma^\infty) = (P(\Sigma^* \times (\Sigma^*)^{\mathrm{op}}), P(\Sigma^\infty))$ *is a completely idempotent semiring-semimodule pair and an ω-idempotent Conway semiring-semimodule pair.*

Thus all of the rational operations are defined on $(P(\Sigma^* \times (\Sigma^*)^{\mathrm{op}}), P(\Sigma^\infty))$. Our aim is to show the following result.

Theorem 6.1. *A language $X \subseteq \Sigma^\infty$ can be generated from the singleton subsets of $\Sigma^* \times (\Sigma^*)^{\mathrm{op}}$ by the rational operations if and only if it is linear.*

In our argument establishing Theorem 6.1 we will make use of Theorem 5.1.

Define $\mathcal{F}$ as the set of all singleton subsets of $\Sigma^* \times (\Sigma^*)^{\mathrm{op}}$. An automaton (α, M, k) over $\mathcal{F}$ is essentially the same thing as a linear grammar with Büchi acceptance condition. To see this, consider a linear grammar $G = (N, \Sigma, P, X_{i_0}, R)$ with $N = \{X_1, \ldots, X_n\}$ and $R = \{X_1, \ldots, X_k\}$. By Remark 1.1, without loss of generality we may assume that each production has a (single) nonterminal on the right side. Define an automaton $\mathbb{A}_G = (\alpha, M, k)$ where for each $1 \leq i, j \leq n$,

$$\alpha_i = \begin{cases} 1 \text{ if } i = i_0 \\ 0 \text{ otherwise} \end{cases}$$

$$M_{ij} = \{(u,v) : X_i \to uX_jv \in P\}$$

Example 6.1. Consider the grammar G_2 defined in Example 1.2 with $X_1 = Y$ and $X_2 = X$. Then the corresponding automaton is

$$\left((\emptyset, \{\epsilon\}), \begin{pmatrix} \{(\epsilon, b)\} & \emptyset \\ \{(\epsilon,\epsilon)\} & \{(a,\epsilon)\} \end{pmatrix}, 1 \right).$$

Proposition 6.2. *For each linear grammar G over Σ with Büchi acceptance condition having no production whose right side is in Σ^* it holds that $L^\infty(G) = |\mathbb{A}_G|$.*

Proof. Let us write the transition matrix M of $\mathbb{A}_G$ in the form

$$M = \begin{pmatrix} A & B \\ C & D \end{pmatrix}$$

where $A \in P(\Sigma^* \times (\Sigma^*)^{\mathrm{op}})^{k \times k}$, $B \in P(\Sigma^* \times (\Sigma^*)^{\mathrm{op}})^{k \times m}$, $C \in P(\Sigma^* \times (\Sigma^*)^{\mathrm{op}})^{m \times k}$ and $D \in P(\Sigma^* \times (\Sigma^*)^{\mathrm{op}})^{m \times m}$, $k + m = n$. Now for each $1 \leq i, j \leq n$, $(A + BD^*C)_{ij}$ is the set of all pairs of finite words (u, v) such that there is a nontrivial derivation tree rooted X_i whose frontier is uX_jv such that except for the indicated vertices, all nonterminals labeling a vertex of the tree belong to the set $N \setminus R$ of non-repeating nonterminals. Thus, for each $1 \leq i \leq k$, $(A + BD^*C)_i^\omega$ is the set of all words $(u_0, v_0)(u_1, v_1) \cdots = u_0 u_1 \ldots\ldots v_1 v_0$ that can be derived from X_i by a derivation tree which has an infinite number of vertices labeled in the set R. Moreover, for each $1 \leq j \leq m$, the jth component of $D^*C(A + BD^*C)^\omega$ is the set of all words $(u_0, v_0)(u_1, v_1) \cdots = u_0 u_1 \ldots\ldots v_1 v_0$ which have a derivation tree rooted X_{k+j} such that at least one nonterminal in R labels an infinite number of vertices of the tree. By the definition of the initial vector, it follows now that the component of the behavior of $\mathbb{A}_G$ which corresponds to the start symbol is exactly the language generated by the grammar G. □

Using Proposition 6.2, we can now complete the proof of Theorem 6.1. The correspondence $G \mapsto \mathbb{A}_G$ creates a bijection (up to a rearrangement of the nonterminals) between grammars and automata $\mathbb{A} = (\alpha, M, k)$ over $\mathcal{F}$. Thus, a language $L \subseteq \Sigma^\infty$ is linear if and only if it is the behavior of some automaton as above. But by Theorem 5.1, the behaviors of such automata are exactly those languages which can be constructed from $\mathcal{F}$ by the rational operations.

References

1. S.L. Bloom and Z. Ésik: *Iteration Theories*, Springer, 1993.
2. Z. Ésik and W. Kuich: A semiring-semimodule generalization of ω-regular languages, parts I. and II. *J. Autom. Lang. Comb.*, 10(2005), 203–242 and 243–264.
3. Z. Ésik and W. Kuich: On iteration semiring-semimodule pairs, *Semigroup Forum*, 75(2007), 129–159.
4. J.S. Golan: *The Theory of Semirings with Applications in Computer Science*, Longman Scientific and Technical, 1993.
5. D. Perrin and J.-É. Pin: *Infinite Words*, Pure and Applied Mathematics, Vol 141, Elsevier, 2004.
6. A. Salomaa: *Formal Languages.* ACM Monograph Series. Academic Press, 1973.
7. Th. Wilke: An algebraic theory for regular languages of finite and infinite words. *Internat. J. Algebra Comput.*, 3(1993), 447–489.

Received: October 19, 2009

Revised: May 14, 2010

EXTENDED TEMPORAL LOGICS ON FINITE TREES*

Z. ÉSIK and Sz. IVÁN
Department of Computer Science,
University of Szeged,
Hungary

Wolper associated a temporal logic to each class of (regular) languages. A different semantics for essentially the same logic was given by Ésik. Both approaches can be extended to trees resulting in families of branching time temporal logics with regular modalities. Here, we compare the two semantics of these branching time temporal logics and use this comparison to derive an algebraic characterization of their expressive power with respect to the Wolper-style semantics in terms of varieties of finite algebras. We also provide a game-theoretic characterization.

1. Introduction

Wolper [13] introduced a proper extension of Linear Temporal Logic by associating a modality to each (regular) language of words. His approach can be extended to (finite, ranked, ordered, variable-free) trees; the resulting family of logics subsumes e.g., the widely researched logic CTL[a] [1] and its modular extension.

A logic on finite trees with the same syntax but equipped with a different semantics was defined in [3]. The relation between the two logics was investigated in [7] for the case of unary trees (words), where also an algebraic characterization has been achieved using results from [4]. Here, we compare the two semantics for trees and use this comparison to derive an algebraic characterization of the expressive power with respect to the Wolper-style semantics in terms of varieties of finite algebras. To this end, we make use of results from [6].

---

*Research supported by grant no. K 75249 from the National Foundation of Hungary for Scientific Research and by the TÁMOP-4.2.2/08/1/2008-0008 program of the Hungarian National Development Agency.

[a]The original CTL is defined over unranked and unordered, typically infinite trees, see also Remark 6.1.

Besides algebraic characterizations (see e.g., [10]), a general tool for studying the expressive power of logics is provided by the Ehrenfeucht-Fraïssé games (see e.g., [12]). Here, we introduce a class of two-player games characterizing Wolper's logic. The game-theoretic characterization, in conjunction with some results of [5], provides an alternative proof of the relation between the two semantics.

2. Trees

When n is a nonnegative integer, $[n]$ denotes the set $\{1,\ldots,n\}$. Thus, $[0]$ is another notation for the empty set $\emptyset$. A *rank type* R is a finite subset of $\mathcal{N} = \{0,1,\ldots\}$ containing 0 and at least one positive integer. A *ranked alphabet* $\Sigma = \bigcup_{n\in R} \Sigma_n$ *(of rank type* R*)* is a disjoint union of finite nonempty sets of *n-ary symbols* Σ_n, $n \in R$.

For the whole paper we fix a rank type R and every alphabet is assumed to have rank type R.

Given a ranked alphabet Σ, the set T_Σ of Σ-terms (or (Σ-)*trees*) is the least set T satisfying the following condition: whenever $n \in R$, $\sigma \in \Sigma_n$ and $t_1,\ldots,t_n \in T$, then $\sigma(t_1,\ldots,t_n)$ is also in T. Since $0 \in R$, T_Σ is not empty. When $\sigma \in \Sigma_0$, we also write just σ to denote the term $\sigma()$.

Any tree $t \in T_\Sigma$ can be viewed as a mapping from a tree domain $\mathrm{dom}(t) \subseteq \mathcal{N}^*$ to Σ as follows: the domain of a tree t is

$$\mathrm{dom}(t) = \begin{cases} \{\epsilon\} & \text{if } t = \sigma \in \Sigma_0, \text{ where } \epsilon \\ & \text{stands for the empty word;} \\ \{\epsilon\} \cup \bigcup_{i\in[n]} \{i \cdot v : v \in \mathrm{dom}(t_i)\} & \text{if } t = \sigma(t_1,\ldots,t_n) \text{ for} \\ & \text{some } n > 0, \sigma \in \Sigma_n, t_i \in T_\Sigma, \end{cases}$$

and for any tree $t = \sigma(t_1,\ldots,t_n)$, the mapping from $\mathrm{dom}(t)$ to Σ, also denoted t is defined as

$$t(x) = \begin{cases} \sigma & \text{if } x = \epsilon; \\ t_i(v) & \text{if } x = i \cdot v \text{ for some } i \in [n] \text{ and } v \in \mathcal{N}^*. \end{cases}$$

An element of $\mathrm{dom}(t)$ is a *node* of t. A node x of $t \in T_\Sigma$ is called an *n-ary node* for some $n \in R$ if $t(x) \in \Sigma_n$. When $t \in T_\Sigma$ and $x \in \mathrm{dom}(t)$, $t|_x$ stands for the *subtree of t rooted at* x, i.e., the unique tree with $\mathrm{dom}(t|_x) = \{u : x \cdot u \in \mathrm{dom}(t)\}$ and $t|_x(u) = t(x \cdot u)$ for any $u \in \mathrm{dom}(t|_x)$.

For better readability, $\mathrm{Root}(t)$ stands for $t(\epsilon)$. When s is a Δ-tree and t is a Σ-tree for some ranked alphabets Δ and Σ, we say that s is a (Δ-)*relabeling* of t if $\mathrm{dom}(s) = \mathrm{dom}(t)$.

A *(Σ-)tree language* is any subset L of T_Σ.

When Σ and Δ are ranked alphabets, a relation $\varrho \subseteq \Sigma \times \Delta$ is called *rank-preserving* if $(\sigma, \delta) \in \varrho$, $\sigma \in \Sigma_n$ implies $\delta \in \Delta_n$ for any $n \in R$, $\sigma \in \Sigma$ and $\delta \in \Delta$. A rank-preserving relation $\varrho \subseteq \Sigma \times \Delta$ *induces a literal substitution*, also denoted ϱ, defined as the relation

$$\{(s,t) \in T_\Sigma \times T_\Delta : \mathrm{dom}(s) = \mathrm{dom}(t), \forall x \in \mathrm{dom}(s) : (s(x), t(x)) \in \varrho\}.$$

A rank-preserving function $h : \Sigma \to \Delta$ (i.e., a function mapping Σ_n into Δ_n for each n) induces a *literal homomorphism*, also denoted h, from T_Σ to T_Δ, where for each $s \in T_\Sigma$, $h(s)$ is the Δ-relabeling of s defined by $(h(s))(x) = h(s(x))$ for all $x \in \mathrm{dom}(s)$. Clearly any literal homomorphism induced by $h : \Sigma \to \Delta$ is a literal substitution induced by the relation $\{(\sigma, h(\sigma)) : \sigma \in \Sigma\}$.

When $\varrho \subseteq A \times B$ is a relation and X is a subset of A, then $\varrho(X)$ stands for the set $\{b \in B : (a,b) \in \varrho \text{ for some } a \in X\}$. When $\mathcal{L}$ is a class of tree languages, let $\mathbf{S}(\mathcal{L})$ ($\mathbf{H}(\mathcal{L})$, $\mathbf{H}^{-1}(\mathcal{L})$, respectively) denote the class of all tree languages of the form $f(L)$ where $L \in \mathcal{L}$ and f is a literal substitution (literal homomorphism, inverse literal homomorphism, respectively). It is clear that the inverse of a literal substitution is also a literal substitution, moreover, for any literal substitution S induced by some relation $\varrho \subseteq \Sigma \times \Delta$ there exist literal homomorphisms $h_1 : \varrho \to \Sigma$ and $h_2 : \varrho \to \Delta$ (here ϱ is viewed as a subset of $\Sigma \times \Delta$) such that $S = h_1^{-1} \circ h_2$, so that $S(L) = h_2(h_1^{-1}(L))$ for any Σ-tree language L. It is also clear that $\mathbf{H}$, $\mathbf{H}^{-1}$ and $\mathbf{S}$ are closure operators. Thus, $\mathbf{S}(\mathcal{L}) = \mathbf{H}(\mathbf{H}^{-1}(\mathcal{L}))$ holds for any class $\mathcal{L}$ of tree languages.

When Σ is a ranked alphabet, $\Sigma(\bullet)$ denotes the ranked alphabet resulting from Σ endowed with a new constant symbol $\bullet$. A Σ-*context* is a $\Sigma(\bullet)$-tree which has exactly one node labeled $\bullet$. When ζ is a Σ-context and t is a Σ-tree, then $\zeta(t)$, or simply ζt denotes the Σ-tree resulting from ζ by substituting t in place of the "hole symbol" $\bullet$. When L is a Σ-tree language and ζ is a Σ-context, the *quotient of* L *with respect to* ζ is the tree language $\zeta^{-1}(L) = \{t \in T_\Sigma : \zeta t \in L\}$. When $\mathcal{L}$ is a class of tree languages, let $\mathbf{Q}(\mathcal{L})$ stand for the class of all quotients of the members of $\mathcal{L}$.

When Σ is a ranked alphabet, $P(\Sigma)$ denotes the *power alphabet* of Σ, where $P(\Sigma)_n = \{(n, D) : D \subseteq \Sigma_n\}$[b]. We also define the rank-preserving relation

$$S_\Sigma = \{((n,D),\sigma) : n \in R, \sigma \in D \subseteq \Sigma_n\} \subseteq P(\Sigma) \times \Sigma.$$

[b]We include n to ensure disjointness of the sets $P(\Sigma)_n$, since otherwise $\emptyset$ would be ambiguous. Alternatively, we could define $P(\Sigma)_n$ as $P(\Sigma_n - \{\emptyset\}) \cup \{\emptyset_n\}$.

3. Temporal logics

Let Σ be a ranked alphabet. The set of FTL-formulae (over Σ) is the least set satisfying the following conditions:

(1) for any $\sigma \in \Sigma$, p_σ is an (atomic) formula (of depth 0);
(2) whenever φ and ψ are formulae (of maximal depth d), then $(\neg\varphi)$ and $(\varphi \vee \psi)$ are also formulae (of depth d);
(3) if $L \subseteq T_\Delta$ for some ranked alphabet Δ and for each $\delta \in \Delta$, φ_δ is a formula (of maximal depth d), then

$$L(\delta \mapsto \varphi_\delta)_{\delta\in\Delta} \tag{1}$$

is also a formula (of depth $d+1$).

We make use of the shorthands $(\varphi \wedge \psi) = (\neg(\neg\varphi \vee \neg\psi))$, $(\varphi \to \psi) = ((\neg\varphi) \vee \psi)$ as usual. We may also omit some parentheses following the usual precedence order of the connectives. Subformulae of a formula are defined as usual.

We define two different semantics. In both semantics, a Σ-tree t *satisfies* the atomic formula p_σ if and only if $\mathrm{Root}(t) = \sigma$. The Boolean connectives are handled as usual. Only the formulae of the form (1) are interpreted differently in the two semantics.

Semantics 1. For this semantics we assume that each ranked alphabet has a fixed lexicographic order. A tree $t \in T_\Sigma$ satisfies a formula $L(\delta \mapsto \varphi_\delta)_{\delta\in\Delta}$ with respect to Semantics 1 if and only if the *characteristic tree* $\widehat{t} \in T_\Delta$ *of* t *determined by the family* $(\varphi_\delta)_{\delta\in\Delta}$ belongs to L.

The tree $\widehat{t}$ is the (unique) Δ-relabeling of t defined as follows: for each $n \in R$ and n-ary node $x \in \mathrm{dom}(t)$, let $\widehat{t}(x) = \delta$ if and only if $\delta \in \Delta_n$ and one of the following conditions holds:

(1) either $t|_x$ satisfies φ_δ with respect to Semantics 1 and δ is the first such symbol of Δ_n;
(2) or $t|_x$ does not satisfy any $\varphi_{\delta'}$ with $\delta' \in \Delta_n$ with respect to Semantics 1 and δ is the last symbol of Δ_n.

Semantics 2. The Σ-tree t satisfies a formula $L(\delta \mapsto \varphi_\delta)_{\delta\in\Delta}$ with respect to Semantics 2 if and only if *there exists* a Δ-relabeling $\widehat{t} \in L$ of t such that for each node $x \in \mathrm{dom}(t)$, $t|_x$ satisfies $\varphi_{\widehat{t}(x)}$ with respect to Semantics 2.

We write $t \models_i \varphi$ if t satisfies φ with respect to Semantics i, $i \in \{1,2\}$. The *language defined by the formula* φ *with respect to Semantics* i, $i \in \{1,2\}$ is $L_{\varphi,i} = \{t \in T_\Sigma : t \models_i \varphi\}$.

Let FTL($\mathcal{L}$) consist of those formulae all of whose subformulae of the form (1) satisfy $L \in \mathcal{L}$. For $i = 1, 2$, let $\mathbf{FTL}_i(\mathcal{L})$ stand for the class of tree languages definable by some FTL($\mathcal{L}$)-formula with respect to Semantics i.

Remark 3.1. We call a formula φ over Σ *deterministic* if for every subformula of φ of the form (1), $n \in R$ and for every tree $t \in T_\Sigma$ with $\text{Root}(t) \in \Sigma_n$ there is *exactly one* symbol $\delta \in \Delta_n$ with $t \models_1 \varphi_\delta$.

It is easy to see that when φ is deterministic, then $L_{\varphi,1}$ is not affected by the respective ordering of the alphabets, moreover, for any class $\mathcal{L}$ of tree languages and formula φ (of some depth d) of FTL($\mathcal{L}$) there exists a deterministic formula ψ (of depth at most d) of FTL($\mathcal{L}$) with $L_{\varphi,1} = L_{\psi,1}$. Thus, $\mathbf{FTL}_1(\mathcal{L})$ does not depend on the chosen respective orderings of the alphabets.

Also observe that when φ is deterministic, then $L_{\varphi,1} = L_{\varphi,2}$. Thus, $\mathbf{FTL}_1(\mathcal{L}) \subseteq \mathbf{FTL}_2(\mathcal{L})$ for any class $\mathcal{L}$ of tree languages.

Example 3.1. Let $R = \{0, 2\}$, Σ be the alphabet with $\Sigma_0 = \{c\}$ and $\Sigma_2 = \{a, b\}$, and let $L \subseteq T_\Sigma$ be the tree language consisting of those Σ-trees having exactly one node labeled a.

Then an FTL($\{L\}$)-formula (still over Σ) is $\varphi = L(\sigma \mapsto \varphi_\sigma)_{\sigma \in \Sigma}$, where

$$\begin{aligned} \varphi_c &= p_c, \\ \varphi_a &= p_a, \\ \varphi_b &= p_a \vee p_b. \end{aligned}$$

Considering the Σ-trees $t_1 = b(a(c,c),\ a(c,c))$ and $t_2 = b(b(c,c),\ b(c,c))$, we get that

- $t_1 \models_2 \varphi$, since $b(a(c,c),\ b(c,c)) \in L$ is a relabeling of t_1 satisfying the conditions of Semantics 2;
- $t_2 \not\models_2 \varphi$, since the tree $b(b(c,c),\ b(c,c)) \notin L$ is the only relabeling t' of t_2 with $t_2|_x \models_2 \varphi_{t'(x)}$ for each node $x \in \text{dom}(t)$.

At the same time, the characteristic tree of t_2 is $b(b(c,c),\ b(c,c)) \notin L$ with respect to the family $(\varphi_\sigma)_{\sigma \in \Sigma}$, and the characteristic tree of t_1 is either t_1 (when $a < b$ in the chosen ordering of Σ) or t_2 (otherwise). Thus, neither t_1 nor t_2 satisfies φ with respect to Semantics 1, hence the two semantics are indeed different.

Moreover, it is easy to see (via a straightforward induction on the depth of the formula) that no FTL($\{L\}$)-formula ψ exists for which $t_1 \models_1 \psi$ and $t_2 \not\models_1 \psi$ both hold, i.e., these two trees are indistinguishable, when one considers FTL($\{L\}$)-formulae with respect to Semantics 1. Hence, $\mathbf{FTL}_1(\{L\})$

is a proper subclass of $\mathbf{FTL}_2(\{L\})$ for the singleton class $\{L\}$ of tree languages.

4. The correspondence

In this section we relate the two semantics defined in the previous section, yielding an algebraic characterization of the classes $\mathbf{FTL}_2(\mathcal{L})$, at least when $\mathcal{L}$ satisfies a natural condition defined later. Let Σ denote a ranked alphabet.

Lemma 4.1. *For any class $\mathcal{L}$ of tree languages, $\mathbf{FTL}_2(\mathcal{L}) \subseteq \mathbf{FTL}_1(\mathbf{S}(\mathcal{L}))$.*

Proof. Let φ be a formula of FTL($\mathcal{L}$). We construct a formula φ' of FTL($\mathbf{S}(\mathcal{L})$) with $L_{\varphi,2} = L_{\varphi',1}$, by induction of the structure of φ.

When φ is an atomic formula, let $\varphi' = \varphi$.

When $\varphi = (\neg\varphi_1)$ or $\varphi = (\varphi_1 \vee \varphi_2)$, let $\varphi' = \neg(\varphi_1')$ or $\varphi' = (\varphi_1' \vee \varphi_2')$, respectively.

Assume that $\varphi = L(\delta \mapsto \varphi_\delta)_{\delta\in\Delta}$. We define a family of formulae indexed by $P(\Delta)$ as follows: for any $n \in R$ and $(n, D) \in P(\Delta)_n$, let $\varphi_{(n,D)}$ stand for the formula

$$\bigwedge_{\delta\in D} \varphi'_\delta \wedge \bigwedge_{\delta\in\Delta_n - D} \neg\varphi'_\delta.$$

Note that the family $(\varphi_{(n,D)})_{(n,D)\in P(\Delta)}$ is deterministic. We will show that for all trees $t \in T_\Sigma$,

$$t \models_2 L(\delta \mapsto \varphi_\delta)_{\delta\in\Delta} \;\Leftrightarrow\; t \models_1 (S_\Delta^{-1}(L))((n, D) \mapsto \varphi_{(n,D)})_{(n,D)\in P(\Delta)}.$$

Indeed,

$$\begin{aligned}
t \models_2 L(\delta \mapsto \varphi_\delta)_{\delta\in\Delta} &\Leftrightarrow \widehat{t} \in L \text{ for some } \Delta\text{-relabeling } \widehat{t} \text{ of } t \\
&\qquad \text{with } t|_x \models_2 \varphi_{\widehat{t}(x)} \text{ for each } x \in \operatorname{dom}(t) \\
&\Leftrightarrow \widehat{t} \in L \text{ for some } \Delta\text{-relabeling } \widehat{t} \text{ of } t \\
&\qquad \text{with } t|_x \models_1 \varphi'_{\widehat{t}(x)} \text{ for each } x \in \operatorname{dom}(t) \\
&\Leftrightarrow S_\Delta(\widehat{t'}) \cap L \neq \emptyset, \text{ where } \widehat{t'} \text{ is the } P(\Delta)\text{-relabeling} \\
&\qquad \text{of } t \text{ defined by } \widehat{t'}(x) = (n, \{\delta \in \Delta_n : t|_x \models_1 \varphi'_\delta\}) \\
&\qquad \text{for each } n\text{-ary node } x \in \operatorname{dom}(t) \\
&\Leftrightarrow t \models_1 (S_\Delta^{-1}(L))((n, D) \mapsto \varphi_{(n,D)})_{(n,D)\in P(\Delta)}.
\end{aligned}$$

Thus, we can define φ' as $(S_\Delta^{-1}(L))((n, D) \mapsto \varphi_{(n,D)})_{(n,D)\in P(\Delta)}$. □

The following facts are proved in [2]:

Lemma 4.2. $\mathbf{FTL}_1$ *is a closure operator on classes of tree languages preserving regularity, i.e., for any classes $\mathcal{L}_1, \mathcal{L}_2$ of tree languages, the following hold:*

(1) $\mathcal{L}_1 \subseteq \mathbf{FTL}_1(\mathcal{L}_1)$;
(2) $\mathbf{FTL}_1(\mathbf{FTL}_1(\mathcal{L}_1)) \subseteq \mathbf{FTL}_1(\mathcal{L}_1)$;
(3) *if additionally* $\mathcal{L}_1 \subseteq \mathcal{L}_2$, *then* $\mathbf{FTL}_1(\mathcal{L}_1) \subseteq \mathbf{FTL}_1(\mathcal{L}_2)$;

moreover, if $\mathcal{L}$ is a class of regular tree languages, then so is $\mathbf{FTL}_1(\mathcal{L})$.

It also holds that $\mathbf{FTL}_1(\mathcal{L})$ is closed under inverse literal homomorphisms and the Boolean operations, and is closed under quotients if and only if $\mathbf{Q}(\mathcal{L}) \subseteq \mathbf{FTL}_1(\mathcal{L})$.

Making use of the above lemma, we can prove the reverse of the containment relation in Lemma 4.1:

Lemma 4.3. *For any class $\mathcal{L}$ of tree languages,* $\mathbf{FTL}_1(\mathbf{S}(\mathcal{L})) \subseteq \mathbf{FTL}_2(\mathcal{L})$.

Proof. Let φ be an FTL($\mathbf{S}(\mathcal{L})$)-formula over Σ.

We construct an FTL($\mathcal{L}$)-formula φ' with $L_{\varphi,1} = L_{\varphi',2}$, by induction on the structure of the formula.

The only nontrivial case is when $\varphi = L(\delta \mapsto \varphi_\delta)_{\delta\in\Delta}$ for some Δ-tree language $L \in \mathbf{S}(\mathcal{L})$. Let us write in more detail $L = S(K)$, for a Γ-tree language $K \in \mathcal{L}$, where S is the literal substitution induced by the rank-preserving relation $\varrho \subseteq \Gamma \times \Delta$. Without loss of generality we may also assume that φ is deterministic.

For each $n \in R$ and $\gamma \in \Gamma_n$, let ψ_γ stand for the formula

$$\bigvee_{\delta\in\Delta_n:(\gamma,\delta)\in\varrho} \varphi'_\delta.$$

Then for any Σ-tree t,

$$\begin{aligned} t \models_1 \varphi &\Leftrightarrow \widehat{t} \in S(K) \text{ for the unique } \Delta\text{-relabeling } \widehat{t} \text{ of } t \\ &\qquad \text{where } t|_x \models_1 \varphi_{\widehat{t}(x)} \text{ for all } x \in \mathrm{dom}(t) \\ &\Leftrightarrow \widehat{t} \in S(K) \text{ for the unique } \Delta\text{-relabeling } \widehat{t} \text{ of } t \\ &\qquad \text{where } t|_x \models_2 \varphi'_{\widehat{t}(x)} \text{ for all } x \in \mathrm{dom}(t) \\ &\Leftrightarrow \widehat{t'} \in K \text{ for some } \Gamma\text{-relabeling } \widehat{t'} \text{ of } t \\ &\qquad \text{where } t|_x \models_2 \psi_{\widehat{t'}(x)} \text{ for all } x \in \mathrm{dom}(t) \\ &\Leftrightarrow t \models_2 K(\gamma \mapsto \psi_\gamma)_{\gamma\in\Gamma}. \end{aligned}$$

Thus, we can define φ' as $K(\gamma \mapsto \psi_\gamma)_{\gamma\in\Gamma}$. □

Corollary 4.1. *For any class $\mathcal{L}$ of tree languages,*

$$\mathbf{FTL}_2(\mathcal{L}) = \mathbf{FTL}_1(\mathbf{S}(\mathcal{L})).$$

Thus, for each $\mathcal{L}$, $\mathcal{L} \subseteq \mathbf{FTL}_1(\mathcal{L}) \subseteq \mathbf{FTL}_2(\mathcal{L})$.

5. Algebraic characterization

Let Σ be a ranked alphabet. A *(Σ-)tree automaton* $\mathbb{A} = (A, \Sigma)$ consists of a nonempty *state set* A, and an associated *elementary operation* $\sigma^{\mathbb{A}} : A^n \to A$ for each $\sigma \in \Sigma_n$, $n \in R$. When $\mathbb{A} = (A, \Sigma)$ is a tree automaton, each tree $t \in T_\Sigma$ *evaluates* to an element $t^{\mathbb{A}} \in A$ as usual. *Homomorphic images, subautomata* are defined as usual.

For any set $A' \subseteq A$, the tree language $L(\mathbb{A}, A')$ is defined as $\{t \in T_\Sigma : t^{\mathbb{A}} \in A'\}$. A tree language $L \subseteq T_\Sigma$ is *recognized by* the tree automaton $\mathbb{A} = (A, \Sigma)$ if $L = L(\mathbb{A}, A')$ for some set $A' \subseteq A$ of final states.

A tree language is called *regular* if it is recognized by some *finite* tree automaton, i.e., a tree automaton having a finite state set. It is well-known that for any Σ-tree language L there exists a *minimal tree automaton* $\mathbb{A}_L$, unique up to isomorphism, such that whenever L is recognized by a tree automaton $\mathbb{B}$, then $\mathbb{A}_L$ is a homomorphic image of a subautomaton of $\mathbb{B}$ (i.e., $\mathbb{A}_L$ *divides* $\mathbb{B}$.) Thus, a tree language is regular if and only if its minimal tree automaton is finite. For more information on tree automata, the reader is referred to [11].

A tree automaton $\mathbb{B} = (B, \Delta)$ is a *renaming* of $\mathbb{A} = (A, \Sigma)$ if $A = B$ and each elementary operation of $\mathbb{B}$ is also an elementary operation of $\mathbb{A}$.

When $\mathbb{A} = (A, \Sigma)$ is a tree automaton, we define its *power automaton* $P(\mathbb{A})$ as $(P(A), P(\Sigma))$ equipped with the following elementary operations:

$$(n, D)^{P(\mathbb{A})}(A_1, \ldots, A_n) = \{\sigma^{\mathbb{A}}(a_1, \ldots, a_n) : \sigma \in D, a_i \in A_i, i \in [n]\}$$

for each $n \in R$, $D \subseteq \Sigma_n$ and $A_1, \ldots, A_n \subseteq A$.

When $\mathbb{A} = (A, \Sigma)$ and $\mathbb{B} = (B, \Delta)$ are tree automata and $\alpha : A \times \Sigma \to \Delta$ is a rank-preserving function, i.e., $\alpha(a, \sigma) \in \Delta_n$ for any $n \in R$ and $\sigma \in \Sigma_n$, then the *Moore product [6] of $\mathbb{A}$ and $\mathbb{B}$ determined by α* is the tree automaton $\mathbb{A} \times_\alpha \mathbb{B} = (A \times B, \Sigma)$ equipped with the elementary operations

$$\sigma^{\mathbb{A} \times_\alpha \mathbb{B}}((a_1, b_1), \ldots, (a_n, b_n)) = (\sigma^{\mathbb{A}}(a_1, \ldots, a_n), \delta^{\mathbb{B}}(b_1, \ldots, b_n)),$$

for each $n \in R$, $\sigma \in \Sigma_n$ and $(a_i, b_i) \in A \times B$, $i \in [n]$, where

$$\delta = \alpha(\sigma^{\mathbb{A}}(a_1, \ldots, a_n), \sigma).$$

We also define the following tree automaton $\mathbb{D}_0 = (\{0, 1\}, \text{Bool})$ over the ranked alphabet Bool, where for each $n \in R$, $\text{Bool}_n = \{\uparrow_n, \downarrow_n\}$ and

$\uparrow_n^{\mathbb{D}_0}$ is the constant function $\{0,1\}^n \to \{0,1\}$ with value 1, and $\downarrow_n^{\mathbb{D}_0}$ is the constant function $\{0,1\}^n \to \{0,1\}$ with value 0.

The following result is proved in [6]:

Theorem 5.1. *Let $\mathcal{L}$ be a class of regular tree languages with $\mathbf{Q}(\mathcal{L}) \subseteq \mathbf{FTL}_1(\mathcal{L})$. Then the following are equivalent for any tree language K:*

(1) *$K \in \mathbf{FTL}_1(\mathcal{L})$;*
(2) *$\mathbb{A}_K$ is contained in the least class of finite tree automata which contains $\mathbb{D}_0$ as well as the minimal tree automata of each member of $\mathcal{L}$, which is closed under taking renamings, Moore products, and divisors.*

Using this theorem and Corollary 4.1, our aim is now to give a related characterization of $\mathbf{FTL}_2(\mathcal{L})$.

Lemma 5.1. *Suppose the tree language L is recognizable by the tree automaton $\mathbb{A} = (A, \Sigma)$. Then any image of L under a literal substitution is recognizable by a renaming of $P(\mathbb{A})$.*

Proof. Let us assume in more detail that L is recognizable by $\mathbb{A}$ with the set $F \subseteq A$ of final states, and assume that $K = S(L)$, where K is a Δ-tree language and S is a literal substitution induced by the rank-preserving relation $\varrho \subseteq \Sigma \times \Delta$.

Let us define the Δ-renaming $\mathbb{B}$ of $P(\mathbb{A})$ determined by

$$\delta^{\mathbb{B}} = (n, \{\sigma \in \Sigma_n : (\sigma, \delta) \in \varrho\})^{P(\mathbb{A})},$$

for each $n \in R$ and $\delta \in \Delta_n$.

It is clear that for any Δ-tree t,

$$t^{\mathbb{B}} = \{\hat{t}^{\mathbb{A}} : \hat{t} \in S_\Delta^{-1}(t)\}.$$

Thus, K is recognized by $\mathbb{B}$ with the set $\{A' \subseteq A : A' \cap F \neq \emptyset\}$ of final states. □

Lemma 5.2. *Let Σ be a ranked alphabet and suppose that $K \subseteq T_{P(\Sigma)}$ is a tree language recognizable by $P(\mathbb{A})$ for a finite tree automaton $\mathbb{A} = (A, \Sigma)$.*

Then K is a Boolean combination of images under literal substitutions of languages recognizable by $\mathbb{A}$.

Proof. Let us consider the literal substitution S_Σ. For any $P(\Sigma)$-tree t and state $a \in A$,

$$\begin{aligned} a \in t^{P(\mathbb{A})} &\Leftrightarrow a = s^{\mathbb{A}} \text{ for some } s \in S_\Sigma(t) \\ &\Leftrightarrow t \in S_\Sigma^{-1}(L_a), \end{aligned}$$

where L_a is a shorthand for $L(\mathbb{A}, \{a\})$.

Thus, for any state $a \in A$, the language recognized by $P(\mathbb{A})$ with the set $A_{\exists a} = \{A' \subseteq A : a \in A'\}$ of final states is an image of a language recognizable by $\mathbb{A}$, under the literal substitution S_Σ^{-1}.

Since for any $A' \subseteq A$,

$$L(P(\mathbb{A}), \{A'\}) = \bigcap_{a \in A'} L(P(\mathbb{A}), A_{\exists a}) - \bigcup_{a \notin A'} L(P(\mathbb{A}), A_{\exists a}),$$

and obviously

$$L(P(\mathbb{A}), \{A_1, \ldots, A_n\}) = \bigcup_{i \in [n]} L(P(\mathbb{A}), \{A_i\}),$$

the lemma is proved. □

From Corollary 4.1, Theorem 5.1 and Lemmas 5.1, 5.2 we immediately get the following characterization:

Theorem 5.2. *Suppose that $\mathcal{L}$ is a class of regular tree languages with $\mathbf{Q}(\mathbf{S}(\mathcal{L})) \subseteq \mathbf{FTL}_2(\mathcal{L})$. Then the following are equivalent for any tree language K:*

(1) $K \in \mathbf{FTL}_2(\mathcal{L})$;
(2) $\mathbb{A}_K$ *is contained in the least class of finite tree automata which contains $\mathbb{D}_0$, the tree automaton $P(\mathbb{A}_L)$ for each $L \in \mathcal{L}$, which is closed under renamings, divisors and Moore products.*

6. Ehrenfeucht-Fraïssé type games

In this section we provide a game-theoretic characterization of FTL with respect to Semantics 2. For a game-theoretic treatment of Semantics 1 see [5]. For the purposes of this section, when $\underline{t} = (t_1, \ldots, t_n)$ is a tuple of trees, we define $\mathrm{dom}(\underline{t})$ as $\bigcup_{i \in [n]} \{i \cdot u : u \in \mathrm{dom}(t_i)\}$, and for any node $i \cdot u \in \mathrm{dom}(\underline{t})$, we define $\underline{t}(i \cdot u)$ as $t_i(u)$ and $\underline{t}|_{i \cdot u}$ as $t_i|_u$.

We say that a formula φ over Σ *separates* two Σ-trees s and t, with respect to Semantics 2, if *exactly one* of the trees satisfies φ with respect to Semantics 2.

Let $\mathcal{L}$ be a class of tree languages, $r \geq 0$ be an integer, and t_0, t_1 be two trees over a ranked alphabet Σ. The *r-round $\mathcal{L}$-Wolper game* on the pair (t_0, t_1) of trees is played between two competing players, Spoiler and Duplicator, according to the following rules:

(1) If $\text{Root}(t_0) \neq \text{Root}(t_1)$, then Spoiler wins. Otherwise, Step 2 follows.
(2) If $r = 0$, then Duplicator wins. Otherwise, Step 3 follows.
(3) Spoiler picks a tree language $L \in \mathcal{L}$ over some ranked alphabet Δ, and two $P(\Delta)$-relabelings, $\widehat{t_0}$ and $\widehat{t_1}$ of t_0 and t_1, respectively, such that exactly one of them is contained in $S_\Delta^{-1}(L)$. If he cannot do so, Duplicator wins; otherwise, Step 4 follows.
(4) Duplicator picks two nodes $x, y \in \text{dom}((t_0, t_1))$ with $(\widehat{t_0}, \widehat{t_1})(x) \neq (\widehat{t_0}, \widehat{t_1})(y)$. If he cannot do so, Spoiler wins; otherwise, an $(r-1)$-round $\mathcal{L}$-Wolper game is played on the pair $((t_0, t_1)|_x, (t_0, t_1)|_y)$ of trees. The player winning the subgame wins the game.

Clearly, for any fixed $\mathcal{L}$, r, t_0 and t_1, exactly one of the players has a winning strategy; let $t_0 \sim_{\mathcal{L}}^r t_1$ denote that Duplicator has a winning strategy in the r-round $\mathcal{L}$-Wolper game on (t_0, t_1). Also, let $t_0 \equiv_{\mathcal{L}}^r t_1$ denote that the trees t_0 and t_1 satisfy the same set of FTL($\mathcal{L}$)-formulae having depth at most r, with respect to Semantics 2.

Example 6.1. Let $R = \{0, 2\}$ and Δ the ranked alphabet with $\Delta_2 = \{\vee, \wedge\}$ and $\Delta_0 = \{\uparrow, \downarrow\}$, and let $L \subseteq T_\Delta$ be the language consisting of exactly those Δ-trees evaluating to 1 according to the standard interpretation of the logical connectives.

Let Σ be the ranked alphabet with $\Sigma_2 = \{f, g\}$ and $\Sigma_0 = \{c, d\}$, t_1 be the Σ-tree $g(f(c, d),\ g(c, d))$ and t_2 be the Σ-tree $g(f(c, c),\ g(d, d))$. (See Figure 1.)

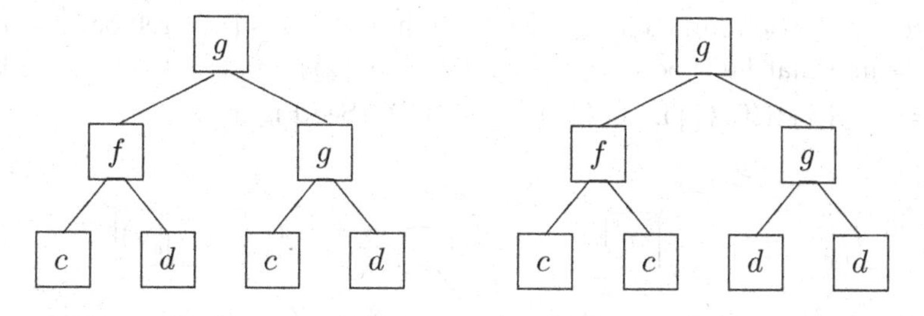

Fig. 1. The trees t_1 and t_2.

A possible 2-round $\{L\}$-Wolper game on (t_1, t_2) is played as follows. Since $\text{Root}(t_1) = \text{Root}(t_2)$ $(= g)$, and there are 2 rounds left, Spoiler has to choose a tree language $K \in \{L\}$. Having no other option, he chooses the Δ-tree language L. Moreover, Spoiler also has to choose two $P(\Delta)$-relabelings, $\widehat{t_1}$ and $\widehat{t_2}$ of t_1 and t_2, respectively, such that exactly one of them is contained in $S_\Delta^{-1}(L)$.

Let Spoiler choose the relabelings

$$\widehat{t_1} = (2,\{\vee,\wedge\})\big((2,\{\vee\})((0,\{\downarrow\}),(0,\{\uparrow\})),\ (2,\{\vee,\wedge\})((0,\{\downarrow\}),(0,\{\uparrow\}))\big),$$
$$\widehat{t_2} = (2,\{\vee,\wedge\})\big((2,\emptyset)((0,\{\downarrow\}),(0,\{\downarrow\})),\ (2,\{\vee\})((0,\{\uparrow\}),(0,\{\uparrow\}))\big).$$

(See Figure 2. For better readability, the arities are omitted from the labels.)

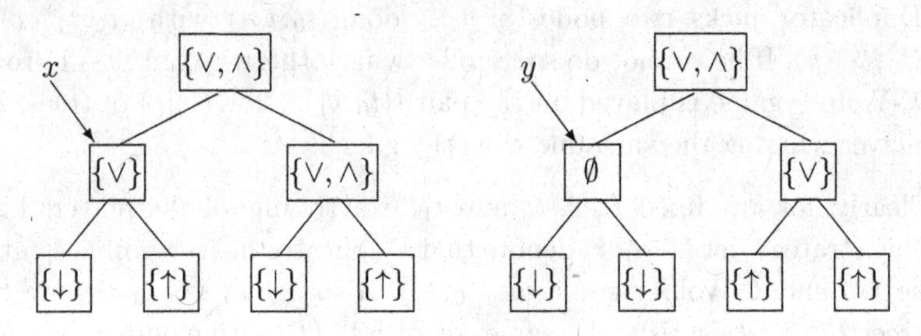

Fig. 2. The relabelings $\widehat{t_1}$ and $\widehat{t_2}$.

Then, $\widehat{t_1} \in S_\Delta^{-1}(L)$, since $\widehat{t_1} \in S_\Delta^{-1}(t)$ for $t = \vee(\vee(\downarrow,\uparrow),\ \vee(\downarrow,\uparrow)) \in L$. On the other hand, $\widehat{t_2} \notin S_\Delta^{-1}(L)$, thus this is a valid move of Spoiler.

Now Duplicator has to choose two nodes, x and y, having different labels in the relabelings. Assume he chooses the nodes x and y indicated on Figure 2.

Then, Spoiler and Duplicator play a 1-round $\{L\}$-Wolper game on the trees $t_1' = f(c,d)$ and $t_2' = f(c,c)$. Since $\mathrm{Root}(t_1') = \mathrm{Root}(t_2')$ $(= f)$, and there is still one round left, Spoiler again has to choose two relabelings. Let us assume that he chooses $\widehat{t_1'} = (2,\{\vee\})\big((0,\{\downarrow\}),\ (0,\{\uparrow\})\big) \in S_\Delta^{-1}(L)$ and $\widehat{t_2'} = (2,\{\vee\})\big((0,\{\downarrow\}),\ (0,\{\downarrow\})\big) \notin S_\Delta^{-1}(L)$. (See Figure 3.)

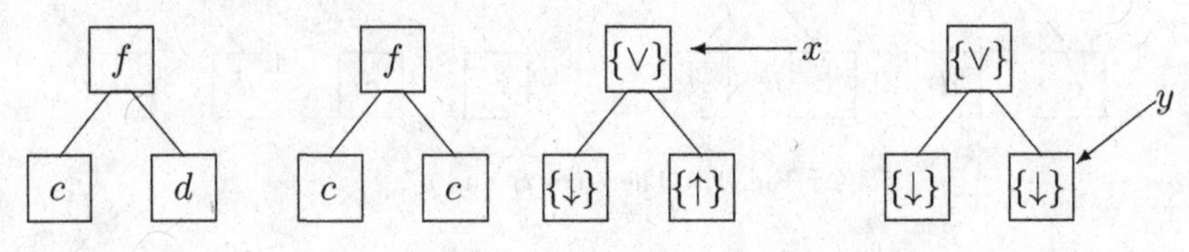

Fig. 3. Finishing the game: the trees $t_1', t_2', \widehat{t_1'}$ and $\widehat{t_2'}$.

Now Duplicator has to choose two nodes of (t_1', t_2') having different labels in $(\widehat{t_1'}, \widehat{t_2'})$. Suppose he chooses the nodes x and y indicated on Figure 3. Then, a zero-round $\{L\}$-Wolper game begins on the trees $f(c,d)$ and c.

Since $f = \mathrm{Root}(f(c,d)) \neq \mathrm{Root}(c) = c$, Spoiler wins this subgame and also the whole game.

Actually, Spoiler has a winning strategy in the 2-round $\{L\}$-Wolper game played on (t_1, t_2) but not in the 1-round game, i.e., $t_1 \sim^1_{\{L\}} t_2$ and $t_1 \not\sim^2_{\{L\}} t_2$. Similarly, one can check that $t_1 \equiv^1_{\{L\}} t_2$ and $t_1 \not\equiv^2_{\{L\}} t_2$. As Corollary 6.1 states, this is not a coincidence.

Lemma 6.1. *For any class $\mathcal{L}$ of tree languages, $r \geq 0$, ranked alphabet Σ and trees $t_0, t_1 \in T_\Sigma$, if $t_0 \sim^r_{\mathcal{L}} t_1$ then $t_0 \equiv^r_{\mathcal{L}} t_1$.*

Proof. We prove the lemma by induction on r. When $r = 0$, the claim is obvious: Duplicator wins in 0 rounds if and only if $\mathrm{Root}(t_0) = \mathrm{Root}(t_1)$.

Suppose that $r > 0$ and that the claim holds for any integer less than r. Assume that $t_0 \not\equiv^r_{\mathcal{L}} t_1$. We have two cases: either $t_0 \not\equiv^{r-1}_{\mathcal{L}} t_1$ or some $\mathrm{FTL}(\mathcal{L})$-formula $L(\delta \mapsto \varphi_\delta)_{\delta \in \Delta}$ of depth r separates t_0 and t_1 (with respect to Semantics 2).

If $t_0 \not\equiv^{r-1}_{\mathcal{L}} t_1$, we can apply the induction hypothesis and conclude that Spoiler already wins the $(r-1)$-round $\mathcal{L}$-Wolper game on (t_0, t_1), thus also wins the r-round game.

Suppose that $t_0 \equiv^{r-1}_{\mathcal{L}} t_1$ and that the formula $\varphi = L(\delta \mapsto \varphi_\delta)_{\delta \in \Delta}$ of depth r separates t_0 and t_1, say, $t_0 \models_2 \varphi$ and $t_1 \not\models_2 \varphi$. We design a winning strategy for Spoiler as follows: let Spoiler choose the tree language L and the $P(\Delta)$-relabeling $\widehat{t_i}$, $i = 0, 1$ defined by $\widehat{t_i}(x) = (n, \{\delta \in \Delta_n : t_i|_x \models_2 \varphi_\delta\})$ for each n-ary node $x \in \mathrm{dom}(t_i)$, $n \in R$.

Since $t_0 \models_2 \varphi$ and $t_1 \not\models_2 \varphi$, by Semantics 2 we get that $\widehat{t_0} \in S^{-1}_\Delta(L)$, while $\widehat{t_1} \notin S^{-1}_\Delta(L)$.

Now if Duplicator cannot choose two nodes having different labels according to the relabelings, Spoiler wins the game. Otherwise, suppose Duplicator picks the node x of t_i and the node y of t_j, $i, j \in \{0, 1\}$ with $(n, D_x) = \widehat{t_i}(x) \neq \widehat{t_j}(y) = (m, D_y)$. We have two subcases:

(1) If $n \neq m$, then the arities, thus the labels of the two nodes are different, hence Spoiler wins the subgame in the next step.
(2) Otherwise $D_x \neq D_y$ and there exists a symbol $\delta \in \Delta_n$ contained in exactly one of the sets D_x and D_y. By the construction of the relabelings, the $\mathrm{FTL}(\mathcal{L})$-formula φ_δ of depth at most $r-1$ separates $t_i|_x$ and $t_j|_y$ (with respect to Semantics 2). Applying the induction hypothesis we get that Spoiler wins the subgame, thus also wins the game. □

Lemma 6.2. *For any class $\mathcal{L}$ of tree languages, $r \geq 0$ and trees $t_0, t_1 \in T_\Sigma$ for some ranked alphabet Σ, if $t_0 \equiv^r_{\mathcal{L}} t_1$, then $t_0 \sim^r_{\mathcal{L}} t_1$.*

Proof. We argue by induction on r. When $r = 0$, Duplicator wins if and only if $\mathrm{Root}(t_0) = \mathrm{Root}(t_1)$.

Suppose that $r > 0$ and that the claim holds for any integer less than r. Assume that $t_0 \not\sim^r_{\mathcal{L}} t_1$.

If Spoiler has a winning strategy in the $(r-1)$-round $\mathcal{L}$-Wolper game on (t_0, t_1), then by the induction hypothesis $t_0 \not\equiv^{r-1}_{\mathcal{L}} t_1$, which obviously implies $t_0 \not\equiv^r_{\mathcal{L}} t_1$.

Suppose Spoiler has a winning strategy in the r-round $\mathcal{L}$-Wolper game on (t_0, t_1) but not in the $(r-1)$-round one. Let us fix one of Spoiler's winning strategies and assume that Spoiler chooses the tree language $L \in \mathcal{L}$ over Δ and the $P(\Delta)$-relabelings $\widehat{t_0}, \widehat{t_1}$ of t_0 and t_1 when playing this winning strategy. For ease of notation, let $\underline{t}$ stand for the pair (t_0, t_1) and $\widehat{\underline{t}}$ for $(\widehat{t_0}, \widehat{t_1})$.

Then for any pair of nodes $x, y \in \mathrm{dom}(\underline{t})$ with $\widehat{\underline{t}}(x) \neq \widehat{\underline{t}}(y)$, Spoiler has a winning strategy in the $(r-1)$-round $\mathcal{L}$-Wolper game on $(\underline{t}|_x, \underline{t}|_y)$. Applying the induction hypothesis, for any such pair x, y of nodes there exists an $\mathrm{FTL}(\mathcal{L})$-formula $\varphi_{x,y}$ of depth at most $r-1$ separating $\underline{t}|_x$ and $\underline{t}|_y$, say, $\underline{t}|_x \models_2 \varphi_{x,y}$ and $\underline{t}|_y \not\models_2 \varphi_{x,y}$.

For each $n \in R$, $\delta \in \Delta_n$, let φ_δ stand for

$$\bigvee_{\substack{x \in \mathrm{dom}(\underline{t}): \\ \delta \in S_\Delta(\widehat{\underline{t}}(x))}} \bigwedge_{\substack{y \in \mathrm{dom}(\underline{t}): \\ \delta \notin S_\Delta(\widehat{\underline{t}}(y))}} \varphi_{x,y}. \tag{2}$$

Note that for any possible pair x, y, the formula $\varphi_{x,y}$ is well-defined, thus so is φ_δ.

Let $z \in \mathrm{dom}(\underline{t})$ be a node of the pair (t_0, t_1) such that $\widehat{\underline{t}}(z) = (n, D)$ for some $n \in R$ and $D \subseteq \Delta_n$.

Assume that $\delta \in D$. We claim that $\underline{t}|_z \models_2 \varphi_\delta$. Indeed, from the definition of the formulae $\varphi_{z,y}$ we have $\underline{t}|_z \models_2 \varphi_{z,y}$ for any node $y \in \mathrm{dom}(\underline{t})$ with $\widehat{\underline{t}}(y) \neq \widehat{\underline{t}}(z)$, and thus for any $y \in \mathrm{dom}(\underline{t})$ with $\delta \notin S_\Delta(\widehat{\underline{t}}(y))$. Thus, $\underline{t}|_z \models_2 \bigwedge_{\substack{y \in \mathrm{dom}(\underline{t}): \\ \delta \notin S_\Delta(\widehat{\underline{t}}(y))}} \varphi_{z,y}$, which implies (by choosing $x = z$) that $\underline{t}|_z \models_2 \varphi_\delta$.

Assume that δ is a symbol that is not in D. We claim that $\underline{t}|_z \not\models_2 \varphi_\delta$. Suppose to the contrary that $\underline{t}|_z \models_2 \varphi_\delta$. Then there exists a node $x \in \mathrm{dom}(\underline{t})$ with $\delta \in S_\Delta(\widehat{\underline{t}}(x))$ such that $\underline{t}|_z \models_2 \bigwedge_{\substack{y \in \mathrm{dom}(\underline{t}): \\ \delta \notin S_\Delta(\widehat{\underline{t}}(y))}} \varphi_{x,y}$. Since z is a node of $\underline{t}$

with $\delta \notin S_\Delta(\widehat{\underline{t}}(y))$, in particular $\underline{t}|_z \models_2 \varphi_{x,z}$, contradicting the definition of the formulae $\varphi_{x,z}$. Thus, $\underline{t}|_z \not\models_2 \varphi_\delta$.

Hence, the FTL($\mathcal{L}$)-formula $\varphi = L(\delta \mapsto \varphi_\delta)_{\delta\in\Delta}$ of depth at most r separates t_0 and t_1, proving $t_0 \not\equiv^r_{\mathcal{L}} t_1$. □

Corollary 6.1. *For any class $\mathcal{L}$ of tree languages and $r \geq 0$, the relations $\equiv^r_{\mathcal{L}}$ and $\sim^r_{\mathcal{L}}$ coincide.*

Using the game-theoretic characterization theorem from [5], the above result proves an alternative formulation of Corollary 4.1:

Corollary 6.2. *Let $\mathcal{L}$ be an arbitrary class of tree languages. Let $\mathcal{L}'$ consist of all languages of the form $S_\Sigma^{-1}(L) \subseteq T_{P(\Sigma)}$, where $L \subseteq T_\Sigma$ is a member of $\mathcal{L}'$. Then,*

$$\mathbf{FTL}_2(\mathcal{L}) \;=\; \mathbf{FTL}_1(\mathcal{L}'),$$

moreover, for any FTL($\mathcal{L}$)*-formula φ of depth r there exists an* FTL($\mathcal{L}'$)*-formula φ' of depth at most r with $L_{\varphi,2} = L_{\varphi',1}$, and vice versa.*

Remark 6.1. An infinite Σ-tree t as a function from a tree domain to Σ is defined as a finite tree except that the domain may be infinite. The results of this section also hold for infinite trees, at least when $\mathcal{L}$ is finite. The reason for this is that if $\mathcal{L}$ is finite, then for each r, and for each ranked alphabet, up to equivalence there are only a finite number of formulas of depth at most r. Thus, (2) is a finite formula even for infinite trees.

References

1. E. A. Emerson, E. M. Clarke. Using branching time temporal logic to synthesize synchronization skeletons. *Science of Computer Programming* 2(3), p. 241–266, 1982.
2. Z. Ésik. An algebraic characterization of temporal logics on finite trees. Parts I, II, III. In: *Proc. 1st International Conference on Algebraic Informatics, 2005*, Aristotle University, Thessaloniki, p. 53–77, 79–99, 101–110, 2005.
3. Z. Ésik. Characterizing CTL-like logics on finite trees. *Theoretical Computer Science* 356, p. 136–152, 2006.
4. Z. Ésik. Extended temporal logic on finite words and wreath products of monoids with distinguished generators. In: *Proc. DLT 02, Kyoto*, LNCS 2450, p. 43–48, Springer, 2003.
5. Z. Ésik and Sz. Iván. Games for temporal logics on trees. In: *Proc. CIAA 2008*, LNCS 5148, Springer, p. 191–200, 2008.
6. Z. Ésik and Sz. Iván. Products of Tree Automata with an Application to Temporal Logic. *Fundamenta Informaticæ* 82, p. 61–78, 2007.

7. Z. Ésik and G. S. Martín. A note on Wolper's logic. In: *Proc. Workshop on Semigroups and Automata*, Lisboa, 2005.
8. G. Grätzer. *Universal Algebra*, Springer, 1979.
9. R. Milner. *Communication and Concurrency*, Prentice Hall International Series in Computer Science, 1989.
10. J-É. Pin. *Varieties of Formal Languages*. North Oxford Academic, 1986.
11. F. Gécseg and M. Steinby. *Tree Automata*. Akadémiai Kiadó, 1984.
12. H. Straubing. *Finite Automata, Formal Logic, and Circuit Complexity*. Birkhauser, 1994.
13. P. Wolper. Temporal logic can be more expressive. *Information and Control* 56, p. 71–99, 1983.

Received: December 15, 2009

Revised: March 29, 2010

THE NUMBER OF DISTINCT 4-CYCLES AND 2-MATCHINGS OF SOME ZERO-DIVISOR GRAPHS

MITSUO KANEMITSU

Department of Mathematics, College of Contemporary Education,
Chubu University, 487-8501 Matsumoto, 1200, Kasugai City, Japan
mkanemit@isc.chubu.ac.jp

Let $\mathbf{Z}$ be the ring of integers. We will write $\mathbf{Z}_n$ for $\mathbf{Z}/n\mathbf{Z}$. The zero-divisor graph of $\mathbf{Z}_n$, denoted $\Gamma(\mathbf{Z}_n)$, is the graph whose vertices are the non-zero zero-divisors of $\mathbf{Z}_n$ with two distinct vertices joined by an edge when the product of the vertices is zero. Let $f(\lambda, \Gamma(\mathbf{Z}_n))$ be the characteristic polynomial of $\Gamma(\mathbf{Z}_n)$. To $n = 4p$ $(n \neq 8)$ and $n = 9p$ $(n \neq 27)$ where p is a prime number, we find the characteristic polynomial $f(\lambda, \Gamma(\mathbf{Z}_n))$, the number of distinct 4-cycles and the number of 2-matchings of $\Gamma(\mathbf{Z}_n)$.

1. Introduction

The concept of zero-divisor graph of a commutative ring was introduced by I.Beck in [2]. In [1], Anderson and Livingston introduced and studied the zero-divisor graph whose vertices are the non-zero zero-divisors. As usual, the ring of integers and integers modulo n will be denoted by $\mathbf{Z}$ and $\mathbf{Z}_n = \{0, 1, 2, \cdots, n-1\}$, respectivelly. The graph, $\Gamma(\mathbf{Z}_n)$ as studied in [1] has as its vertices the set $Z(\mathbf{Z}_n)^* = Z(\mathbf{Z}_n) - \{0\}$, where $Z(\mathbf{Z}_n)$ is the set of zero divisors of $\mathbf{Z}_n$. Two distinct vertices x, y are adjacent (connected by an edge) when $xy = 0$ in $\mathbf{Z}_n$. In this paper, we give the number of distinct 4-cycles and 2-matchings of graphs $\Gamma(\mathbf{Z}_n)$, where $n = 4p$ (p is a prime number $(\neq 2)$) and $n = 9p$ where p is a prime number $(\neq 3)$.

Let G be a graph and let A be the adjacent matrix of G.

Let us suppose that the characteristic polynomial of a graph G is

$$f(\lambda, G) = \lambda^n + C_1\lambda^{n-1} + C_2\lambda^{n-2} + C_3\lambda^{n-3} + C_4\lambda^{n-4} + \cdots + C_n.$$

It is proved in the theory matrices that all coefficients can be expressed in terms of the principal minors of A, where a principal minor is the determinant of a submatrix obtained by taking a subset of the rows and the same subset of the columns.

Let $\varphi(n)$ be the Euler function. Then we have that $f(\lambda, \Gamma(\mathbf{Z}_n)) = \lambda^{\varphi(n)} + C_1\lambda^{\varphi(n)-1} + C_2\lambda^{\varphi(n)-2} + C_3\lambda^{\varphi(n)-3} + C_4\lambda^{\varphi(n)-4} + \cdots + C_{\varphi(n)}$ is a characteristic polynomial of $\Gamma(\mathbf{Z}_n)$.

The number of distinct 4-cycles and the number of 2-matchings of $\Gamma(\mathbf{Z}_n)$ will be denoted by n_C and n_M, respectivelly. It is well known that $C_1 = 0$. Also, $-C_2$ is the number of edges of $\Gamma(\mathbf{Z}_n)$. $-C_3$ is twice the number of triangles in $\Gamma(\mathbf{Z}_n)$. It follows that $C_4 = n_M - 2n_C$ ([3]).

2. Results

In this section, we give some examples and some results for the number of distinct 4-cycles and the number of 2-matchings in some zero divisor graphs.

We start the following Lemma.

Lemma 2.1. ([4]) *Let G be a simple graph having the order n and let $(d_1, d_2, \cdots, d_n)$ be the degree sequence of G. Then the number of 2-matchings of G is*

$$n_M = \tfrac{1}{8}(\textstyle\sum_{i=1}^n d_i)^2 - \tfrac{1}{2}\sum_{i=1}^n d_i^2 + \tfrac{1}{4}\sum_{i=1}^n d_i$$

where d_i is a degree of a vertex v_i $(i = 1, 2, \cdots, n)$ of G.

For a complete graph K_n with order n, we have that the characteristic polynomial $f(\lambda, K_n) = (\lambda - (n-1))(\lambda+1)^{n-1}$ and have that $n_M = \frac{1}{8}((n-1)n)^2 - \frac{1}{2}((n-1)^2 n) + \frac{1}{4}(n-1)n = \frac{1}{8}n(n-1)(n-2)(n-3)$. Also, we have that $n_C = 3 \times {}_nC_4 = \frac{1}{8}n(n-1)(n-2)(n-3)$. Hence $C_4 = n_M - 2n_C = -\frac{1}{8}n(n-1)(n-2)(n-3)$. Thus we obtain that the characteristic polynomial $f(\lambda, K_n) = \lambda^n - \frac{1}{2}n(n-1)\lambda^{n-2} - \frac{1}{3}n(n-1)(n-2)\lambda^{n-3} - \frac{1}{8}n(n-1)(n-2)(n-3)\lambda^{n-4} + g(\lambda)$, where $g(\lambda)$ is a polynomial with respect to λ.

Example 2.2. Let p be a prime number. Then the order of the zero-divisor graph $\Gamma(\mathbf{Z}_{p^2})$ is $p^2 - \varphi(p^2) - 1 = p - 1$. The vertices set $V(\Gamma(\mathbf{Z}_{p^2}))$ is $\{p, 2p, 3p, \cdots, (p-2)p, (p-1)p\}$. Therefore $\Gamma(\mathbf{Z}_{p^2})$ is a complete graph K_{p-1}.

Lemma 2.3. *Let $K_{m,n}$ be a complete bipartite graph with vertex sets having m and n elements, respectively. Then we have that the following satatements.*

(1) *The characteristic polynomial of $K_{m,n}$ is $f(\lambda, K_{m,n}) = \lambda^{m+n} - mn\lambda^{m+n-2}$.*

(2) $n_M = \frac{1}{2}mn(m-1)(n-1), n_C = \frac{1}{4}mn(n-1)(m-1)$.

Proof. The number of vertices in $K_{m,n}$ is $m+n$. We have that the size of $K_{m,n}$ equals to mn. Hence $C_2 = -mn$. Also, $C_3 = 0$. By Lemma 2.1, $n_M = \frac{1}{8}(n\times m+m\times n)^2 - \frac{1}{2}(n^2\times m+m^2\times n)+\frac{1}{4}(n\times m+m\times n) = \frac{1}{2}mn(m-1)(n-1)$. Moreover, we have that $n_C = {}_mC_2 \times {}_nC_2 = \frac{1}{4}mn(m-1)(n-1)$. Therefore $C_4 = n_M - 2n_C = 0$. Hence the characteristic polynomial $f(\lambda, K_{m,n}) = \lambda^{m+n} - mn\lambda^{m+n-2}$. ■

Corollary 2.4. *Let $n = pq$ ($p < q, p$ and q are prime numbers). Then we have that the following statements.*

(1) $\Gamma(\mathbf{Z}_{pq}) = K_{p-1,q-1}$.

(2) $f(\lambda, \Gamma(\mathbf{Z}_{pq})) = \lambda^{p+q-2} - (p-1)(q-1)\lambda^{p+q-4}$.

(3) $n_M = \frac{1}{2}(p-1)(q-1)(p-2)(q-2), n_C = \frac{1}{4}(p-1)(p-2)(q-1)(q-2)$.

Theorem 2.5. *Let $n = 4p$, where p ($\neq 2$) is a prime number. Then the following statements hold.*

(1) *The characteristic polynomial $f(\lambda, \Gamma(\mathbf{Z}_{4p}))$ of $\Gamma(\mathbf{Z}_{4p})$ is the following.*

$$f(\lambda, \Gamma(\mathbf{Z}_{4p})) = \lambda^{2p+1} - 4(p-1)\lambda^{2p-1} + 2(p-1)^2\lambda^{2p-3}.$$

(2) $n_M = (p-1)(5p-8), n_C = \frac{3}{2}(p-1)(p-2)$.

Proof. The prime divisors of $4p$ are $1, 2, 4, p, 2p$ and $4p$ ($= 0$). Let $[2p] = \{2p\}, [p] = \{p, 2p, 3p\} - [2p] = \{p, 3p\}, [4] = \{4a \mid a = 1, 2, 3, \cdots, p-1\}$ and $[2] = \{2a \mid a = 1, 2, \cdots, 2p-1\} - [4] - [2p]\}$.

We have that $||[2p]|| = 1, ||[p]|| = 2, ||[4]|| = p-1, ||[2]|| = p-1$.

Let $f(\lambda, \Gamma(\mathbf{Z}_{4p})) = |\lambda E - A| = \lambda^n + C_1\lambda^{n-1} + C_2\lambda^{n-2} + \cdots + C_n$, where E is the identity matrix and A is the adjacent matrix of $\Gamma(\mathbf{Z}_{4p})$.

(a) The number of vertices of $\Gamma(\mathbf{Z}_{4p})$ i.e. the order of $\Gamma(\mathbf{Z}_{4p})$.

The number of vertices in the zero-divisor graph $\Gamma(\mathbf{Z}_{4p})$ equals to $n - \varphi(4p) - 1$, where $\varphi(4p)$ is the Euler function. We notice that $n - \varphi(4p) - 1 = 4p - 2(p-1) - 1 = 2p+1$. Hence the order of $\Gamma(\mathbf{Z}_{4p})$ is $2p+1$. Hence the degree of $f(\lambda, \Gamma(\mathbf{Z}_{4p}))$ is $2p+1$.

(b) The number of edges of $\Gamma(\mathbf{Z}_{4p})$, i.e. the size of $\Gamma(\mathbf{Z}_{4p})$.

The edges of $\Gamma(\mathbf{Z}_{4p})$ are the following $a - b$.

(i) $a - b$ ($a \in [2p], b \in [4]$), that is, $2p - b$. The number of this type's edges equals to $p-1$.

(ii) $a - b$ ($a \in [2p], b \in [2]$), i.e. $2p - b$. The number of this type's edges equals to $p-1$.

(iii) $a - b$ ($a \in [4], b \in [p]$). The number of this type's edges equals to $2(p-1)$.

Therefore the number of edges is $1 \times (p-1) + 1 \times (p-1) + (p-1) \times 2 = 4(p-1)$. Hence $C_2 = -4(p-1)$.

(c) The number of triangles is 0. Hence $C_3 = 0$.

(d) The distinct 4-cycles of $\Gamma(\mathbf{Z}_{4p})$ are the following $a - b - c - d - a$;

(i) $a - b - c - d - a$ $(a \in [2p], b, d \in [4], c \in [p])$.

(ii) $a - b - c - d - a$ $(a, c \in [4], b, d \in [p])$. Hence we have that

$n_C = {}_{p-1}C_2 \times 2 + {}_{p-1}C_2 \times {}_2C_2 = (p-1)(p-2) + \frac{1}{2}(p-1)(p-2) = \frac{3}{2}(p-1)(p-2)$.

By Lemma 2.1, we have that $n_M = \frac{1}{8}(1 \times (p-1) + 2(p-1) \times 1 + (p-1) \times 2 + 3 \times (p-1))^2 - \frac{1}{2}(1^2 \times (p-1) + (2(p-1))^2 \times 1 + 3^2 \times (p-1) + (p-1)^2 \times 2) + \frac{1}{4}(1 \times (p-1) + 2(p-1) \times 1 + (p-1) \times 2 + (p-1) \times 3)) = (p-1)(5p-8)$.

Hence $C_4 = n_M - 2n_C = (p-1)(5p-8) - 2 \times \frac{3}{2}(p-1)(p-2) = 2(p-1)^2$.

If $i \geq 5$, then $C_i = 0$. In fact, those coefficients can be expressed in terms of the i-th degree principal minors of A, where a principal minor is the determinant of a submatrix obtained by taking a subset of the rows and the same subset of the columns. Therefore five or moreover vertices contain the same type vertices and so minor determinant equal to 0.

Thus we have that $f(\lambda, \Gamma(\mathbf{Z}_{4p})) = \lambda^{2p+1} - 4(p-1)\lambda^{2p-1} + 2(p-1)^2\lambda^{2p-3}$. The proof is complete. ■

Example 2.6. Let $n = 28 = 4 \times 7$. The prime divisors of 28 are $1, 2, 4, 7, 14, 28 \ (= 0)$. Put

$[14] = \{14\}, [7] = \{7, 21\}, [4] = \{4, 8, 12, 16, 20, 24\}$

and

$[2] = \{2, 6, 10, 18, 22, 26\}$.

The edges of $\Gamma(\mathbf{Z}_{28})$ are the following edges $a - b$:

(i) $14 - a$ $(a \in [2], [4])$. The number of this type's edges is $12 (= |[2]| + |[4]|)$.

(ii) $a - b$ $(a \in [4], b \in [7])$. The number of this type's edges is $12 (= |[7]| \times |[4]|)$.

The order of $\Gamma(\mathbf{Z}_{28})$ equals to $15 \ (= 28 - \varphi(28) - 1)$. The size of $\Gamma(\mathbf{Z}_{28})$ equals to 24. Hence $C_2 = -24$. Also, the number of triangles are 0. Therefore $C_3 = 0$. Moreover, $n_M = \frac{1}{8}(6 \times 2 + 3 \times 6 + 12 \times 1 + 1 \times 6)^2 - \frac{1}{2}(6^2 \times 2 + 3^2 \times 6 + 12^2 \times 1 + 1^2 \times 6) + \frac{1}{4}(6 \times 2 + 3 \times 6 + 12 \times 1 + 1 \times 6) = 162$. Also, $n_C = {}_6C_2 + 2 \times {}_6C_2 = 45$. Since $C_4 = n_M - 2n_C = 162 - 2 \times 45 = 72$, and $0 = C_5 = C_6 = \cdots$. Therefore we have that $f(\lambda, \Gamma(\mathbf{Z}_{28})) = \lambda^{15} - 24\lambda^{13} + 72\lambda^{11}$.

Theorem 2.7. *Let $n = 9p$, where $p \ (\neq 3)$ is a prime number. Then the following statements hold.*

(1) *The characteristic polynomial* $f(\lambda, \Gamma(\mathbf{Z}_{9p}))$ *of* $\Gamma(\mathbf{Z}_{9p})$ *is the following.*

$f(\lambda, \Gamma(\mathbf{Z}_{9p})) = \lambda^{3p+5} - (12p-11)\lambda^{3p+3} - 6(p-1)\lambda^{3p+2} + 6(p-1)(4p-3)\lambda^{3p+1} + a\lambda^{3p}$, *where* a *is an integer.*

(2) $n_M = 12(p-1)(5p-7), n_C = (p-1)(18p-33)$.

Proof. The prime divisors of $9p$ are $1, 3, 9, p, 3p, 9p\ (=0)$. Let $[3p] = \{3p, 6p\}, [p] = \{rp \mid r = 1, 2, \cdots, 8\} - [3p] = \{p, 2p, 4p, 5p, 7p, 8p\}, [3^2] = \{r \cdot 3^2 \mid r = 1, 2, \cdots, p-1\}$ and $[3] = \{r \cdot 3 \mid r = 1, 2, \cdots, 3p-1\} - [3p] - [3^2]$. Thus, we have that $|[3p]| = 2, |[p]| = 6, |[3^2]| = p-1$ and $|[3]| = 2(p-1)$.

Let $f(\lambda, \Gamma(\mathbf{Z}_{9p})) = |\lambda E - A| = \lambda^n + C_1\lambda^{n-1} + C_2\lambda^{n-2} + \cdots + C_n$, where E is the identity matrix and A is the adjacent matrix of $\Gamma(\mathbf{Z}_{9p})$.

(a) The number of vertices of $\Gamma(\mathbf{Z}_{9p})$ i.e. the order of $\Gamma(\mathbf{Z}_{9p})$.

The number of vertices in the zero-divisor graph $\Gamma(\mathbf{Z}_{9p})$ equals to $n - \varphi(9p) - 1$, where $\varphi(9p)$ is the Euler function. We notice that $n - \varphi(9p) - 1 = 3p+5$. Hence the order of $\Gamma(\mathbf{Z}_{9p})$ is $3p+5$. Hence the degree of $f(\lambda, \Gamma(\mathbf{Z}_{9p}))$ is $3p+5$.

(b) The number of edges in $\Gamma(\mathbf{Z}_{9p})$, i.e. the size of $\Gamma(\mathbf{Z}_{9p})$.

The edges of $\Gamma(\mathbf{Z}_{9p})$ are the following $a - b$.

(i) $3p - 6p$. The number of this type's is only one.

(ii) $a - b\ (a \in [3p], b \in [3])$. The number of this type's edges is $4(p-1)$.

(iii) $a - b\ (a \in [3p], b \in [3^2])$. The number of this type's edges is $2(p-1)$.

(iv) $a - b\ (a \in [3^2], b \in [p])$. The number of this type's is $6(p-1)$.

Therefore the number of edges equals to $2 \times 2(p-1) + 1 + 2(p-1) + 6(p-1) = 12p - 11$. Hence $C_2 = -(12p-11)$.

(c) The number of triangles are the following $a - b - c - a$.

(i) $a - b - c - a\ (a \in [3]; b, c \in [3p])$. The number of this type's triangles is $2(p-1)$.

(ii) $a - 3p - 6p - a\ (a \in [3^2])$. The number of this type's triangles is $p-1$.

Hence the number of distinct 3-cycles equals to $2(p-1) + (p-1)\ (= 3(p-1))$. Thus $C_3 = -2 \times 3(p-1) = -6(p-1)$.

(d) The distinct 4-cycles of $\Gamma(\mathbf{Z}_{4p})$ are the following $a - b - c - d - a$;

(i) $a - b - c - d - a\ (a, c \in [3p], b, d \in [3])$. The number of this type's distinct 4-cycles is ${}_{2(p-1)}C_2\ (= (p-1)(2p-3))$.

(ii) $a - b - c - d - a\ (a, c \in [3p], b \in [3], d \in [3^2])$. The number of this type's distinct 4-cycles is $2(p-1)^2 (= 2(p-1) \times (p-1))$.

(iii) $a - b - c - d - a\ (a, c \in [3p], b, d \in [3^2])$. The number of this type's distinct 4-cycles is $\frac{1}{2}(p-1)(p-2)\ (= {}_{p-1}C_2)$.

(iv) $a - b - c - d - a$ ($a \in [3p], b, d \in [3^2], c \in [p]$). The number of this type's distinct 4-cycles is $6(p-1)(p-2)$ $(= 6 \times 2 \times {}_{p-1}C_2)$.

(v) $a - b - c - d - a$ ($a, c \in [p], b, d \in [3^2]$). The number of this type's distinct 4-cycles is $\frac{15}{2}(p-1)(p-2)$ $(= {}_6C_2 \times {}_{p-1}C_2)$.

Thus the number of distinct 4-cycles equals to ${}_{2(p-1)}C_2 + (p-1) \times 2(p-1) + {}_6C_2 \times {}_{p-1}C_2 + 6 \times 2 \times {}_{p-1}C_2$. Therefore $n_C = (p-1)(2p-3) + 2(p-1)^2 + \frac{15}{2}(p-1)(p-2) + \frac{1}{2}(p-1)(p-2) + 6(p-1)(p-2) = (p-1)(18p-33)$.

Also, we obtain that $n_M = 12(p-1)(5p-7)$.

Since $C_4 = n_M - 2n_C$, we have that $C_4 = 12(p-1)(5p-7) - 2(p-1)(18p-33) = 6(p-1)(4p-3)$.

If $i \geq 6$, we have that $C_i = 0$ using the principal minors of degree $i \geq 6$.

Thus we have that $f(\lambda, \Gamma(\mathbf{Z}_{9p})) = \lambda^{3p+5} - (12p-11)\lambda^{3p+3} - 6(p-1)\lambda^{3p+2} + 6(p-1)(4p-3)\lambda^{3p+1} + a\lambda^{3p}$. The proof is complete. ∎

Example 2.8. The characteristic polynomial $f(\lambda, \Gamma(\mathbf{Z}_{18}))$ of $\Gamma(\mathbf{Z}_{18})$ is $\lambda^{11} - 13\lambda^9 - 6\lambda^8 + 30\lambda^7 + 24\lambda^6$ by using the Mathematica. In the above Theorem 2.7, we see that $a = 24$ in this case.

Example 2.9. $\Gamma(\mathbf{Z}_{45})$ has the following vertices;

$[3] = \{3, 6, 12, 21, 24, 33, 39, 42\}, [15] = \{15, 30\}, [9] = \{9, 18, 27, 36\}$ and $[5] = \{5, 10, 20, 25, 35, 40\}$. The edges of $\Gamma(\mathbf{Z}_{45})$ are the following $a - b$:

(i) $a - b$ ($a \in [3], b \in [15]$). The number of this type's edges 8×2.

(ii) $a - b$ ($a, b \in [15]$), i.e. $15 - 30$. The mumber of this type's edges is only one.

(iii) $a - b$ ($a \in [15], b \in [9]$). The number of this type's edges is 2×4.

(iv) $a - b$ ($a \in [9]), b \in [5]$). The number of this type's edges is 4×6.

Hence the number of all edges is 49. The triangles of $\Gamma(\mathbf{Z}_{45})$ are the following $a - b - c - a$:

(i) $a - 15 - 30 - a$ ($a \in [3]$). The number of this type's triangles is 8.

(ii) $a - 15 - 30 - a$ ($a \in [9]$). The number of this type's triangles is 4.

The number of triangles of $\Gamma(\mathbf{Z}_{45})$ is 12. Next, the distinct 4-cycles are the following $a - b - c - d - a$:

(i) $a - 15 - b - 30 - a$ ($a \in [3], b \in [9]$). The number of this type's distinct 4-cycles is 32.

(ii) $a - 15 - b - 30 - a$ ($a, b \in [3], a \neq b$). The number of this type's distinct 4-cycles is 28 $(= {}_8C_2)$.

(iii) $15 - a - 30 - b - 15$ ($a, b \in [9]), a \neq b$). The number of this type's distinct 4-cycles is 6 $(= {}_4C_2)$.

(iv) $15 - a - b - c - 15$ ($a, c \in [9], a \neq b, b \in [5]$). The number of this type's distinct 4-cycles is 72 $(= {}_2C_1 \times {}_4C_2 \times {}_6C_1)$.

(v) $a - b - c - d - a$ $(a, c \in [9]$ $(a \neq c), b, d \in [5]$ $(b \neq d))$. The number of this type's distinct 4-cycles is 90 $(= {}_4C_2 \times {}_6C_2)$.

Hence, the number of distinct 4-cycles equals to $n_C = 228$. Also, we have that $n_M = 864$. Thus $C_4 = n_M - 2n_C = 864 - 456 = 408$. Therefore $f(\lambda, \Gamma(\mathbf{Z}_{45})) = \lambda^{20} - 49\lambda^{18} - 24\lambda^{17} + 408\lambda^{16} + a\lambda^{15}$, where a is an integer.

References

1. D.F.Anderson and P.S.Livingston, The zero-divisor graph of a commutative ring, *J. Algebra* **217** (1999), 434-447.
2. I.Beck, Coloring of commutative rings, *J. Algebra* **116** (1988), 208-226.
3. N.Biggs, *Algebraic Graph Theory* (second edition), Cambridge University Press 1993.
4. N.Biggs, Y.Jin and M.Kanemitsu, Beck's graphs associated with $\mathbf{Z}_n$ and their characteristic polynomials, *to appear in International J. Applied Math. and Statistics.*

Received: August 17, 2009

Revised: September 6, 2009

ON NORMAL FORM GRAMMARS AND THEIR SIZE*

A. KELEMENOVá[1,2], L. CIENCIALOVá[1] and L. CIENCIALA[1]

[1] *Institute of Computer Science,*
Silesian University in Opava, Czech Republic,
[2] *Department of Computer Science, Catholic University,*
Ružomberok, Slovakia
E-mail: {*alica.kelemenova,lucie.ciencialova,ludek.cienciala*}*@fpf.slu.cz*

In this paper normal forms for context-free grammars, namely position restricted grammars, are treated and their influence to the size of the description of languages are presented. We discuss and compare several types of position restricted grammars and minimal size of grammars, expressed by the number of rules, needed to generate a language. Several techniques for the transformation of given grammar to an equivalent grammar in required form are used to reach upper bounds of the possible increase of the size complexity of languages related to different types of the position restricted grammars.

Keywords: Normal form grammar, position restricted grammar, size complexity.

1. Introduction

Study of the normal forms of grammars for (the context-free) languages started in the late fifties of the last century. Ideas leading to introduce them were following: To find such restrictions of the form of rules of context-free grammars, which do not change their generative power, in order to handle easily the parsing procedures or, in order to reduce the (complexity of) proofs of theorems for (context-free) languages and to make them more transparent. Classical examples of such description of languages are Chomsky normal form grammars, introduced in 1959 and Greibach normal form grammars, introduced in 1965. In the first one we look for as short as possible rules. Rules with two nonterminals are sufficient in the right side of rules completed by rules with single terminal on their the right sides. Cru-

*This research is partially supported by projects VEGA 1/0692/08 (A. Kelemenová), IGS 37/2009, GAČR 201/09/P075 (L. Ciencialová) and by research plan MSM 4781305903 (L. Cienciala).

cial for the Greibach normal form are terminal symbols in front on the right side of all rules. Both aspects, i.e. short rules with terminal start symbols of the right side characterize Greibach binary form.

Several modifications or generalizations of these basic normal forms appeared during the decades. For example more (than one) terminal symbols were required in the front of the right side of production, or more (than two) nonterminals, but bounded size, were allowed (k normal form). Terminal symbols/string were required on the end of the right side of the rules (reverse Greibach normal form), or terminals appeared both in front of and at the end of the right sides of the rules (double Greibach normal form). Neighboring nonterminals are not allowed in rules of the operator grammars. Nice recapitulation of the normal forms for context-free grammars is presented in[6] chapter 4.

After series of these results the fascinating uniform view to the normal form grammars was presented by M. Blattner and S. Ginsburg in 1977[1] and in1982.[2] Main idea was to fix the position of terminals and nonterminals in rules of grammars. Position restricted grammars represent the collection of the normal form grammars for context-free languages determined by k tuples ($k \geq 3$) of natural numbers. k tuple $(m_1, m_2, \ldots, m_k)$ determines rules with right side containing terminal strings of length $m_1, m_2, \ldots, m_k$ separated by single nonterminals. More formally: According to[1] a grammar is called position restricted grammar of type $(m_1, m_2, \ldots, m_k)$, where $k \geq 3$ and each m_i is a nonnegative integer if each rule of grammar has either the form $A_0 \rightarrow w_1 A_1 w_2 \ldots w_{k-1} A_{k-1} w_k$, where each A_i is a nonterminal and each w_i is a terminal word of the length m_i, or $A \rightarrow aB$ or $A \rightarrow a$.

From this point of view Chomsky normal form grammars belong to $(0,0,0)$ position restricted grammars, Greibach binary form coincides with $(1,0,0)$ position restricted grammars. In[2] the authors slightly modified the terminal rules. They replaced the right linear rules by rules $A \rightarrow w$, where w is a terminal string with no bound to its length. As the next step the length set of terminal rules was taken under the consideration. Only terminal rules allowed in supernormal-forms[11] are rules $A \rightarrow w$ with $|w|$ in the length set of the generated language. This means that the language generated by a grammar with rule $A \rightarrow w$, where w is a terminal string has to produce some word of the length $|w|$. The supernormal-form was presented for position restricted grammars specified by three-tuples (m_1, m_2, m_3) and possible generalization for arbitrary k tuple was mentioned.

Special normal form of grammars influences significantly the size of grammars needed to generate languages and consequently the size of languages determined by the size of minimal grammar of given type, which describes the language. See[3–5,7] for previous results on grammatical complexity of languages.

In the present paper the size of languages will be discussed for their description by position restricted grammars of type (m_1, m_2, m_3). We use original right linear rules to terminate the derivation. The length of each rule of the position restricted grammars of given type is bounded by a constant and so the total length of the grammars is linearly bounded by the number of rules of grammars. Therefore we choose the number of rules to characterize the size of grammars and consequently also the size of languages. In more detail, number of production rules in grammar G, denoted by *Prod* G will characterize the size of the grammar and the size of the language L with respect to the grammars of type $t = (m_1, m_2, m_3)$ will be determined by the number of rules of minimal grammar of type t, which describes L, and will be denoted by $Prod_t L$.

We will compare $Prod_t L$ and $Prod_{t'} L$ for pairs of types $t = (m_1, m_2, m_3)$ and $t' = (m'_1, m'_2, m'_3)$ and for context-free languages L. In order to find function f, as good as possible, which satisfies $Prod_{t'} L \leq f(Prod_t L)$ for all context-free languages we will analyze various techniques of transformation for the grammars of type t to an equivalent grammars of type t'.

In this paper we discuss types with at most one nonzero component. We will consider:

(i) matrix algorithm in the case when nonzero components are in the different positions in t and t';
(ii) simulation of one derivation step or several derivation steps in t by corresponding derivation steps in t'.

Polynomial bounds will be achieved, in all considered cases, where the degree of the polynomial depends on t and t'. Optimality of the presented upper bound polynomial will be not discussed in the present paper. In the conclusion we analyze consequences of the presented results for other types of position restricted grammars and we recapitulate open cases.

2. Preliminaries and formalism for language descriptions

We assume that the reader is familiar with the theory of context-free grammars and languages. First, we fix the notations used in the paper.

A position restricted grammar $G = (N, T, P, S)$ of type $t = (m_1, \ldots, m_k)$, where $k \geq 3$ and $(m_1, \ldots, m_k)$ is a vector of natural numbers, is a context-free grammar with rules of the following forms:

(1) $A \to w_1 A_1 w_2 A_2 \dots w_{k-1} A_{k-1} w_k$, called a t rule,
(2) $A \to aB$,
(3) $A \to a$,

where $A_i, A, B \in N$, $w_i \in T^*$, $|w_i| = m_i$, $1 \le i \le k$ and $a \in T$.

In this paper the complexity of the grammar G will be characterized by its size and based on the number of its production rules. We denote it $Prod\ G$, i.e. $Prod\ G = |P|$.

The size of the context-free language L with respect to position restricted grammars of type t, denoted $Prod_t\ L$, is given by

$$Prod_t\ L = \min\{Prod_t\ G \mid L(G) = L,\ G \text{ is of type } t\}.$$

Let $G = (N, T, P, S)$ be a context-free grammar, $N = \{X_i \mid 0 \le i \le n\}$ and let $X_i \to P_{i,1} \mid P_{i,2} \mid \cdots \mid P_{i,m_i}$ be all rules of G with left side X_i. With the set P one can associate the set of the equations

$$X_i = P_{i,1} + P_{i,2} + \cdots + P_{i,m_i}, \quad i = 0, \dots, n,$$

which can be expressed also in matrix form

$$\vec{a} = \vec{a} \cdot \mathcal{D} + \vec{b},$$

where $\vec{a}$ is the vector $(X_1, \dots, X_n)$ of all nonterminals, $\vec{b}$ is the vector whose elements b_i are sums of the right sides of rules for X_i beginning with terminal symbols and $\mathcal{D}$ is a matrix, an element $D_{i,j}$ in the i-th row and j-th column of $\mathcal{D}$ is the sum of the right sides of the rules for X_i starting with X_j.

We illustrate the above mentioned representation in the next example.

Example 2.1. Let $G = (\{A, B\}, \{a, b\}, P, A)$ be a context-free grammar. A set of rules P consists of the following rules:

$$\begin{array}{ll} A \to AB & B \to BA \\ A \to AA & B \to AA \\ A \to aB & B \to b \\ A \to a & \end{array}$$

Matrix representation of the grammar G has form:

$$(A, B) = (A, B) \cdot \begin{pmatrix} A + B & A \\ \emptyset & A \end{pmatrix} + (aB + a,\ b)$$

One of the methods used to transform grammars to Greibach normal form is the matrix algorithm. Starting with grammar G represented by

$$\vec{a} = \vec{a} \cdot \mathcal{D} + \vec{b},$$

it produces grammar H described by equations

$$\vec{a} = \vec{b} \cdot \mathcal{Y} + \vec{b} \qquad \mathcal{Y} = \mathcal{D}' \cdot \mathcal{Y} + \mathcal{D}',$$

where $\mathcal{Y}$ is an $n \times n$ matrix of new nonterminals $Y_{i,j}$ and matrix $\mathcal{D}'$ is constructed from $\mathcal{D}$ substituting each nonterminal, which stays as the first symbol in the component $D_{i,j}, 1 \leq i, j \leq n$ by corresponding sum determined by $\vec{a} = \vec{b} \cdot \mathcal{Y} + \vec{b}$. For details of this transformation we refer[6] pp. 125–131.

Next example presents a Greibach normal form grammar constructed using matrix algorithm for grammar G from Example 2.1.

Example 2.2. Matrix algorithm transforms grammar G to grammar G' specified by

$$(A, B) = (aB + a, b) \cdot \begin{pmatrix} Y_{1,1} & Y_{1,2} \\ Y_{2,1} & Y_{2,2} \end{pmatrix} + (aB + a, b)$$

$$\begin{pmatrix} Y_{1,1} & Y_{1,2} \\ Y_{2,1} & Y_{2,2} \end{pmatrix} = \begin{pmatrix} D'_{1,1} & D'_{1,2} \\ D'_{2,1} & D'_{2,2} \end{pmatrix} \cdot \begin{pmatrix} Y_{1,1} & Y_{1,2} \\ Y_{2,1} & Y_{2,2} \end{pmatrix} + \begin{pmatrix} D'_{1,1} & D'_{1,2} \\ D'_{2,1} & D'_{2,2} \end{pmatrix},$$

where

$$\begin{aligned} D'_{1,1} &= aBY_{1,1} + aY_{1,1} + bY_{2,1} + aB + a + aBY_{1,2} + aY_{1,2} + bY_{2,2} + b \\ D'_{1,2} &= aBY_{1,1} + aY_{1,1} + bY_{2,1} + aB + a \\ D'_{2,1} &= \emptyset \\ D'_{2,2} &= aBY_{1,1} + aY_{1,1} + bY_{2,1} + aB + a \end{aligned}$$

Note that $\mathit{Prod}\ G = 7$ and $\mathit{Prod}\ G' = 9 + 19 \cdot 3 = 66$.

Similarly, we can express grammar G by reverse matrix form

$$\vec{a} = \mathcal{D} \cdot \vec{a} + \vec{b},$$

where $\vec{a}$ is a column vector $(X_1, \ldots, X_n)$ of all nonterminals, $\vec{b}$ is a column vector whose elements b_i are sums of the right sides of rules for X_i ending with terminal symbols and $\mathcal{D}$ is a matrix, an element $D_{i,j}$ in the i-th row and j-th column of $\mathcal{D}$ is the sum of the right sides of the rules for X_i ending with

X_j. Reverse matrix form can be transformed to reverse Greibach normal form

$$\vec{a} = \mathcal{Y} \cdot \vec{b} + \vec{b} \qquad \mathcal{Y} = \mathcal{Y} \cdot \mathcal{D}' + \mathcal{D}',$$

where $\mathcal{Y}$ is an $n \times n$ matrix of new nonterminals $Y_{i,j}$ and matrix $\mathcal{D}'$ is constructed from $\mathcal{D}$ substituting each nonterminal, which stays as the last symbol in the component $D_{i,j}, 1 \leq i,j \leq n$ by corresponding sum determined by $\vec{a} = \mathcal{Y} \cdot \vec{b} + \vec{b}$.

3. Matrix algorithm and size estimation of languages

Matrix transformation gives bases for all results presented in this section. Following Theorem 4.9.1 in[6] (see also[7]), matrix transformation gives cubic bound for the size of produced grammars. The matrix transformation associates with grammar in Chomsky normal form an equivalent grammar in Greibach binary form. This gives immediately

Theorem 3.1. *Let L be a context-free language. Then*

$$Prod_{(1,0,0)}L \leq cProd^3_{(0,0,0)}L$$

for some c.

The matrix transformation associates with grammar of type $(0,1,0)$ immediately an equivalent grammar in Greibach binary form. All components of $\mathcal{D}$ start with the terminal symbols so $\mathcal{D}' = \mathcal{D}$. Therefore quadratic bound is sufficient in this case.

Theorem 3.2. *Let L be a context-free language. Then*

$$Prod_{(1,0,0)}L \leq cProd^2_{(0,1,0)}L$$

for some c.

Proof. Let L be a context-free language, $Prod_{(0,1,0)}L = p$ and let $G = (N, T, P, S)$ be a minimal position restricted grammar of the type $(0,1,0)$, which generates L. The set of rules of G of the form $A \to BaC$, $A \to aB$ and $A \to a$ has matrix form $\vec{a} = \vec{a} \cdot \mathcal{D} + \vec{b}$, where the elements of the matrix $\mathcal{D}$ are in TN and the vector $\vec{b}$ has the elements in $TN \cup T$. Grammar H determined in matrix form by

$$\vec{a} = \vec{b} \cdot \mathcal{Y} + \vec{b}, \quad \mathcal{Y} = \mathcal{D} \cdot \mathcal{Y} + \mathcal{D},$$

is of the type $(1,0,0)$ and $ProdH \leq (n+1) \cdot p \leq cp^2$. This together with $Prod_{(1,0,0)}L \leq ProdH$ proves the theorem. □

In the following theorem we compare complexity of the type $(0, m-1, 0)$ with respect to the type $(0, 0, m)$ grammars.

Theorem 3.3. *Let L be a context-free language and $m \geq 2$. Then*

$$Prod_{(0,m-1,0)}L \leq cProd^4_{(0,0,m)}L$$

for some c.

Proof. Let L be a context-free language, $m \geq 2$ and $Prod_{(0,0,m)}L = p$. Let $G = (N, T, P, S)$ be a minimal position restricted grammar of the type $(0, 0, m)$ for L with the set of p_1 rules of the form $A \to BCa_1 \dots a_m$, p_2 rules of the form $A \to aB$ and p_3 rules of the form $A \to a$ and with corresponding matrix representation

$$\vec{a} = \vec{a} \cdot \mathcal{D} + \vec{b}.$$

The components of the matrix $\mathcal{D}$ are (sums of) elements of NT^m and the components of the vector $\vec{b}$ are (sums of) elements in $TN \cup T$. Moreover there are p_1 elements in the components of $\mathcal{D}$ and $p_2 + p_3$ elements in components of $\vec{b}$. We analyze an equivalent grammar H determined by equations:

$$\vec{a} = \vec{b} \cdot \mathcal{Y} + \vec{b} \qquad \mathcal{Y} = \mathcal{D} \cdot \mathcal{Y} + \mathcal{D},$$

where matrix $\mathcal{Y}$ is an $n \times n$ matrix of new nonterminals $Y_{i,j}$, $1 \leq i, j \leq n$, and n is the number of nonterminals of grammar G.

(1) $\vec{a} = \vec{b} \cdot \mathcal{Y} + \vec{b}$
represents $n \cdot p_3$ rules of form $A \to aY$, p_2 rules of form $A \to aB$ and p_3 rules of form $A \to a$. Remaining np_2 rules of form $A \to aBY$ have to be transformed further to the type $(0, m-1, 0)$ rules. We replace each rule of form $A \to aBY$ by at most $3p$ rules, where we substitute nonterminal B according to corresponding rules in P. Strings in [ ] denote new nonterminals.

$$\begin{array}{lll}
A \to a[bY], & [bY] \to bY & \text{for } B \to b \text{ in } P \\
A \to a[bCY], & [bCY] \to b[CY] & \text{for } B \to bC \text{ in } P, \\
A \to [aCD]w[bY], & [aCD] \to a[CD],\ [bY] \to bY & \text{for } B \to CDwb \text{ in } P, \\
 & & w \in T^{m-1},\ a, b \in T.
\end{array}$$

Total number of rules in part (1) is bounded by $3p \cdot n \cdot p_2 + p_2 + (n+1) \cdot p_3$, which gives cubic bound with respect to p.

(Rules for new nonterminals of the form $[CD], [CY]$ where C, D are nonterminals of G will be constructed in part (3).)

(2) $\mathcal{Y} = \mathcal{D} \cdot \mathcal{Y} + \mathcal{D}$
represents rules of H for new nonterminals Y, which are of form $Y \to BwaZ$ and $Y \to Bwa$, $w \in T^{m-1}$, $a \in T$ and $B \in N; Y, Z \in N'$.
We slightly modify them to obtain rules of type $(0, m-1, 0)$:

$$\begin{array}{lll} Y \to Bw[a], & [a] \to a & \text{for } Y \to Bwa \text{ and} \\ Y \to Bw[aC], & [aC] \to aC & \text{for } Y \to BwaC. \end{array}$$

The matrix $\mathcal{D}$ represents at most p_1 rules so there are at most $2(n+1)\cdot p_1$ rules for nonterminals in $\mathcal{Y}$ constructed in the part (2).

(3) Rules for new nonterminals $[AB]$, $A \in N$, $B \in N \cup \{Y_{i,j} \mid 1 \leq i, j \leq n\}$ will be determined by rules of G for A:

$$\begin{array}{ll} [AB] \to aB & \text{for } A \to a \in P, \\ [AB] \to a[BC] & \text{for } A \to aB \in P, \\ [AB] \to [CD]w[aB]\ [aB] \to aB & \text{for } A \to CDwa \in P. \end{array}$$

n nonterminals in G and n^2+n nonterminals in $N \cup \{Y_{i,j} \mid 1 \leq i, j \leq n\}$ produces $n^2 + n^3$ nonterminals of the form $[AB]$. By construction in part (3) we obtain at most $2p(n^2 + n^3)$ rules.

By (1) - (3) and for $n \leq p$ it holds $Prod\ H \leq cp^4$ for some $c > 0$. We proved $Prod_{(0,m-1,0)}L \leq cProd^4_{(0,0,m)}L$. □

Theorem 3.4. *Let L be a context-free language. Then*

$$Prod_{(0,0,1)}L \leq cProd^4_{(0,1,0)}L$$
$$Prod_{(0,1,0)}L \leq cProd^4_{(1,0,0)}L$$

for some c.

Proof. To prove $Prod_{(0,0,1)}L \leq cProd^4_{(0,1,0)}L$ we assume that L is a context-free language, $Prod_{(0,1,0)}L = p$ and $G = (N, T, P, S)$, a minimal position restricted grammar of type $(0, 1, 0)$ which generates L has p_1 rules of form $A \to BaC$, p_2 rules of form $A \to aB$ and p_3 rules of form $A \to a$. Note that each rule contains one terminal symbol, so $|T| \leq p$. Reverse matrix form for G is

$$\vec{a} = \mathcal{D} \cdot \vec{a} + \vec{b},$$

where elements of $\mathcal{D}$ are in $NT \cup T$ and vector $\vec{b}$ has elements from T.

Grammar H determined in reverse Greibach normal form given by

$$\vec{a} = \mathcal{Y} \cdot \vec{b} + \vec{b} \qquad \mathcal{Y} = \mathcal{Y} \cdot \mathcal{D} + \mathcal{D},$$

where $\mathcal{Y}$ is $n \times n$ matrix of new nonterminals $Y_{i,j}$, $1 \leq i, j \leq n$, n is the number of nonterminals of grammar G gives

(1) p_3 rules of form $A \to a$, np_1 rules of form $Y \to YAb$ and p_2 rules of type $Y \to a$, which are type $(0,0,1)$ rules and
(2) np_3 rules of form $A \to Ya, np_2$ rules of form $Y \to Ya$, p_1 rules of form $Y \to Aa$, which have to be transformed further using rules of H to obtain rules of type $(0,0,1)$ grammar.
   (a) Each of the p_1 rules $Y \to Aa$ will be replaced by at most p_3 rules $A \to b$ and at most $n \cdot p_3$ rules $A \to Yb$ which add to the set H rules

$$p_1p_3 \text{ rules } Y \to b[a], \quad \text{and} \quad np_1p_3 \text{ rules } Y \to Y[b]a.$$

   (b) Each of the np_2 rules $Y \to Ya$ and np_3 rules $A \to Ya$ will be transformed by rules from H for Y from part I which add to the set H rules

$$\begin{array}{lll} Y \to Y[b]a, & A \to Y[b]a & \text{by at most } np_3 \text{ rules } Y \to Yb, \\ Y \to b[a], & A \to b[a] & \text{by at most } p_3 \text{ rules } \quad Y \to b, \\ Y \to A[b]a, & A \to A[b]a & \text{by at most } np_2 \text{ rules } Y \to Ab, \\ Y \to Y[Ab]a, & A \to Y[Ab]a & \text{by at most } np_1 \text{ rules } Y \to YAb. \end{array}$$

   At most n^2p^2 rules were added to H in this step.

Additionally we complete H with rules for $[Ab], A \in N$ and for $[a], a \in T$:

$$[Ab] \to a[b] \text{ (at most } p_1p_3 \text{ rules)}, \quad [Ab] \to Y[a]b \text{ (at most } np_1p_3 \text{ rules)},$$
$$[a] \to a \quad \text{(at most } p \text{ rules)}.$$

Number of rules of H is bounded by cp^4 for some constant c. So we conclude with $Prod_{(0,0,1)}L \leq cProd^4_{(0,1,0)}L$.

The second result $Prod_{(0,1,0)}L \leq cProd^4_{(1,0,0)}L$ can be proven in the same way. Only difference is that elements of matrix $\mathcal{D}$ are in $TN \cup T$ for $(1,0,0)$ type rules $A \to aBC$ which leads to type $(0,1,0)$ rules instead of type $(0,0,1)$ rules everywhere in the construction. □

4. Partition of the derivations and size estimation of languages

Type division of the position restricted grammars were studied in,[10] type t' divides the type t if each rule of a grammar of type t can be replaced by some derivation steps in the grammar of type t', or equivalently each derivation step in a grammar of type t can be replaced by some derivation tree of type t' grammar.

More formally: The type $t' = (m'_1, \ldots, m'_{r+1})$ divides the type $t = (m_1, \ldots, m_{n+1})$ if there is a grammar $G = (N, a \cup A, P, S)$ such that $L(G) = \{a^{m_1} A a^{m_2} A \ldots a^{m_{n+1}}\}$ and each production of G is of the form $X \rightarrow a^{m'_1} X_1 a^{m'_2} X_2 \ldots X_r a^{m'_{r+1}}$ or $X \rightarrow aZ$ or $X \rightarrow a$ and every non-terminal occurs exactly once in the right hand of side of the productions P.

Simulation of the derivation in t by a derivation in t' consists simply in the subsequent simulations of individual derivation steps. Evidently, linear bound $Prod_{t'} L \leq cProd_t L$ holds in this case.[10] Adapting this result for types given by 3-tuples we get

Theorem 4.1. *Let L be a context-free language and $0 \leq n \leq m$. Then*

$$Prod_{(0,0,0)} L \leq cProd_{(m,0,0)} L \qquad Prod_{(n,0,0)} L \leq cProd_{(m,0,0)} L$$
$$Prod_{(0,0,0)} L \leq cProd_{(0,m,0)} L \qquad Prod_{(0,n,0)} L \leq cProd_{(0,m,0)} L$$
$$Prod_{(0,0,0)} L \leq cProd_{(0,0,m)} L$$

where c is a constant for fixed m, n.

In what follows we will decompose derivation trees in a grammar of type t to subtrees in such a way that each chosen subtree can be replaced by some derivation steps (a derivation subtree) in type t' grammar. So again simulation of the derivation in type t grammar will consist in composition of derivation subtrees of the type t' grammar.

We illustrate this treatment on the transformation of grammars of the type $(m, 0, 0)$ to an equivalent grammar of type $(m + 1, 0, 0)$.

Theorem 4.2. *Let L be a context-free language and $m \geq 1$. Then*

$$Prod_{(m+1,0,0)} L \leq cProd^4_{(m,0,0)} L$$

for some c.

Proof. Let L be a context-free language and $m \geq 1$. Let $Prod_{(m,0,0)} L = p$ and $G = (N, T, P, S)$ be a minimal position restricted grammar of the type

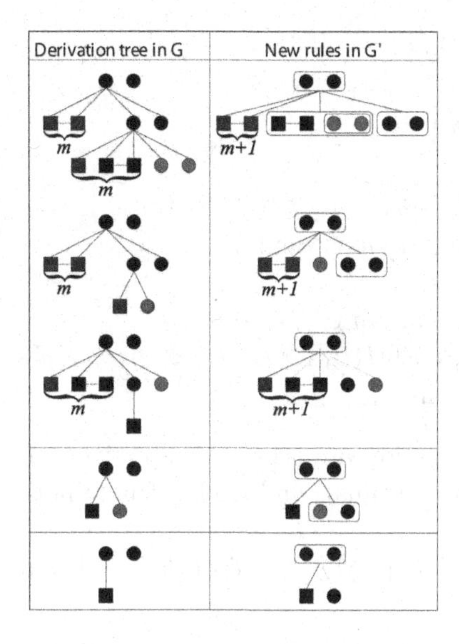

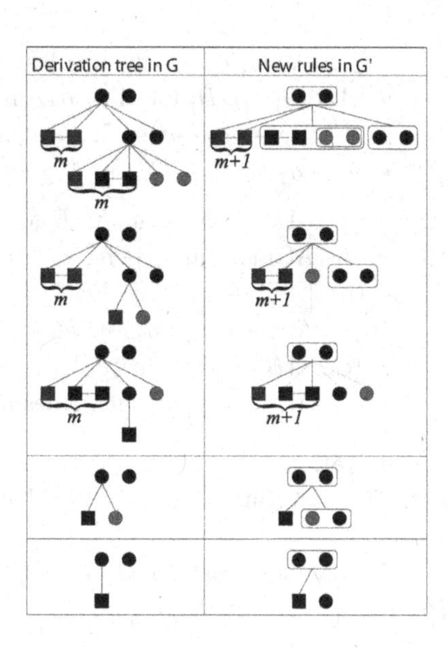

Fig. 1. The transformation of G to G'

$(m, 0, 0)$ for L. Assume that G has p_1 rules of the form $A \to a_1 a_2 \ldots a_m BC$, p_2 rules of the form $A \to aB$ and p_3 rules of the form $A \to a$; $p_1+p_2+p_3 = p$.

To produce rules of $(m+1, 0, 0)$ grammar G' we replace two consecutive derivation steps in G starting with rule of type $A \to a_1 a_2 \ldots a_m BC$ by corresponding derivation in G':

(1) $A \Rightarrow^2 a_1 a_2 \ldots a_m bC$ in G will be realized by $m+1$ regular rules in G'. At most $(m+1)p_1 \cdot p_3$ such rules are in G'.
(2) $A \Rightarrow^2 a_1 a_2 \ldots a_m bDC$ in G gives $A \to a_1 a_2 \ldots a_m bDC$ in G'. At most $p_1 \cdot p_2$ such rules are in G'.
(3) $A \Rightarrow^2 a_1 \ldots a_m b_1 b_2 \ldots b_m DEC$ in G gives $A \to a_1 \ldots a_m b_1 [b_2 \ldots b_m D][EC]$ in G'. At most p_1^2 such rules are in G' with at most p_1 nonterminals $[b_2 \ldots b_m D]$, and p_1^2 nonterminals $[EC]$.
(4) Each nonterminal $[b_2 \ldots b_m D]$ will be rewritten by $m-1$ regular rules to string $b_2 \ldots b_m D$.
(5) Each nonterminal $[AB], A, B \in N$ has in G' rules simulating one or two derivation steps in G. Namely:
   - $[AB] \to aB$ for $A \to a$ in G.

At most $p_3 \cdot n^2$ such rules are in G'.

- $[AB] \to a[DB]$ for $A \to aD$ in G.
  At most $p_2 \cdot n^2$ such rules are in G' and
- $A \to a_1a_2 \dots a_mCD$ in G gives:
  (a) $[AB] \to a_1 \dots a_m bDB$ for $C \to b$.
  At most $p_1 \cdot p_3 \cdot n^2$ such rules are in G' and
  (b) $[AB] \to a_1 \dots a_m bE[DB]$ for $C \to bE$
  At most $p_1 \cdot p_2 \cdot n^2$ such rules are in G'
  (c) $[AB] \to a_1 \dots a_m b_1[b_2 \dots b_m[EF]][DB]$ for $C \to b_1b_2 \dots b_mEF$.
  At most $p_1^2 n^2$ such rules are in G'.

Derivation trees in presented construction are schematically given on Fig. 1, where squares are used to denote terminals and circles denote nonterminals.

We conclude that $Prod\ G' \leq f(p)$ for polynomial f of degree 4 and $Prod_{(m+1,0,0)}L \leq cProd^4_{(m,0,0)}L$. □

Theorem 4.3. *Let L be a context-free language. Then*

$$Prod_{(0,0,0)}L \leq cProd_{(0,0,1)}L$$

$$Prod_{(0,0,1)}L \leq cProd^2_{(0,0,2)}L$$

$$Prod_{(0,0,m-1)}L \leq cProd^3_{(0,0,m)}L, m \geq 3$$

for some c.

Proof. Let L be a context-free language and $m \geq 1$. Let $Prod_{(0,0,m)}L = p$ and $G = (N, T, P, S)$ be a minimal position restricted grammar of the type $(0, 0, m)$ for L. Assume that G has p_1 rules of the form $A \to BCa_1a_2 \dots a_m$, p_2 rules of the form $A \to aB$ and p_3 rules of the form $A \to a$; $p_1+p_2+p_3 = p$.

To produce $(0, 0, m-1)$ grammar G' equivalent to G we replace each derivation step in G determined by one of the rules $A \to BCa_1a_2 \dots a_m$, by corresponding derivations in G' starting with rule $A \to B[Ca_1]a_2 \dots a_m$, where $[Ca_1]$ is new nonterminal.
Rules for $[Ca_1]$ are determined by rules of G as follows:

(a) $m = 1$:
$[Ca_1] \to C[a_1], [a_1] \to a_1$
G' has at most $2p_1 + p$ rules ie. $Prod_{(0,0,0)}L \leq cProd_{(0,0,1)}L$.

(b) $m = 2$:

$[Ca_1] \to b[a_1], [a_1] \to a_1$ for $C \to b \in P$.

At most $p_3p_1 + p_1$ rules.

$[Ca_1] \to [b]Da_1, [b] \to b$ for $C \to bD \in P$.

At most $p_1p_2 + p_2$ rules.

$[Ca_1] \to D[Eb_1b_2]a_1, [Eb_1b_2] \to E[b_1]b_2$ for $C \to DEb_1b_2$ in G.

At most $p_1p_1 + p_1$ rules.

G' has at most $p(3 + p_1)$ rules, ie. $Prod_{(0,0,1)}L \leq cProd^2_{(0,0,2)}L$.

(c) $m \geq 3$.

Rules for $[Eb_1b_2 \dots b_i]$, $1 \leq i \leq m-1$ determined by rules of G:

(c1) $[Eb_1b_2 \dots b_i] \to b[b_1b_2 \dots b_i]$ for $E \to b \in P$.

At most p_3 rules for each $[Eb_1b_2 \dots b_i]$ and p_1 new nonterminals $[b_1b_2 \dots b_i]$.

(c2) $[Eb_1b_2 \dots b_i] \to b[Db_1b_2 \dots b_i]$ for $E \to bD \in P$, $i < m-1$.

$[Eb_1b_2 \dots b_{m-1}] \to [b]Db_1b_2 \dots b_{m-1}$ for $E \to bD$.

At most p_2 rules for each $[Eb_1b_2 \dots b_i]$, $i < m-1$ and at most p_2 rules for each $[Eb_1b_2 \dots b_{m-1}]$. Moreover p_1p_2 new nonterminals $[Db_1b_2 \dots b_i]$ for each $[Eb_1b_2 \dots b_i]$ and $i < m-1$.

(c3) $[Eb_1 \dots b_i] \to D[Fc_1 \dots c_{i+1}]c_{i+2} \dots c_mb_1 \dots b_i$ for $E \to DFc_1 \dots c_m \in P$, $i < m-1$.

$[Eb_1 \dots b_{m-1}] \to D[Fc_1 \dots c_m]b_1 \dots b_{m-1}$, and

$[Fc_1 \dots c_m] \to F[c_1]c_2 \dots c_m$ for $E \to DFc_1 \dots c_m \in P$.

At most p_1 rules and p_1 new nonterminals $[Fc_1c_2c_3 \dots c_{i+1}]$ for each $[Eb_1b_2 \dots b_i]$ and $i < m-1$.

(c4) We add to G' i regular rules for $[a_1 \dots a_i], i \leq m-1$ for each rule $A \to BCa_1a_2 \dots a_m$ ie. less than $(m-1)p_1$ rules.

In parts c1) – c3) we constructed at most $(p_1 + p_1p_2)p$ rules for each $1 \leq i \leq m-2$ and $2pp_1$ rules for $i = m-1$. *Prod* G' for constructed G' is bounded by cp^3 for some c.

$Prod_{(0,0,m-1)}L \leq cProd^3_{(0,0,m)}L$ for all $m \geq 3$. □

5. Conclusion

In this paper we have presented several algorithms, which transform vice versa two types of position restricted grammars, determined by (m_1, m_2, m_3), where $m_i \neq 0$ for at most one i. The results presented in the paper can be summarized as follows.

Let m, n be two natural numbers. Then there are polynomials p such that

(i) $Prod_{(m,0,0)}L \leq p(Prod_{(n,0,0)}L)$

(ii) $Prod_{(m,0,0)}L \leq p(Prod_{(0,n,0)}L)$
(iii) $Prod_{(m,0,0)}L \leq p(Prod_{(0,0,n)}L)$
Moreover let $m < n$. Then
(iv) $Prod_{(m,0,0)}L \leq cProd_{(n,0,0)}L$
(v) $Prod_{(0,m,0)}L \leq cProd_{(0,n,0)}L$ and
(vi) $Prod_{(0,m,0)}L \leq p(Prod_{(0,0,n)}L)$
for some constant c and polynomial p.

Polynomial bounds for remaining pairs of types as well as the optimality of the order of polynomials can be the subject of further study.

References

1. Blattner, M., Ginsburg, S., *Canonical forms of context-free grammars and position restricted grammar forms*, Lecture Notes in Computer Science 56, (Springer Verlag, Berlin, 1977), pp. 49–53 .
2. Blattner, M., Ginsburg, S., *Position restricted grammar forms and grammars*, Theoretical Computer Science 17, (1982), pp. 1–27.
3. Csuhaj-Varjú, E., Kelemenová, A., *Descriptional complexity of context-free grammar forms*, Theoretical Computer Science 112,(1993), pp. 277–289.
4. Gruska, J., *Descriptional complexity of context-free languages*, Proc. of Mathematical Foundations of Computer Science MFCS'73, (High Tatras, 1973), pp. 71–83.
5. Gruska, J., *Descriptional complexity (of languages). A short survey*, Proc. of Mathematical Foundations of Computer Science MFCS'76, Lecture Notes in Computer Science 45, (Springer Verlag, Berlin, 1976), pp. 65–80.
6. Harrison, M. A., *Introduction to formal language theory*, (Addison Wesley P. C., Reading, Massachusets, 1978).
7. Kelemenová, A., *Complexity of normal form grammars*, Theoretical Computer Science 28, (1984), pp. 299–314.
8. Kelemenová, A., *Grammatical levels of the position restricted grammars*, Proc. of Mathematical Foundations of Computer Science MFCS'81, Lecture Notes in Computer Science 118, (Springer Verlag, Berlin, 1981), pp. 347–359.
9. Kelemenová, A., *Minimal position restricted grammars. Relations between complexity measures*, Mathematical Models in Computer Systems, (Akadémiai Kiadó, Budapest, 1981), pp. 159–169.
10. Kelemenová, A., *Type division of position restricted grammars and sizes of languages*, Colloquia Mathematica Societatis János Bolyai, 42. Algebra, Combinatorics and Logic in Computer Science, (Györ, Hungary, 1983), pp. 515–522.
11. Maurer, H. A., Salomaa, A., Wood, D., *A supernormal-form theorem for context-free grammars*, JACM, (1983), pp. 95–102.

Received: June 15, 2009

Revised: April 25, 2010

GRÖBNER BASES ON ALGEBRAS BASED ON WELL-ORDERED SEMIGROUPS*

YUJI KOBAYASHI

Department of Information Science,
Toho University, Funabashi 274–8510, Japan
E-mail: kobayasi@is.sci.toho-u.ac.jp

We develop the theory of Gröbner bases on an algebra based on a well-ordered semigroup inspired by the discussions in Farkas et al.,[3,4] where the authors study multiplicative bases in an axiomatic way. We consider a reflexive semigroup with 0 equipped with a suitable well-order, and use it as a base of an algebra over a commutative ring, on which we develop a Gröbner basis theory.

Our framework is considered to be fairly general and unifies the existing Gröbner basis theories on several types of algebras (ref.[1,6–10]). We discuss a Gröbner basis theory from a view point of rewriting systems. We study behaviors of critical pairs in our situation and give a so-called critical pair theorem. We need to consider z-elements as well as usual critical pairs come from overlapping applications of rules.

Keywords: Gröbner basis; well-ordered semigroup; rewriting system; critical pair; z-element.

1. Well-ordered reflexive semigroups

Let $S = B \cup \{0\}$ be a semigroup with zero element 0. A semigroup S is *well-ordered* if B has a well-order $>$, which is compatible in the following sense: For $a, b, c, d \in B$,

(i) $a > b, ca \neq 0, cb \neq 0 \;\Rightarrow\; ca > cb$,
(ii) $a > b, ac \neq 0, bc \neq 0 \;\Rightarrow\; ac > bc$,
(iii) $a > b, c > d, ac \neq 0, bd \neq 0 \;\Rightarrow\; ac > bd$.

A semigroup S is called *reflexive* if for any $a \in B$ there are $e, f \in B$ such that $a = eaf$. If B is a monoid, S is reflexive.

*This work was partially supported by Grant-in-Aid for Scientific Research (No. 21540048).

In the rest of this section $S = B \cup \{0\}$ is a well-ordered reflexive semigroup with 0.

Lemma 1.1. *For $a, b, c, d \in S$, we have*
(1) $ca = cb \neq 0$ *implies* $a = b$,
(2) $ac = bc \neq 0$ *implies* $a = b$,
(3) $0 \neq ca > cb \neq 0$ *implies* $a > b$,
(4) $0 \neq ac > bc \neq 0$ *implies* $a > b$.
(5) $ac = bd \neq 0$ *and* $a > b$ *imply* $d > c$.

Proof. (1) If $a \neq b$ but $ca = cb \neq 0$, then either $a > b$ or $b > a$ holds. So, either $ca > cb$ or $cb > ca$ holds by (i), a contradiction. The other assertions can be proved similarly. □

Lemma 1.2. *For $a, b \in B$, if $ab^n \neq 0$ (resp. $b^n a \neq 0$) for all $n > 0$, then $ab \geq a$ (resp. $ba \geq a$).*

Proof. If $a > ab$ and $ab^n \neq 0$ for all n, then we would have an infinite sequence

$$a > ab > ab^2 > \cdots .$$

But, this contradicts that B is well ordered. □

Proposition 1.1. *Any element of S is either idempotent or nilpotent or of infinite order.*

Proof. If $a \in B$ is not nilpotent, $a^2 \geq a$ by Lemma 1.2. If this a is not an idempotent, $a^2 > a$. So, we have an infinite sequence

$$a < a^2 < a^3 < \cdots ,$$

that is, a is of infinite order. □

Lemma 1.3. *For any $a \in B$, there is a unique pair $(e, f) \in B \times B$ such that $a = eaf$.*

Proof. Suppose that there are two pairs $(e, f), (e', f') \in B \times B$ such that $a = eaf = e'af'$. Then, $e^n af^n = a \neq 0$ for all $n > 0$. Hence, by Lemma 1.2, we see

$$a = eaf \geq ea \geq a.$$

It follows that $a = ea$. Similarly, we have $a = e'a$ and $af = af' = a$. By cancellativity (Lemma 1.1), we conclude that $e = e'$ and $f = f'$. □

Lemma 1.4. *Let (e, f) be a unique pair in Lemma 1.3. We have*
(1) *e and f are idempotents.*
(2) *$ea = af = a$.*

Proof. The statement in (2) is already proved above. Because

$$a = eaf = e^2af^2,$$

we have $e = e^2$ and $f = f^2$ by cancellativity. □

Thus, we have

Proposition 1.2. *For any $a \in B$, there is a unique pair (e, f) of idempotents such that $a = ea = af$.*

In the above lemma, e (resp. f) is called the *source* (resp. *terminal*) of a and denoted by $\sigma(a)$ (resp. $\tau(a)$). Two elements $a, b \in B$ are *parallel* and written as $a\|b$, if $\sigma(a) = \sigma(b)$ and $\tau(a) = \tau(b)$.

Let $E(B)$ be the set of idempotents in B. Idempotents in B are orthogonal to each other as stated in the following lemma.

Proposition 1.3. *$ef = 0$ for any distinct $e, f \in E(B)$.*

Proof. Let e and f be distinct idempotents in B. Assume that $ef \neq 0$. If $e > f$, then

$$ef = e^2f > ef^2 = ef,$$

a contradiction. Similarly, $f > e$ is impossible. □

Corollary 1.1. *For any $a, b \in B$, $ab \neq 0$ implies $\tau(a) = \sigma(b)$.*

Proof. If $\tau(a) \neq \sigma(b)$, then $ab = a\tau(a)\sigma(b)b = 0$ by Proposition 1.3. □

Corollary 1.2. *For $a, b \in B$ and $e \in E(B)$, $ae \neq 0$ implies $e = \tau(a)$, $eb \neq 0$ implies $e = \sigma(b)$ and $aeb \neq 0$ implies $e = \tau(a) = \sigma(b)$. Moreover, for $a, b, c \in B$, $ca = a$ implies $\sigma(a) = c$, $ac = a$ implies $\tau(a) = c$ and $acb = ab \neq 0$ implies $c = \tau(a) = \sigma(b)$.*

For idempotents e and f in B, ${}_eB$ (resp. B_f) denotes the set of elements of B with source e (resp. terminal f). Moreover, set ${}_eB_f = {}_eB \cap B_f$. We see that ${}_eB = eB \setminus \{0\}$, $B_f = Bf \setminus \{0\}$ and ${}_eB_f = eBf \setminus \{0\}$.

Example 1.1. The semigroups $(S, >)$ given below are well-ordered reflexive semigroups with 0.

(1) Let Σ^* be the free monoid generated by an alphabet Σ, a set of symbols (or variables). Let $>$ the *length-lexicographic order* on Σ^* defined as follows. Let $>$ be any total order on Σ. For two words $u = a_1a_2\cdots a_m$ and $v = b_1b_2\cdots b_n$, $u > v$ if only if (i) $m > n$ or (ii) $m = n$ and u is lexicographically larger than v with respect the order $>$ on Σ. Let $S = \Sigma^* \cup \{0\}$.

(2) Let $\mathrm{Ab}(\Sigma)$ be the free abelian monoid generated by an alphabet $\Sigma = \{a_1, a_2, \ldots, a_k\}$. For elements $u = a_1^{m_1}a_2^{m_2}\cdots a_k^{m_k}$ and $v = a_1^{n_1}a_2^{n_2}\cdots a_k^{n_k}$ of $\mathrm{Ab}(\Sigma)$, $u > v$ if and only if $(m_1, m_2, \ldots, m_k)$ is lexicographically larger than $(n_1, n_2, \ldots, n_k)$. Let $S = \mathrm{Ab}(\Sigma) \cup \{0\}$.

(3) Let $\Gamma = (\Gamma^0, \Gamma^1)$ be a quiver with a set Γ^0 of vertices and a set Γ^1 of edges. Let Γ^* be the set of all paths in Γ. Consider the following semigroup operation $\circ$ on $S = \Gamma^* \cup \{0\}$. For two paths p and q, $p \circ q$ is the path obtained by concatenating them at the end vertex v of p, if v coincides with the initial vertex of q, and $p \circ q = 0$ otherwise. We can define a compatible well-order $>$ on Γ^* as follows. Let $p, q \in \Gamma^*$. If $|p| > |q|$, then $p > q$, where $|p|$ and $|q|$ are the lengths of p and q respectively. If $|p| = |q| = 0$, that is $p, q \in \Gamma^0$, then compare them in a linear ordered given beforehand on Γ^0. If $|p| = |q| > 0$, compare them in the lexicographic order with respect to a linear order given beforehand on Γ^1.

(4) Let $n \geq 2$ and let $S = \{1 > a > a^2 > \cdots > a^{n-1}, a^n = 0\}$.

(5) Let $n \geq 2$ and let $S = \{1 < a < a^2 < \cdots < a^{n-1}, a^n = 0\}$.

(6) Let $S = \{a < ab < ba < b\} \cup \{0\}$ with $aba = a, bab = b, aa = bb = 0$.

If $a||b$, $a > b$ and $cad = 0$ imply $cbd = 0$ for any $a, b, c, d \in B$, the semigroup S is called *normally ordered.* If $\tau(a) = \sigma(b)$ implies $ab \neq 0$ for any $a, b \in B$, S is called *coherent.*

Proposition 1.4. *If S is coherent, it is normally ordered.*

Proof. If $cbd \neq 0$ for $b, c, d \in B$, then $\tau(c) = \sigma(b)$ and $\tau(b) = \sigma(d)$ by Corollary 1.1. Thus, if S is coherent, $cad \neq 0$ for $a \in B$ with $a||b$. □

The semigroups in (1), (2), (3) and (6) in Example 1.1 are coherent, but the semigroup in (4) is not coherent though it is normally ordered. The semigroup in (5) is not even normal.

2. Factors and appearances

If $a = bcd$ for $a, b, c, d \in B$, c is a *factor* of a. If in particular $b = \sigma(c)$, that is, $a = cd$, c is a *left factor*, and if $d = \tau(c)$, c is a *right factor.*

Proposition 2.1. *Any $a \in B$ has only a finite number of (left ,right) factors.*

Proof. If $a \in B$ has an infinite number of left factors, it has an infinite increasing sequence $a_1 < a_2 < \cdots < a_n < \cdots$ of left factors, because B is well ordered. Let $a = a_n b_n$ with $b_n \in B$. Then by Lemma 1.1, (5), we have an infinite decreasing sequence $b_1 > b_2 > \cdots > b_n > \cdots$, a contradiction. Similarly, B does not have an infinite number of right factors. Since a factor of a is a left factor of a right factor of a, there is only a finite number of factors. □

Corollary 2.1. *The set of triples $(a_1, a_2, a_3) \in B \times B \times B$ such that $a = a_1 a_2 a_3$ is finite for any $a \in B$.*

A factor of an idempotent in B is called *idempotential.* An element of B that is not idempotential is *nonidempotential.*

The semigroups in (1), (2), (4) and (5) in Example 1.1 have a unique idempotent 1 and have no other idempotential elements. In the semigroup in (3), Γ^0 is the set of idempotents, and there are no idempotential elements other than idempotents. In the semigroup in (6), all (nonzero) elements are idempotential, and among them, ab and ba are idempotents.

Lemma 2.1. *For any idempotential element $a \in B$, there is an element $b \in B$ such that ab and ba are idempotents.*

Proof. Since a is a factor of an idempotent, $a'aa''$ is an idempotent for some $a', a'' \in B$, that is, $a'aa''a'aa'' = a'aa''$. By Corollary 1.2, $aa''a' = \sigma(a)$ and $a''a'a = \tau(a)$. Letting $b = a''a'$, ab and ba are idempotents. □

An element $b \in B$ is an *associate* of an element $a \in B$, if $b = eaf$ for some idempotential elements e and f, and we write as $a \sim b$. It is easy to see that the relation $\sim$ is an equivalence relation on B. A (left, right) factor of $a \in B$ is *proper* if it is not idempotential nor an associate of a. An element $x \in B$ is a *prime* if it is not idempotential and has no proper factor.

In the semigroups in (1) and (2), Σ is the set of primes, while Γ^1 is the set of primes of the semigroup in (3). In the semigroups in (4) and (5), a is an only prime.

Proposition 2.2. *Any nonidempotential element a in B is a product of finite number of primes.*

Proof. If the assertion were not true, then for any arbitrary large k we would have a sequence $a_1, a_2, \ldots, a_k$ of nonidempotential elements of B such that a_1, a_1a_2, $a_1a_2a_3$, $\ldots$ are factors of a. By Proposition 2.1, $a_1 \cdots a_m = a_1 \cdots a_n$ for some $m < n \leq k$, and hence, $a_{m+1} \cdots a_n$ is an idempotent by Corollary 1.2. But this is a contradiction because a_n is not idempotential. □

It is not unique to decompose $a \in B$ into a product of primes in general, but the length of the decomposition is bounded. Let $\ell(a)$ denote this bound;

$$\ell(a) = \max\{\, n \mid a = p_1, \ldots p_n,\ p_i \text{ are primes} \,\}.$$

In particular, we define $\ell(a) = 0$ if a is idempotential. By definition

$$\ell(ab) \geq \ell(a) + \ell(b)$$

for $a, b \in B$ such that $ab \neq 0$. Hence, if b is a proper factor of a, then

$$\ell(a) > \ell(b).$$

Elements a and b in B are *left coprime* (resp. *right coprime*) if they have no nonidempotential common left (resp. right) factor. They are *coprime*, if they are left and right coprime. Clearly, for any $a, b \in B$, there are $c, d \in B$ such that $a = ca'd$, $b = cb'd$, and a' and b' are coprime.

For pairs $(a, b), (a', b') \in B \times B$, we order them as

$$(a, b) > (a', b') \ \Leftrightarrow\ a > a' \text{ or } (a = a' \text{ and } b' > b).$$

A pair (a, b) is *nonempty* if $axb \neq 0$ for some $x \in B$. Pairs $(a, b), (a', b')$ are *equivalent* if $axb = a'xb'$ for any $x \in B$. The equivalence class of a nonempty pair is a finite set by Corollary 2.1 and is called a *context*. The context of a pair (a, b) is denoted by $C(a, b)$. We always represent a context by its maximal pair unless stated otherwise, so when we refer to a context $C(a, b)$, it implicitly means that (a, b) is maximal (with respect to the order defined above) in the context.

When B is the free monoid Σ^* generated by Σ, pairs $(a, b), (a', b')$ are equivalent if and only if $a = a'$ and $b = b'$. Thus there is no ambiguity. When B is the free abelian monoid $\mathrm{Ab}(\Sigma)$ generated by Σ, pairs $(a, b), (a', b')$ are equivalent if and only if $ab = a'b'$. For example, for $a, b \in \Sigma$, $(ab, 1), (a, b), (b, a), (1, ab)$ are equivalent to each other and consist one context. Among them $(ab, 1)$ is the maximal representative.

We order contexts by the maximal representatives, that is, for contexts $c = C(a,b), c' = C(a',b')$,

$$c > c' \Leftrightarrow (a,b) > (a',b').$$

Remark that if (a,b) or (a',b') is not maximal representatives above, then $c > c'$ if and only if there is a pair $(d,e) \in C(a,b)$ such that $(d,e) > (d',e')$ for any pair $(d',e') \in C(a',b')$.

If (a,b) and (a',b') are equivalent, then (da,be) and $(da',b'e)$ are equivalent for any $d,e \in B$ (as far as da, be, da' and $b'e$ are nonzero), so for a context $c = C(a,b)$ we can define the context $d \cdot c \cdot e$ which is the equivalent class of (da,be).

Contexts c and c' are *coprime* if a and a' are left coprime and b and b' are right coprime for any $(a,b) \in c$ and $(a',b') \in c'$. For any pair c, c' of contexts, there are $a, b \in B$ such that $c = a \cdot c_1 \cdot b$, $c' = a \cdot c_1' \cdot b$ and c_1 and c_1' are coprime.

Let U be a subset of B. If an element $x \in B$ is decomposed as $x = aub$ with $a, b \in B$ and $u \in U$, the triple (a,u,b) is called an *appearance* of U in x. We do not distinguish appearances with the same context, and $A(a,u,b)$ denotes the appearance with the context $C(a,b)$. For an appearance $A = A(a,u,b)$ of U in $x = aub$ and $d, e \in B$, $d \cdot A \cdot e$ denotes the appearance $A(da,u,be)$ of U in dxe. Two appearances $A(a,u,b)$ and $A(a',u',b')$ are *coprime* if so are the contexts $C(a,b)$ and $C(a',b')$. For any appearances A and A', there are coprime appearances $A'', A^\dagger$ and $a, b \in B$ such that $A = a \cdot A'' \cdot b$ and $A' = a \cdot A^\dagger \cdot b$.

For two appearances $A(a,u,b)$ and $A(a',u',b')$ of U in x, we order them as

$$A(a,u,b) > A(a',u',b') \;\Leftrightarrow\; C(a,b) > C(a',b').$$

Clearly, this order is a total order, and hence by Corollary 2.1, the set of all appearances of U in a forms a finite chain. Let

$$A_0 > A_1 > \cdots > A_n \tag{1}$$

be the chain of appearances of U in a. The first A_0 is the *rightmost appearance*, and A_{i-1} *appears at the immediate right of* A_i for $i = 1, \ldots, n$. Two appearances A and A' are *adjacent* if either A is at the immediate right of A' or A' is at the immediate right of A.

The following technical result will be used in the proof of the main results in Section 5.

Proposition 2.3. *For any distinct appearances A and A' of U in $x \in B$, there is a sequence of appearances A_i ($n \geq 1, i = 0, ..., n$) of U in x such that*

$A_0 = A$, $A_n = A'$, and for every $i = 1, \dots n$, there are elements $a_i, b_i \in B$ and appearances A''_i, $A^\dagger_i$ of U in a factor of x with $A_{i-1} = a_i \cdot A''_i \cdot b_i$, $A_i = a_i \cdot A^\dagger_i \cdot b_i$ such that A''_i and $A^\dagger_i$ are coprime and adjacent.

Proof. Suppose $A > A'$, and consider the chain (1) of appearances between A and A', where $A_0 = A$, $A_n = A'$. We proceed by double induction on $\ell = \ell(x)$ and n. If $\ell = 0$, that is, x is idempotential, then A_{i-1} and A_i are coprime for every i, and (1) itself is the desired sequence. Suppose that $\ell > 0$. By induction hypothesis there is a desired sequence of appearances $A'_0, \dots, A'_m$ between A_1 and A_n. If A_0 and A_1 are coprime, we have the desired sequence $A = A_0, A_1 = A'_0, \dots, A'_m$. Otherwise, there are $a, b \in B$ one of which is nonidempotential such that $A_0 = a \cdot A''_0 \cdot b$, $A_1 = a \cdot A''_1 \cdot b$ with appearances A''_0 and A''_1 of U in a proper factor x' of x ($x = ax'b$). Since $\ell(x') < \ell(x)$, by induction hypothesis there is a desired sequence of appearances $A^\dagger_0, \dots A^\dagger_k$ between A''_0 and A''_1 of U in x'. Therefore, we have the desired sequence $a \cdot A^\dagger_0 \cdot b, \dots, a \cdot A^\dagger_k \cdot b, A'_1, \dots, A'_m$ of appearances between A and A'. □

Two appearances A and A' of U in x are *disjoint* if for some $(a, u, b) \in A$ and $(a', u', b') \in A'$ (i) $a = a'u'c$ for some left factor c of b', or (ii) $a' = auc$ for some left factor c of b. In case (i), $b' = cub$ for the right factor c of a and $x = a'u'cub$, and in case (ii) $b = cu'b'$ for the right factor c of a' and $x = aucu'b'$.

3. Rewriting on algebras

In this section $S = B \cup \{0\}$ is a well-ordered reflexive semigroup with 0, and K is a commutative ring with 1.

Let $F = K \cdot B$ be the free K-module generated by B. Then, F has an algebra structure with the product induced from the semigroup operation of S. An element f of F is uniquely written as a finite sum

$$f = \sum_{i=1}^{n} k_i x_i \tag{2}$$

with $k_i \in K \setminus \{0\}$ and x_i are distinct elements in B. If $x_1 > x_i$ for all $i = 2, \dots n$, $k_1 x_1$ is the *leading term* of f and k_1 is the *leading coefficient* of f, which are denoted by $\mathrm{lt}(f)$ and $\mathrm{lc}(f)$, respectively. We also set $\mathrm{rt}(f) = f - \mathrm{lt}(f)$.

The well-order $>$ on B is extended to a partial order $\succ$ on F as follows. First, define $f \succ 0$ for any nonzero $f \in F$. Let f, g be nonzero elements of

F with $\mathrm{lt}(f) = kx$ and $\mathrm{lt}(g) = k'x'$, where $k, k' \in K$ and $x, x' \in B$. Then, $f \succ g$ if and only if (i) $x > x'$ or (ii) $x = x'$ and $\mathrm{rt}(f) \succ \mathrm{rt}(g)$. Since $>$ is a well-order, $\succ$ is also well founded, that is, there is no infinite decreasing sequence $f_1 \succ f_2 \succ \cdots \succ f_n \succ \cdots$ in F.

The element f in (2) is *uniform* if $\sigma(x_i) = \sigma(x_j)$ and $\tau(x_i) = \tau(x_j)$ for all i, j, and for this uniform f we define the source $\sigma(f) = \sigma(x_i)$ and the terminal $\tau(f) = \tau(x_i)$. Two uniform elements f and g are *parallel* and written as $f||g$, if $\sigma(f) = \sigma(g)$ and $\tau(f) = \tau(g)$. Any element of F is uniquely written as a sum of uniform elements.

For $e, e' \in E(B)$, eF, Fe' and eFe' are the subalgebras of F spanned by ${}_eB$, $B_{e'}$ and ${}_eB_{e'}$ over K, respectively. We have

$$F = \bigoplus_{e \in E(B)} eF = \bigoplus_{e' \in E(B)} Fe' = \bigoplus_{e, e' \in E(B)} eFe'.$$

A *rewriting rule* on F is a pair $r = (u, v)$ with $u \in B$ and $v \in F$ such that $u \succ v$ (that is, $u > u'$ for every term $k \cdot u'$ of v) and $u - v$ is uniform (that is, v is uniform and $u||v$). The rule r is written as $u \to v$.

A rule $r = (u \to v)$ is *normal* if $xuy = 0$ implies $xvy = 0$ for any $x, y \in B$. If S is normally ordered, any rule is normal.

A *rewriting system* R on F is a (not necessarily finite) set of rewriting rules on F. R is *normal* if every rule in R is normal. If f has a nonzero term $k \cdot x$ ($k \in K, x \in B$) and $x = x'ux''$ with $x', x'' \in B$ and $r = (u \to v) \in R$, then applying the rule r upon this term, f is rewritten to $g = kx'(v - u)x'' + f$. In this situation we write as $f \to_r g$ or $f \to_R g$. We call the relation $\to_R$ the *one-step reduction* by R (or modulo R). The reflexive transitive closure and the reflexive symmetric transitive closure of the relation $\to_R$ is denoted by $\to_R^*$ and $\leftrightarrow_R^*$, respectively.

Lemma 3.1. *Let R be a rewriting system on F. For any $f, g, f', g' \in F$ and $k, \ell \in K$, if $f \leftrightarrow_R^* f'$ and $g \leftrightarrow_R^* g'$, then*

$$kf + \ell g \leftrightarrow_R^* kf' + \ell g'.$$

Proof. It is clear that $f \leftrightarrow_R^* f'$ implies $kf \leftrightarrow_R^* kf'$ for any $k \in K$. Hence it suffices to show that $f \leftrightarrow_R^* f'$ implies $f + g \leftrightarrow_R^* f' + g$ for any $g \in F$. First, suppose that kx ($k \in K \setminus \{0\}$, $x \in F$) is a term of $f, x \to_R t$, and $f' = f - k(x - t)$. If g has no term of the form ℓx ($\ell \in K \setminus \{0\}$), then $f + g \to_R f' + g$, and of course $f + g \leftrightarrow_R^* f' + g$ holds. If g has a term ℓx ($\ell \in K \setminus \{0\}$, then $f' + g \to_R f' + g'$, where $g' = g - \ell(x - t)$. Here, if $k + \ell \neq 0$, $f + g \to_R f' + g'$, and if $k + \ell = 0$, $f + g = f' + g'$. In either case we have $f + g \leftrightarrow_R^* f' + g$. The general case can be proved by induction on the number of steps in the reduction $f \leftrightarrow_R^* f'$. □

Set

$$I_0(R) = \{f \in F \mid f \leftrightarrow_R^* 0\}.$$

By Lemma 3.1, we have

Corollary 3.1. *$I_0(R)$ is a K-submodule of F and $\leftrightarrow_R^*$ is equal to the K-module congruence modulo $I_0(R)$.*

If $f \to_R g$, that is, $f = k \cdot xuy + f'$, and $g = k \cdot xvy + f'$, where $k \in K \setminus \{0\}$, $x, y \in B$ and $u \to v \in R$, then, $xuy > xvy$ by compatibility of $>$, and $f \succ g$ by the definition of $\succ$. Hence, there is no infinite sequence

$$f_1 \to_R f_2 \to_R \cdots \to_R f_n \to_R \cdots$$

of reductions in F, because $\succ$ is well founded. Therefore, R is *noetherian* (*terminating*).

We write $f \downarrow_R g$ for $f, g \in F$, if f and g have a common descendent, that is, there is $h \in F$ such that $f \to_R^* h$ and $g \to_R^* h$. In this case we also write as $h \in f \downarrow_R g$.

A rewriting system R is *confluent*, if $f \downarrow_R g$ holds for any $f, g, h \in F$ such that $h \to_R^* f$ and $h \to_R^* g$. In general, noetherian confluent system is called *complete*, but in our situation a confluent system is complete.

An element $f \in F$ is *R-reducible*, if a rule from R is applicable to f, otherwise, it is *R-irreducible.* An element x of B is R-irreducible if so is as an element of F. The set $\mathrm{Irr}(R)$ of R-irreducible elements of B is given by

$$\mathrm{Irr}(R) = B \setminus B \cdot \mathrm{Left}(R) \cdot B,$$

where $\mathrm{Left}(R) = \{u \mid u \to v \in R\}$, and $f \in F$ is irreducible if and only if f is a K-linear combination of elements of $\mathrm{Irr}(R)$. An R-irreducible element f' such that $f \to_R^* f'$ is a *normal form* of f. Because R is noetherian, any $f \in F$ has at least one normal form.

The following is a basic result on complete rewriting systems and the proof is standard and omitted (see[2,5]).

Theorem 3.1. *Let R be a complete rewriting system on F, and let ρ' be the canonical surjection from F to the quotient K-module $F/I_0(R)$. Then, ρ' is injective on $\mathrm{Irr}(R)$ and $\rho'(\mathrm{Irr}(G))$ forms a free K-base of $F/I_0(R)$. Any f has a unique normal form $\hat{f}$, and we have*

$$\hat{f} = \hat{g} \Leftrightarrow f \downarrow_R g \Leftrightarrow f \leftrightarrow_R^* g \Leftrightarrow f - g \to_R^* 0 \Leftrightarrow \rho'(f) = \rho'(g)$$

for any $f, g \in F$. In particular, we have

$$I_0(R) = \{f \in F \mid \hat{f} = 0\} = \{f \in F \mid f \to_R^* 0\}.$$

The following result is also standard.

Proposition 3.1. *For a rewriting system R on F, the following statements are equivalent.*

(1) *R is complete.*
(2) *$f \to_R^* 0$ for all $f \in I_0(R)$.*
(3) *Any nonzero element of $I_0(R)$ is R-reducible.*
(4) *Every element in F has a unique normal form.*

A system R is *reduced* if for any rule $r = (u \to v) \in R$, u and v are $(R \setminus \{r\})$-irreducible. Two systems R and R' on F are *equivalent* if $\leftrightarrow_R^* = \leftrightarrow_{R'}^*$, or equivalently, $I_0(R) = I_0(R')$.

Lemma 3.2. *Let R be a complete rewriting system on F and let $r = (u \to v) \in R$.*

(1) *Suppose that $u = u_1 u' u_2$ for some $u_1, u_2 \in B$ and some rule $r' = (u' \to v') \in R$ distinct from r. Then, $R' = (R \setminus \{r\})$ is a complete system equivalent to R.*

(2) *Let v' be an element of F such that $v \to_R^* v'$. Then, $R' = (R \setminus \{r\}) \cup \{r'\}$ is a complete system equivalent to R, where $r' = (u \to v')$.*

Proof. (1) It suffices to show that $f \to_{R'}^* 0$ for any $f \in I_0(R)$. On the contrary, assume that there is $f \in I_0(R)$ such that f cannot be reduced to 0 modulo R', and choose a minimal such element f with respect to $\succ$. Since $f \in I_0(R)$ and R is complete, we have $f \to_R^* 0$. Suppose that $f \to_s f_1$ and $f_1 \to_R^* 0$ for some rule $s \in R$. By the minimality of f, we see $f_1 \to_{R'}^* 0$. If $s \in R'$, then $f \to_{R'}^* 0$. This contradiction implies that $s = r$. Hence, f has a term $kxuy$ with $k \in K \setminus \{0\}$ and $x, y \in B$. Since $u = u_1 u' u_2$, the rule r' can be applied to f to get $f' = f - kxu_1(u' - v')u_2 y$. Because $f' \in I_0(R)$ and $f \succ f'$, we see $f' \to_{R'}^* 0$ by the minimality of f. But this implies $f \to_{R'}^* 0$, a contradiction.

(2) Similar to the proof of (1). □

Proposition 3.2. *For any complete rewriting system R on F, there is a reduced complete system R' equivalent to R. If R is finite, so is R'.*

Proof. If there are distinct rules $r = u \to v$ and $r' = u' \to v'$ in R such that u' is a factor of u, then remove r from R. If there is a rule $r = u \to v \in R$

such that v is R-reducible, replace it by a rule $u \to \hat{v}$, where $\hat{v}$ is the normal form of v modulo R. In either case the system obtained is complete and equivalent to R by Lemma 3.2. Repeat this procedure until the system becomes reduced. If R is finite, the procedure stops in a finite number of steps and gives a finite reduced system R'. If R is infinite, we obtain a reduced system R' as a limit in our process. □

4. Gröbner bases on algebras

As proved in Section 3, $I_0(R)$ is a K-submodule of F, but, in general, $I_0(R)$ is not an ideal of F and $\leftrightarrow_R^*$ is not the congruence modulo an ideal. To fill this gap, define

$$Z(R) = \{xvy \mid x, y \in B, u \to v \in R, xuy = 0\}.$$

Set

$$G_R = \{u - v \mid u \to v \in R\},$$

and let $I(R)$ be the (two-sided) ideal generated by G_R.

Lemma 4.1. *We have $G_R \subset I_0(R)$, and $Z(R) \subset I(R)$.*

Proof. Since $u - v \to_R 0$ for $u - v \in G_R$, we have $G_R \subset I_0(R)$. If $xuy = 0$ for $u \to v \in R$, then $xvy = x(u - v)y \in I(R)$. This implies $Z(R) \subset I(R)$. □

Lemma 4.2. *$I_0(R) \subset I(R)$ and the relation $\leftrightarrow_R^*$ is included in the congruence modulo $I(R)$.*

Proof. If g is obtained from f by an application of a rule $u \to v$ of R, that is, f has a term kx with $k \in K \setminus \{0\}$ and $x \in B$ such that $x = x'ux''$ and $g = f - kx'(u - v)x''$. Then, $f - g = kx'(u - v)x''$ is in $I(R)$. Thus, we can show (by induction) that $f \leftrightarrow_R^* g$ implies $f \equiv g \pmod{I(R)}$. In particular, $I_0(R) \subset I(R)$. □

Proposition 4.1. *Let R be a rewriting system on F. The following statements are equivalent.*

(1) *$I_0(R) = I(R)$.*
(2) *The relation $\leftrightarrow_R^*$ coincides with the congruence modulo $I(R)$.*
(3) *$I_0(R)$ is an ideal of F.*
(4) *$Z(R) \subset I_0(R)$.*

Proof. The equivalence of (1) and (2) follows from Corollary 3.1 and Lemma 4.2. Implication (1) $\Rightarrow$ (3) is trivial. The converse is also true because $G_R \subset I_0(R)$ by Lemma 4.1. Implication (1) $\Rightarrow$ (4) is true because $Z(R) \subset I(R)$ by Lemma 4.1. Suppose that (4) holds. To show (1) it suffices to prove that $x(u-v)y \in I_0(R)$ for any $x, y \in B$ and $u \to v \in R$. If $xuy \neq 0$, then $x(u-v)y \to_R 0$. If $xuy = 0$, then $x(u-v)y = xvy \in Z(R)$. In either case we find $x(u-v)y \in I_0(R)$. □

Corollary 4.1. *If $Z(R) \subset I_0(R)$, then $f \leftrightarrow_R^* g$ implies $xfy \leftrightarrow_R^* xgy$ for any $f, g \in F$ and $x, y \in B$.*

When R is normal, $Z(R) = \{0\}$. Thus, we have

Corollary 4.2. *If R is a normal rewriting system on F, then $I_0(R) = I(R)$, and $\leftrightarrow_R^*$ is equal to the congruence modulo $I(R)$.*

Let G be a set of monic uniform elements of F. We associate a rewriting system R_G on F by

$$R_G = \{\mathrm{lt}(g) \to -\mathrm{rt}(g) \mid g \in G\}.$$

We sometimes confuse G with the associated rewriting system R_G. We write $g = u - v \in G$, implicitly assuming that $u = \mathrm{lt}(g)$ and $v = -\mathrm{rt}(g)$. We write simply $\to_G$, $\to_G^*$ and $\leftrightarrow_G^*$ instead of $\to_{R_G}$, $\to_{R_G}^*$ and $\leftrightarrow_{R_G}^*$, respectively. We say that f is G-(ir)reducible, if it is R_G-(ir)reducible, and $\mathrm{Left}(G)$ and $\mathrm{Irr}(G)$ denote $\mathrm{Left}(R_G)$ and $\mathrm{Irr}(R_G)$, respectively. We set $I_0(G) = I_0(R_G)$, $I(G) = I(R_G)$ and $Z(G) = Z(R_G)$. G is normal, if it is normal, and G is reduced if R_G is reduced.

A subset G of F is called a *Gröbner basis*, if

(i) every elements of G is monic and uniform,

(ii) the associated system R_G is complete, and

(iii) one of the statements in Proposition 4.1 holds.

If G is normal, we can omit the condition (iii).

If G is a Gröbner basis, then by (iii) $I_0(R_G)$ is equal to the ideal $I(G)$ of F generated by G, so G is called a Gröbner basis of the ideal $I(G)$. The quotient algebra $A = F/I(G)$ is said to be the algebra defined by a Gröbner basis G.

Proposition 4.2. *Let I be an ideal of F and let G be a set of monic uniform elements of an ideal I. The following statements are equivalent.*

(1) *G is a Gröbner basis of I*

(2) *$f \to_G^* 0$ for every $f \in I$.*

(3) *Any nonzero element of I is G-reducible.*

Proof. (1) $\Rightarrow$ (2): Since R_G is complete, $f \to_G^* 0$ for any $f \in I_0(G) = I$.

Conversely, (2) implies $I_0(G) = I$, and R_G is complete by Proposition 3.1.

(2) $\Leftrightarrow$ (3): obvious. □

If G is a Gröbner basis of an ideal I, then $I = I_0(G) = I(G)$, and $\leftrightarrow_G^*$ is equal to the congruence modulo I. Thus, Theorem 3.1 becomes

Theorem 4.1. *Let G be a Gröbner basis of an ideal I of F. Let $A = F/I$ be the quotient algebra of F by I and let $\rho : F \to A$ be the canonical surjection. Then, ρ is injective on* Irr(G) *and $\rho(\mathrm{Irr}(G))$ forms a free K-base of $A = F/I$. Any f has a unique normal form $\hat{f}$, and we have*

$$\hat{f} = \hat{g} \Leftrightarrow f \downarrow_G g \Leftrightarrow f \leftrightarrow_G^* g \Leftrightarrow f - g \to_G^* 0 \Leftrightarrow \rho(f) = \rho(g)$$

for any $f, g \in F$. In particular, we have

$$I = \{f \in F \mid \hat{f} = 0\} = \{f \in F \mid f \to_G^* 0\}.$$

By Proposition 3.2, we have

Proposition 4.3. *For any Gröbner basis G of an ideal I, there is a reduced Gröbner basis G' of an ideal I. If G is finite, so is G'.*

5. Critical pair theorem

In this section we consider conditions for a system to be complete. A rewriting system R on F is *locally confluent* if $f \downarrow_R g$ holds for any $f, g, h \in F$ such that $h \to_R f, h \to_R g$. As is well known (see[5]), a noetherian system is complete if it is locally confluent. Actually, more precise result stated in the following lemma is useful.

Lemma 5.1. *Let R be a rewriting system on F and let $h \in F$. If $f \downarrow_R g$ holds for any $f, g, h' \in F$ such that $h' \to_R f$, $h' \to_R g$ and ($h' = h$ or $h' \prec h$), then h has a unique normal form.*

Let R be a reduced rewriting system on F. Consider two rules $u \to v$ and $u' \to v'$ in R. Let $w \in B$ and suppose that both the lefthand sides u and u' of the rules appear in w, that is,

$$w = xuy = x'u'y' \tag{3}$$

for some $x, x', y, y' \in B$. This situation is called *critical*, if the appearances $A = A(x, u, y)$ and $A' = A(x', u', y')$ are not disjoint, A is at the immediate right of A', and the contexts $C(x, y)$ and $C(x', y')$ are coprime. For the appearances in (3) of u and u' in w, we have two reductions $w \rightarrow_R xvy$ and $w \rightarrow_R x'v'y'$. The pair $(xvy, x'v'y')$ is a *critical pair* if the situation is critical. The pair is *resolvable* if $xvy \downarrow_R x'v'y'$ holds.

First we discuss normal systems.

Lemma 5.2. *If R is a normal rewriting system on F, then for $f, g \in F$ and for $x, y \in B$, $f \rightarrow_R^* g$ implies $xfy \rightarrow_R^* xgy$.*

Proof. We proceed by induction on the number of steps in the reduction from f to g. Let $f = k \cdot x'uy' + f'$ with $k \in K \setminus \{0\}$, $f' \in F$, $u \rightarrow v \in R, x', y' \in B$, and $f \rightarrow_R f_1 \rightarrow_R^* g$, where $f_1 = kx'vy' + f'$. By induction hypothesis $xf_1y \rightarrow_R^* xgy$. If $xx'uy'y \neq 0$, then $xfy \rightarrow_R kxx'vy'y + xf'y = xf_1y$. If $xx'uy'y = 0$, then $xx'vy'y = 0$ because R is normal. Hence, $xfy = xf'y = xf_1y$. In either case we have $xfy \rightarrow_R^* xgy$. □

Lemma 5.3. *If $r = (u \rightarrow v)$ is a normal rule, then for $f \in F$ and $y \in B$, $fuy \rightarrow_r^* fvy$ and $yuf \rightarrow_r^* yvf$.*

Proof. Let $f = k_1x_1 + \cdots + k_nx_n$ with $k_1, \ldots, k_n \in K \setminus \{0\}, x_1, \ldots, x_n \in B$ and $x_n > \cdots > x_1$. If $x_iuy = 0$, then $x_ivy = 0$ because r is normal. Hence $fuy = f'uy$ and $fvy = f'vy$ where $f' = f - k_ix_iy$, and we can neglect such a term in f. So, we may suppose that $x_iuy \neq 0$ for all $i = 1, \ldots n$. Then, applying the rule r on the term k_1x_1uy we have $fuy \rightarrow_r f_1$, where $f_1 = k_1x_1vy + k_2x_2uy + \cdots + k_nx_nuy$. Since every term in k_1x_1vy is less than x_2uy, we can apply r to the term k_2x_2uy of f_1 to get $f_1 \rightarrow_r k_1x_1vy + k_2x_2vy + k_3x_3uy + \cdots + k_nx_nuy$. Repeating this we have the reduction $fuy \rightarrow_r^* fvy$. □

Theorem 5.1. *A normal reduced rewriting system R on F is complete if and only if all the critical pairs are resolvable. A set G of monic uniform normal elements of F is a Gröbner basis if all the critical pairs are resolvable.*

Proof. It suffices to show that R is locally confluent under the condition that all the critical pairs are resolvable. Let $f, g, h \in F$ and suppose that $h \rightarrow_R f$ and $h \rightarrow_R g$. We shall show that $f \downarrow_R g$ by induction on h with respect to $\succ$. Due to Lemma 5.1 the induction hypothesis implies that any h' such that $h \succ h'$ has a unique normal form.

Since $h \rightarrow_R f$ and $h \rightarrow_R g$, h has terms $k{\cdot}w$ and $k'{\cdot}w'$ with $k, k' \in K\backslash\{0\}$ and $w, w' \in B$ such that $w = xuy, w' = x'u'y'$, $x, y \in B$, $u \rightarrow v, u' \rightarrow v' \in R$, $f = h - k \cdot x(u-v)y$ and $g = h - k' \cdot x'(u'-v')y'$.

(a) If $w \neq w'$, then $f = k \cdot xvy + k' \cdot w' + h'$ and $g = k \cdot w + k' \cdot x'v'y' + h'$, where $h' = h - k \cdot w - k' \cdot w'$. Here, if $w \succ w'$ (the case $w' \succ w$ is symmetric), then $k \cdot w$ is a term of g and $g \rightarrow_R g'$, where $g' = k \cdot xvy + k' \cdot x'v'y' + h'$. If xvy has no term of the form $\ell \cdot w'$, then $f \rightarrow_R g'$. If $\ell \cdot w'$ is a term of xvy, then $f = (k\ell + k')x'u'y' + h''$ and $g' = k\ell \cdot x'u'y' + k' \cdot x'v'y' + h''$, where $h'' = k \cdot xvy - \ell \cdot w' + h'$. Thus, $(k\ell + k')x'v'y' + h'' \in f \downarrow_R g'$. In either case we see $f \downarrow_R g$.

(b) If $w = w'$, then $k = k'$, $w = xuy = x'u'y'$, $f = k \cdot xvy + h'$ and $g = k \cdot x'v'y' + h'$, where $h' = h - k \cdot w$.

Here, if h' is R-reducible, that is, h' has a term $\ell \cdot w''$ such that $w'' = x''u''y''$ with $x'', y'' \in B$ and $u'' \rightarrow v'' \in R$, then, $h = k \cdot w + \ell \cdot w'' + h''$ with $h'' \in F$ and $h \rightarrow_R h_1$, where $h_1 = k{\cdot}w + \ell{\cdot}x''v''y'' + h''$. By the result in case (a), there exist $f_1, g_1 \in F$ such that $f_1 \in f \downarrow_R h_1$ and $g_1 \in g \downarrow_R h_1$. Since $h \succ h_1$, h_1 has a unique normal form $\overline{h}_1$, which is in $f_1 \downarrow_R g_1$. Consequently, we see $f \downarrow_R g$.

If h' is R-irreducible, $f \downarrow_R g$ follows from $xvy \downarrow_R x'v'y'$. So, below we suppose that $h = w = xuy = x'u'y'$, $f = xvy$ and $g = x'v'y'$.

(c) First suppose that the appearances u and u' in w are disjoint, that is, there is $z \in B$ such that $x = x'u'z$ and $y' = zuy$ (the case $x' = xuz$ is similar). Then, $f = x'u'zvy$ and $g = x'v'zuy$. Hence, $x'v'zvy \in f \downarrow_R g$ by Lemma 5.3.

(d) We may suppose that the appearances $A = A(x, u, y)$ and $A' = A(x', u', y')$ are distinct. In fact, if $C(x, y) = C(x', y')$, then $x'uy' = xuy = x'u'y'$. Thus, $u = u'$, and it implies $v = v'$ because G is reduced. Now let $A_0, A_1, \ldots, A_n$ be a sequence of appearances of Left(R) in w between $A =$ and $A' =$ in Proposition 2.3. We shall prove $xvy \downarrow_R x'v'y'$ by induction on n. There are $a, b \in B$ such that $A = a \cdot A''_0 \cdot b$, $A_1 = a \cdot A''_1 \cdot b$ and A''_0 and A''_1 are coprime and adjacent. Let $A''_0 = A(x_0, u, y_0)$, $A''_1 = A(x_1, u_1, y_1)$ with $u_1 \rightarrow v_1 \in R$. By induction hypothesis, $ax_1v_1y_1b \downarrow_R x'v'y'$.

Here, if A''_0 and A''_1 are disjoint, $x_0vy_0 \downarrow_R x_1v_1y_1$ by (c) above. If A''_0 and A''_1 are not disjoint, then $(x_0vy_0, x_1v_1y_1)$ is a critical pair, and it is resolvable by assumption. In either case $ax_0vy_0b \downarrow_R ax_1v_1y_1b$ by Lemma 5.2. Since $xuy \succ ax_1v_1y_1b$, by induction hypothesis $ax_1v_1y_1b$ has a unique normal form which is in $xvy \downarrow_R x'v'y'$.

The proof is complete. □

Lemma 5.4. *Suppose that $f \in F$ has a unique normal form $\bar{f}$. If $g \to_R^* g'$ for $g, g' \in F$ and g' is R-irreducible, then $f + g \to_R^* \bar{f} + g'$.*

Proof. We proceed by induction on the number of steps in the reduction $g \to_R^* g'$. If $g = g'$, the assertion is clear. Suppose that g is not equal to g' and has a term $k \cdot xuy$ with $k \in K \setminus \{0\}$, $x, y \in B$ and $u \to v \in R$ such that $g \to g_1 \to_R^* g'$, where $g_1 = g - k \cdot x(u - v)y$. By induction hypothesis $f + g_1 \to_R^* \bar{f} + g'$. If f has no term of the form $k' \cdot xuy$ ($k' \in K \setminus \{0\}$), then $f + g \to_R f + g_1$, and we see $f + g \to_R^* \bar{f} + g'$.

If f has a term $k' \cdot xuy$, then $f \to_R f_1$, where $f_1 = f - k'x(u-v)y$. Here, if $k + k' = 0$, then $f + g = f_1 + g_1$, and if $k + k' \neq 0$, then $f + g \to_R f_1 + g_1$. Since $\bar{f}$ is also the unique normal form of f_1 and $g_1 \to_R^* g'$, we have $f_1 + g_1 \to_R^* \bar{f} + g'$ by induction hypothesis. Thus, in either case $f + g \to_R^* \bar{f} + g'$. □

If a rule $u \to v \in R$ or an element $u - v \in G$ is not normal, that is, $xuy = 0$ but $xvy \neq 0$, the element xvy is called a *z-element*, that is, xvy is a nonzero element of $Z(R)$ (or $Z(G)$). A z-element z is *resolvable* if $z \to_R^* 0$ (or $z \to_G^* 0$). It is *uniquely resolvable* if 0 is its unique normal form.

Lemma 5.5. *Suppose that all the z-elements in $Z(R)$ are uniquely resolvable.*

(1) *If $f \downarrow_R g$, then $xfy \downarrow_R xgy$ for any $x, y \in B$.*

(2) *For $u \to v \in R$, $f \in F$ and $y \in B$, $fuy \downarrow_R fvy$ and $yuf \downarrow_R yvf$.*

Proof. (1) Suppose $f = f_0 \to_R f_1 \to_R \cdots \to_R f_m = h$ and $g = g_0 \to_R g_1 \to_R \cdots \to_R g_n = h$, where h is R-irreducible. We shall prove $xfy \downarrow_R xgy$ by induction on $m + n$. Suppose that $m > 0$ and f has a term $k \cdot x'uy'$ with $k \in K \setminus \{0\}$, $x', y' \in B$, $u \to v \in R$ and $f_1 = f - k \cdot x'(u - v)y'$. By induction hypothesis, $xf_1y \downarrow_R xgy$. If $xx'uy'y \neq 0$, then $xfy \to_R xf_1y$, and hence $xfy \downarrow_R xgy$. If $xx'uy'y = 0$, then $xfy = xf_1y - k \cdot xx'vy'y$ and $xx'vy'y$ is in $Z(R)$. Since $-k \cdot xx'vy'y$ has the unique normal form 0, $xfy \to_R^* \bar{h}$ by Lemma 5.4, where $\bar{h}$ is an R-irreducible element in $xf_1y \downarrow_R xgy$. Hence, $\bar{h} \in xfy \downarrow_R xgy$.

(2) Let $f = k_1x_1 + \cdots + k_nx_n$ with $k_1, \cdots k_n \in K \setminus \{0\}$ and $x_1, \cdots, x_n \in B$. If $x_iuy \neq 0$ for all $i = 1, \ldots, n$, then we can show $fuy \to_R^* fvy$ as in the proof of Lemma 5.3. So assume that f has a term $k \cdot x$ such that $xuy = 0$. By induction we can assume $f'uy \downarrow f'vy$, where $f' = f - kx$. Since xvy is uniquely resolvable, $fvy = f'vy + kxvy \to_R^* h$ by Lemma 5.4, where h is an R-irreducible element in $f'uy \downarrow f'vy$. Since $fuy = f'uy$, it follows that $h \in fuy \downarrow fvy$. □

In the following theorem, G is not necessarily normal. If all the z-elements are resolvable and R_G is confluent, then G is a Gröbner basis. Thus, a similar proof to the proof of Theorem 5.1 is available, where Lemma 5.5 plays roles of Lemmas 5.2 and 5.3. A critical pair for R_G is a *critical pair for G*.

Theorem 5.2. *A set G of monic uniform elements of F is a Gröbner basis if and only if all the critical pairs are resolvable and all the z-elements are uniquely resolvable.*

A z-element xvy with $x, y \in B$, $xuy = 0$ and $u \to v \in R$ is *critical*, if x and y are G-irreducible, $x''uy'' \neq 0$ for any right factor x'' of x' and any left factor y'' of y', one of which is proper, where (x', y') is any pair in $C(x, y)$. We suspect that the following improved statement is true: a set of monic uniform elements of F is a Gröbner basis if all the critical pairs and all the critical z-elements are resolvable. Though we do not have a proof of it, we at least expect that it would hold under some suitable assumptions on S or G.

References

1. T. Becker and V. Weispfenning, *Gröbner bases*, Springer, 1993.
2. R.V. Book and F. Otto, *String rewriting systems*, Springer, 1993.
3. D.R. Farkas, C.D. Feustel and E.L. Green, *Synergy in the theories of Gröbner bases and path algebras*, Can. J. Math. **45** (1993), 727–739.
4. E.D. Green, *Multiplicative bases, Gröbner bases, and right Gröbner bases*, J. Symbolic Comp. **29** (2000), 601–623.
5. G. Fuet, *Confluent reductions: abstract properties and applications to term rewriting systems*, J. ACM **27** (1980), 797–821.
6. Y. Kobayashi, *Gröbner bases of associative algebras and the Hochschild cohomology*, Trans. Amer. Math. Soc. **375** (2005), 1095–1124.
7. Y. Kobayashi, *Gröbner bases on path algebras and the Hochschild cohomology algebras*, Sci. Math. Japonicae **64** (2006), 411–437.
8. K. Madlener and B. Reinert, *Relating rewriting techniques on monoids and rings, congruencies on monoids and ideals in monoid rings* Theoret. Comp. Sci. **208** (1998), 3–31.
9. T. Mora, *Gröbner bases for noncommutative polynomial rings*, In: AAECC3, Lect. Not. Comp. Sci. **229**, 253–262, Springer 1986.
10. T. Mora, *An introduction to commutative and noncommutative Gröbner bases*, Theoret. Comp. Sci. **134** (1994), 131–173.

Received: August 17, 2009

Revised: Septembwe 1, 2010

CONCURRENT FINITE AUTOMATA AND RELATED LANGUAGE CLASSES (AN OVERVIEW)

MANFRED KUDLEK and GEORG ZETZSCHE
Department Informatik MIN-Fakultät, Universität Hamburg
email: {kudlek,3zetzsch}@informatik.uni-hamburg.de

1. Introduction

In classical Turing machines the control is given by a finite automaton. It is an interesting idea to use as control a Petri net in order to introduce concurrency into automata theory, or automata into Petri net theory. This leads to machines with the possibility of creating an arbitrary number of heads on the tape. The heads are represented by tokens of the Petri net pointing to positions on the tape (or vice versa). The heads can only be distinguished if they are associated to different places of the Petri net, or point to different tape positions. The transitions are labelled by symbols of the tape alphabet. A computation step of such a concurrent machine can be performed only if all heads involved are on the same tape position and their corresponding tokens are located in the places forming the pre-condition of one of the Petri net's transitions. This model and some results for concurrent Turing machines have been considered in.[1,3]

This model can be adapted in a straightforward manner to the simpler model of finite automata.[2,8] Since finite automata allow the input word only to be read sequentially from left to right, the tape heads corresponding to the tokens put into the places of the post-condition of a transition will point to the tape position immediately to the right of the previous one, or – in the case of a λ-move – to the same position.

Whereas finite automata accept by reaching a final state, having read the entire input, concurrent finite automata (CFA) accept by reaching a final configuration of the Petri net, having visited all positions of the input at least once. Several possibilities of final configurations are possible, finite sets of such, a singleton (either all tokens of which pointing to different positions or all to the rightmost one), or a singleton with only one token pointing to the rightmost position of the input. For convenience we use an end marker # to allow the recognition of the end of the input.

Furthermore, besides λ moves also erasing of tokens is possible. This allows for each of the possibilties for final marking four classes of accepted languages. However, it can be shown that the simplest possibility for final marking suffices, resulting in normal forms, and that two classes coincide, resulting in the following classes: no λ moves and no erasing transitions, no λ moves but erasing transitions allowed, and λ moves and/or erasing transition allowed. The corresponding language classes are denoted by $\mathcal{C}_0$, $\mathcal{C}'_0$, and $\mathcal{C}_0^\lambda = \mathcal{C}_0'^\lambda$, respectively.

Figure 1 shows an example of a CFA accepting the set $\{a^n b^n \mid n \geq 0\}$. The initial marking of this CFA is $\langle 0 \rangle$, final marking $\langle 3 \rangle$. it starts in 0 and counts the a's in 1. When 0 encounters b one token is put into 2, and then tokens in 1 are reduced whenever a b is encountered, until the end marker # is found.

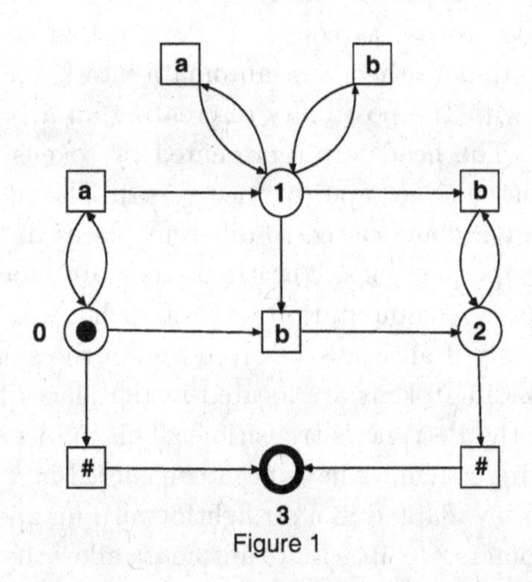

Figure 1

To facilitate the description of CFA and the proofs, the theory of multisets is used. With this it can be shown that it suffices that a CFA works in a leftmost parallel manner, i.e. that in a computation step all tokens point to a leftmost position on the tape and not to any left of it, and that transitions are allowed to fire in parallel, such that after this step all tokens either point to the next position, or to the same for a λ move.

It turns out that the language classes defined in such a way contain corresponding Petri net languages, the regular sets and the context-free languages, but are contained in the class of context-sensitive and recursively enumerable sets. Figure 2 shows the relations among these classes, some of them still open if proper or not. $\mathcal{L}_0$, $\mathcal{L}_0^{\blacktriangle}$, and $\mathcal{L}_0^{\lambda}$ denote Petri net language classes as defined in.[6] Actually, $\mathcal{C}_0' \subseteq NTIMESPACE(n^2, n)$.

2. Definitions

In the sequel we use the concept and notations of multisets.

First we recover the definition of Petri nets and languages defined by them.

Definition 2.1. *(P/T net)*

A place/transition Petri net (*P/T net*) is a quadruple $N = (P, T, g, \mu_0)$, where P is a finite set of places, T a finite set of transitions, g a mapping $g : T \rightarrow \mathbb{N}^{|P|} \times \mathbb{N}^{|P|}$ assigning to each $t \in T$ a pair $g(t) = (\alpha, \beta)$, and $\mu_0 \in \mathbb{N}^{|P|}$ the initial marking.

A transition $t \in T$ with $g(t) = (\alpha, \beta)$ is enabled at marking μ, if $\alpha \sqsubseteq \mu$. In this case t can *fire* (be *applied* on) marking μ, yielding the new marking $\mu' = (\mu \ominus \alpha) \oplus \beta$ where $\mu, \mu' \in \mathbb{N}^{|P|}$.

Definition 2.2. *(Petri net languages)*

Petri net languages are defined by a P/T net $N = (P, T, g, \mu_0)$ and a transition labelling function $\sigma : T \rightarrow \Sigma$ or $\sigma : T \rightarrow \Sigma \cup \{\lambda\}$ with a finite alphabet Σ.

For a transition sequence $\tau \in T^*$ σ is extended to T^* and $\sigma(\tau) \in \Sigma^*$ denotes the canonical extension of σ. For a sequence $\tau = t_1 \cdots t_n$ there is a sequence of markings $\mu_0, \mu_1, \cdots, \mu_n$ corresponding to τ with t_j leading from μ_j to μ_{j+1} for $0 \leq j < n$. To define a language several generating conditions are possible, e.g.

1. By deadlock (of the P/T net N).
2. By a final set of markings $\mathcal{F} \subseteq P^{\oplus} = \mathbb{N}^{|P|}$. This set should be recursive. In this case $\mu_n \in \mathcal{F}$ must hold for $\sigma(\tau)$ to be included in the language. Special cases are
   (a). $\mathcal{F} = \mathbb{N}^{|P|}$
   (b). $\mathcal{F}$ finite
   (c). $|\mathcal{F}| = 1$ singleton
   (d). $\mathcal{F} = \{\mu_f\}$ and $|\mu_f| = 1$.

In the sequel, following,[5,11] we consider variants 2(c) and 2(d) only. In addition, it has been shown there that $|\mu_f| = |\mu_0| = 1$ suffices, too.

Furthermore, only P/T nets with transitions t with $g(t) = (\alpha, \beta)$ where $\alpha \neq \mathbf{0}$ will be considered.

Definition 2.3. (*Petri net language classes*)
The following classes of Petri net languages can be defined:[5,11]
$\mathcal{L}_0^\lambda$ Petri net languages generated by Petri nets with final marking
$\mathcal{L}_0$ Petri net languages generated by λ-free Petri nets with final marking $\mu_f \neq \mu_0$, i.e. $\forall t \in T \;:\; \sigma(t) \neq \lambda$
$\mathcal{CSS} = \mathcal{L}_0^\blacktriangle$ with $\mathbf{X}^\blacktriangle = \mathbf{X} \cup \{L \cup \{\lambda\} \mid L \in \mathbf{X}\}$ for any language class $\mathbf{X}$.
Trivially, $\mathcal{L}_0 \subset \mathcal{CSS} = \mathcal{L}_0^\blacktriangle \subseteq \mathcal{L}_0^\lambda$.
Remark: If $\mu_0 \neq \mu_f$ is removed in the definition then $\mathcal{L}_0 = \mathcal{L}_0^\blacktriangle = \mathcal{CSS}$.

The next definitions introduce concurrent finite automata (CFA).

Definition 2.4. (*Concurrent finite automaton*)
A Concurrent Finite Automaton (CFA) is a triple $C = (N, \Sigma, \sigma)$ where $N = (P, T, g, \mu_0)$ is a P/T net, Σ is a finite tape alphabet for input, and $\sigma : T \to \Sigma \cup \{\lambda\}$ is a transition labelling function. Only P/T nets having transitions $t \in T$ with $g(t) = (\alpha, \beta)$ where $\alpha \neq \mathbf{0}$ are allowed. Transitions t with $\sigma(t) = \lambda$ are called λ-transitions, and such with $g(t) = (\alpha, \mathbf{0})$ erasing transitions.

An expression $(\nu(0), x(1), \nu(1), \cdots, x(m), \nu(m), x(m+1), \nu(m+1))$ denotes a *configuration* of a concurrent finite automaton C, where $\nu(i) \in \mathbb{N}^{|P|}$ for $0 \leq i \leq m+1$, $x(i) \in \Sigma$ for $1 \leq i \leq m$, and $x(m+1) = \# \notin \Sigma$, the *end marker* of the word $w = x(1) \cdots x(m)$ on the tape. In most cases it will be abbreviated by $\nu(0)x(1)\nu(1) \cdots x(m)\nu(m)x(m+1)\nu(m+1)$. Thus $\nu \;:\; \mathbb{N} \to \mathbb{N}^{|P|}$ is a function attaching to each tape position a multiset. $\bigoplus_{i=0}^{m+1} \nu(i)$ represents the marking of the Petri net N. The initial configuration is given by $\mu_0 x(1)\mathbf{0} \cdots \mathbf{0}x(m)\mathbf{0}\#\mathbf{0}$. # is introduced to have a uniform notation also for other kinds of automata where end markers are necessary.

Each token in a place of the P/T net represents a head on the tape. The multiset $\nu(i)$ represents the fact that there are heads of corresponding multiplicities on position i of the tape. In each step a transition of the P/T net N fires, taking away tokens from input places and corresponding heads from position i on the tape, and putting tokens on output places and corresponding heads on position $i+1$, or i if λ-transitions are concerned. Thus heads can move only to the right, or stay on the same position.

In the simplest application of a transition the CFA works *sequentially*. Only tokens corresponding to one tape position i can be taken away, and new tokens put have to correspond to tape position $i+1$, or i in case of

λ-transitions. Then the successor configuration after applying a transition t with $g(t) = (\alpha, \beta)$ and $\sigma(t) = x_i$ (or $\sigma(t) = \lambda$) is given by

$\nu'(i) = \nu(i) \ominus \alpha$, $\nu'(i+1) = \nu(i+1) \oplus \beta$ and $\nu'(j) = \nu(j)$ for $j \neq i$ and $j \neq i+1$,

or $\nu'(i) = \nu(i) \ominus \alpha \oplus \beta$ and $\nu'(j) = \nu(j)$ for $j \neq i$,

respectively.

Several acceptance modes, analogous to Petri nets, can be defined.

1. Acceptance by deadlock (of the P/T net N).

2. Acceptance by a final set of markings $\mathcal{F} \subseteq P^{\oplus} = \mathbb{N}^{|P|}$. Again, this set should be recursive. Special cases are

(a). $\mathcal{F}$ finite

(b). $|\mathcal{F}| = 1$ singleton

(c). $\mathcal{F} = \{\mu_f\}$ with final configuration $\mathbf{0}x_1 \cdots \mathbf{0}x_m\mathbf{0}\#\mu_f$

(d). $\mathcal{F} = \{\mu_f\}$ with final configuration $\mathbf{0}x_1 \cdots \mathbf{0}x_m\mathbf{0}\#\mu_f$ and $|\mu_f| = 1$.

In any case, reaching a final marking, at least one head has to be on position after #, i.e. the entire input word has to be read (2(b)). Another possibility is that all heads represented by a final marking have to be on position after # (2(c), 2(d)).

Definition 2.5. (*Language classes*)

$\mathcal{C}_{01}^{\lambda}$ languages accepted in mode 2(b)

$\mathcal{C}_{02}^{\lambda}$ languages accepted in mode 2(c)

$\mathcal{C}_{03}^{\lambda}$ languages accepted in mode 2(d)

$\mathcal{C}_{01}$ languages accepted by λ-free CFA in mode 2(b)

$\mathcal{C}_{02}$ languages accepted by λ-free CFA in mode 2(c)

$\mathcal{C}_{03}$ languages accepted by λ-free CFA in mode 2(d)

Trivially, $\mathcal{C}_{03} \subseteq \mathcal{C}_{02} \subseteq \mathcal{C}_{01}$ and $\mathcal{C}_{03}^{\lambda} \subseteq \mathcal{C}_{02}^{\lambda} \subseteq \mathcal{C}_{01}^{\lambda}$, as well as $\mathcal{C}_{0i}^{\blacktriangle} \subseteq \mathcal{C}_{0i}^{\lambda}$ for $1 \leq i \leq 3$.

Language classes for which transitions t with $g(t) = (\alpha, \mathbf{0})$ are allowed, will be denoted by $\mathcal{C}'_{0i}$ or $\mathcal{C'}^{\lambda}_{0i}$.

Similarly, $\mathcal{C}'_{03} \subseteq \mathcal{C}'_{02} \subseteq \mathcal{C}'_{01}$ and $\mathcal{C}'^{\lambda}_{03} \subseteq \mathcal{C}'^{\lambda}_{02} \subseteq \mathcal{C}'^{\lambda}_{01}$, as well as $\mathcal{C}'^{\blacktriangle}_{0i} \subseteq \mathcal{C}'^{\lambda}_{0i}$ for $1 \leq i \leq 3$.

The classes of *regular*, *context-free*, *context-sensitive*, and *recursively enumerable* languages are denoted by **REG**, **CF**, **CS**, anf **RE**, respectively.

So far the application of transitions has been defined in a sequential manner. However, it is also possible to apply transitions in a parallel (concurrent) way.

Definition 2.6. (*Parallel transitions*)

Consider the set $T = \{t_1, \cdots, t_{|T|}\}$ with $g(t_j) = (\alpha_j, \beta_j) \in \mathbb{N}^{|P|} \times \mathbb{N}^{|P|}$ for $1 \leq j \leq |T|$. For a multiset $\tau = \bigoplus_{j=1}^{|T|} \tau(t_j)t_j \in T^{\oplus} = \mathbb{N}^{|T|}$ of transitions define $g(\tau) = (\alpha_\tau, \beta_\tau)$ by $\alpha_\tau = \bigoplus_{j=1}^{|T|} \tau(t_j)\alpha_j$ and $\beta_\tau = \bigoplus_{j=1}^{|T|} \tau(t_j)\beta_j$.

In a distributed parallel application one has $\tau_i = \bigoplus_{j=1}^{|T|} \tau_i(t_j)t_j$ and $\tau = \bigoplus_{i=1}^{m} \tau_i$. This means that if $\nu(0)x(0) \cdots \nu(m)x(m+1)\nu(m+1)$ is a configuration, then the application of τ is defined in the following way, called a step in:[10]

$\sigma(t_j) = x(i)$ for t_j with $\tau_i(t_j) > 0$, $\alpha_{\tau_i} \sqsubseteq \nu(i)$, $\nu'(i) = \nu(i) \ominus \alpha_{\tau_i}$,
$\nu'(i+1) = \nu(i+1) \oplus \beta_{\tau_i}$,
and in the case of λ-transitions (at some tape position)
$\sigma(t_i) = \lambda$ for t_j with $\tau_i(t_j) > 0$, $\alpha_{\tau_i} \sqsubseteq \nu(i)$, $\nu'(i) = \nu(i) \ominus \alpha_{\tau_i} \oplus \beta_{\tau_i}$.

A parallel application at one tape position i only, is defined as follows $\tau = \bigoplus_{j=1}^{|T|} \tau(t_j)t_j$ where $\sigma(t_j) = x(i)$ for t_j with $\tau(t_j) > 0$.

Note that both modes contain applications of single (sequential) transitions as special case.

In the last case leftmost application is possible, too. This means that the transitions are applied for the position with smallest i first, as long as possible.

Trivially, in this case parallel and sequential application are equivalent since a sequence of applications of single transitions can be also achieved by one parallel application, and vice versa.

3. Normal Forms

It can be shown that concurrent finite automata and their modes of processing can be transformed into normal forms. For proofs see.[2]

Lemma 3.1. *CFA having one initial place p_0 with initial marking $\langle p_0 \rangle$ and one final marking $\langle p_f \rangle$ suffice.*

Lemma 3.2. *For the cases $\mathcal{C}_{02}^{\lambda}$, $\mathcal{C}_{02}$, $\mathcal{C}_{02}'^{\lambda}$, and $\mathcal{C}_{02}'$ one final place suffices, i.e. $\mathcal{C}_{02}^{\lambda} \subseteq \mathcal{C}_{03}^{\lambda}$, $\mathcal{C}_{02} \subseteq \mathcal{C}_{03}$, $\mathcal{C}_{02}'^{\lambda} \subseteq \mathcal{C}_{03}'^{\lambda}$, and $\mathcal{C}_{02}' \subseteq \mathcal{C}_{03}'$.*

Lemma 3.3. *$\mathcal{C}_{01}^{\lambda} \subseteq \mathcal{C}_{02}^{\lambda}$, $\mathcal{C}_{01} \subseteq \mathcal{C}_{02}$, $\mathcal{C}_{01}'^{\lambda} \subseteq \mathcal{C}_{02}'^{\lambda}$, and $\mathcal{C}_{01}' \subseteq \mathcal{C}_{02}'$.*

From the previous lemmata we get the following theorem, yielding only four implicitly defined language classes, denoted by $\mathcal{C}_0^{\lambda}$, $\mathcal{C}_0$, $\mathcal{C}_0'^{\lambda}$, and $\mathcal{C}_0'$, respectively.

Theorem 3.1. $\mathcal{C}_{01}^{\lambda} = \mathcal{C}_{02}^{\lambda} = \mathcal{C}_{03}^{\lambda} = \mathcal{C}_0^{\lambda}$, $\mathcal{C}_{01} = \mathcal{C}_{02} = \mathcal{C}_{03} = \mathcal{C}_0$, *as well as* $\mathcal{C}'^{\lambda}_{01} = \mathcal{C}'^{\lambda}_{02} = \mathcal{C}'^{\lambda}_{03} = \mathcal{C}'^{\lambda}_{0}$, $\mathcal{C}'_{01} = \mathcal{C}'_{02} = \mathcal{C}'_{03} = \mathcal{C}'_{0}$. *Furthermore, one initial and one final place suffice.*

Another result is

Theorem 3.2. $\mathcal{C}^{\blacktriangle}_{0} = \mathcal{C}_0$, $\mathcal{C}^{\lambda\blacktriangle}_{0} = \mathcal{C}^{\lambda}_{0}$, $\mathcal{C}'^{\blacktriangle}_{0} = \mathcal{C}'_{0}$, $\mathcal{C}'^{\lambda\blacktriangle}_{0} = \mathcal{C}'^{\lambda}_{0}$.

Finally, it can be shown that for $\mathcal{C}_0$, $\mathcal{C}^{\lambda}_{0}$, $\mathcal{C}'_{0}$, and $\mathcal{C}'^{\lambda}_{0}$ the leftmost maximal parallel mode of application of transitions suffices.

Theorem 3.3. *Languages from all languages classes* $\mathcal{C}_0$, $\mathcal{C}^{\lambda}_{0}$, $\mathcal{C}'_{0}$, *and* $\mathcal{C}'^{\lambda}_{0}$ *can also be accepted if transitions are applied in a leftmost maximal parallel manner, emptying all places with tokens pointing to the current tape position.*

4. Relations to Other Language Classes

In this section we present relations to Petri net language classes and to the Chomsky hierarchy. Detailed proofs can be found in.[2]

Since finite automata can be seen as special Petri nets one gets

Theorem 4.1. $\mathbf{REG} \subseteq \mathcal{C}^{\blacktriangle}_{0} = \mathcal{C}_0$.

From the example in Figure 1 follows

Theorem 4.2. $\mathbf{REG} \subset \mathcal{C}_0$.

Theorem 4.3. *All Petri net languages are accepted by CFA with final marking condition and the final set corresponding to that of the Petri net, i.e.*

$\mathcal{L}^{\lambda}_{0} \subseteq \mathcal{C}^{\lambda}_{0}$, $\mathcal{L}_0 \subseteq \mathcal{C}_0$, $\mathcal{CSS} \subseteq \mathcal{C}^{\blacktriangle}_{0} = \mathcal{C}_0$.

Theorem 4.4. $\mathcal{C}'_{0} \subseteq \mathcal{C}^{\lambda}_{0}$ *and* $\mathcal{C}^{\lambda}_{0} = \mathcal{C}'^{\lambda}_{0}$.
This is shown by replacing erasing transition by λ-transirions.

Theorem 4.5. $\mathcal{C}'_{0} \subseteq \mathbf{NTimeSpace(n^2, n)}$.

This can be shown by simulating a CFA by a LBA with its tape partitioned into 6 tracks. The total number of tokens in any reachable configuration is bounded by mn^k where n is the input length, needing only linear space for storage.

This implies

Corollary 4.1. $\mathcal{C}'_{0} \subset \mathbf{CS}$.

5. Characterization and Decidability Results

The following two theorems exhibit characterizations of context-free and recursively enumerable languages using CFA and codings or homomorphisms, shown in.[8]

Theorem 5.1. *For every context-free language L, there is a coding h and a non-erasing λ-free CFA C such that $L = h(L(C))$. Thus,* $\mathbf{CF} \subseteq \mathcal{H}^{cod}(\mathcal{C}_0)$.

This is shown by a CFA whose Petri net has two places, and every computation of it corresponds to a leftmost derivation of the context-free grammar.

Theorem 5.2. *For every recursively enumerable language L, there is a (possibly erasing) homomorphism h and a non-erasing λ-free CFA C such that $L = h(L(C))$. Thereby, h and C are effectively constructible. Therefore,* $\mathbf{RE} = \hat{\mathcal{H}}(\mathcal{C}_0)$.

For this the CFA simulates a 2-counter machine, storing the counters in two places of the Petri net.

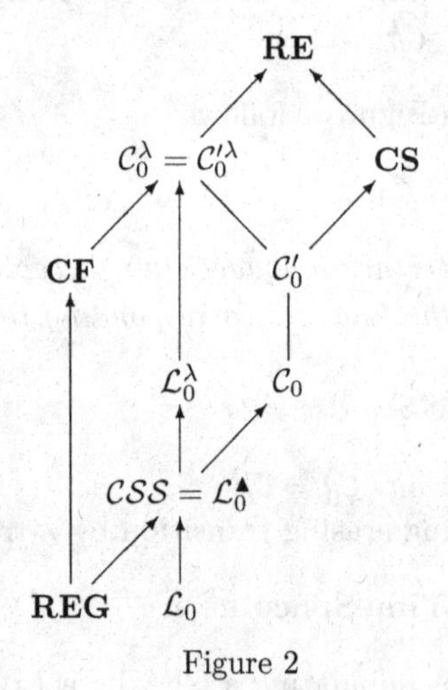

Figure 2

The next theorems show the undecibality of the emptiness problem and the decidability of the word problem for CFA, a fact similar to LBA.

From Theorem 5.2 follows

Corollary 5.1. *The emptiness problem for CFA is undecidable.*

The decidability of the reachability problem for Petri nets implies

Theorem 5.3. *The word problem for CFA is decidable.*

From this follows.

Corollary 5.2. $\mathcal{C}_0^\lambda \subset$ **RE**.

The results can be summarized as a language class hierarchy given in Figure 2 above where there are still two open problems, namely whether $\mathcal{C}_0 \subseteq \mathcal{C}_0'$ and $\mathcal{C}_0' \subseteq \mathcal{C}_0^\lambda$ are proper.

6. Closure Properties

In this section we present some closure properties of language classes defined by CFA. Detailed proofs can be found in.[8]

From Theorem 5.2 follows

Theorem 6.1. $\mathcal{C}_0$, $\mathcal{C}_0'$, $\mathcal{C}_0^\lambda$ *are not closed under arbitrary homomorphisms.*

Theorem 6.2. $\mathcal{C}_0^\lambda$ *is closed under non-erasing homomorphisms.*

This is shown by decomposing every non-erasing homomorphism into simple ones.

As a direct consequence from Theorems 5.1 and 6.2 follows

Corollary 6.1. **CF** $\subseteq \mathcal{C}_0^\lambda$.

By some more technical proofs the following closure properties can also be shown:

Theorem 6.3. $\mathcal{C}_0$, $\mathcal{C}_0'$ *and* $\mathcal{C}_0^\lambda$ *are closed unter union.*

Theorem 6.4. $\mathcal{C}_0$, $\mathcal{C}_0'$ *and* $\mathcal{C}_0^\lambda$ *are closed under intersection.*

Corollary 6.2. $\mathcal{C}_0$, $\mathcal{C}_0'$, $\mathcal{C}_0^\lambda$ *are closed under intersection with regular sets.*

Theorem 6.5. $\mathcal{C}_0$ *and* $\mathcal{C}_0^\lambda$ *are closed unter concatenation.*

Theorem 6.6. $\mathcal{C}_0$ *and* $\mathcal{C}_0'$ *are closed under inverse homomorphisms.*

All results are summarized in the following diagram. Here $\cap R$, $L_1 \cdot L_2$, $\cup$, $\cap$, h^{-1}, h, λ-free h stand for intersection with regular languages, concatenation, union, intersection, inverse homomorphism, arbitrary homomorphism and non-erasing homomorphism, respectively.

Operator	$\mathcal{C}_0$	$\mathcal{C}'_0$	$\mathcal{C}^\lambda_0$
$\cap R$	+	+	+
$\cdot$	+	?	+
$\cup$	+	+	+
$\cap$	+	+	+
h^{-1}	+	+	?
h	−	−	−
λ-free h	?	?	+

Figure 3

References

1. B. Farwer, M. Kudlek, H. Rölke : *Petri-Net-Controlled Machine Models.* FBI-Bericht 274, Hamburg, 2006.
2. B. Farwer, M. Jantzen, M. Kudlek, H. Rölke, G. Zetzsche : *Petri Net Controlled Finite Automata.* FI, vol. 85 (1-4), pp. 111-121, 2008.
3. B. Farwer, M. Kudlek, H. Rölke : *Concurrent Turing Machines.* FI vol. 79 (3-4), pp. 303-317, 2007.
4. S. Greibach : *Remarks on Blind and partially Blind Multicounter Machines.* TCS **7**, pp. 311-324, 1978.
5. M. Hack : *Petri Net Languages.* MIT, Project MAC, Computation Structures Group Memo 124, 1975.
6. M. Jantzen : *On the Hierarchy of Petri Net Languages.* RAIRO **13**, no. 1, pp. 19-30, 1979.
7. M. Jantzen : *Synchronization Operations and Formal Languages.* In : *Fifth Conference of Program Designers.* ed. A. Iványi, Eötvös Loránd University, Faculty of Natural Sciences, Budapest, pp. 15-26, 1989.
8. M. Jantzen, M. Kudlek, G. Zetzsche : *Language Classes Defined by Concurrent Finite Automata.* FI, vol. 85 (1-4), pp. 267-280, 2008.
9. E. W. Mayr : *An Algorithm for the General Petri Net Reachability Problem.* Proc. 13th Ann. ACM STOC, pp. 238-246, 1981; SIAM Journ. of Computation **13**, pp. 441-460, 1984.
10. M. Mukund : *Petri Nets and Step Transition Systems.* IJFCS **3**, no. 4, pp. 443-478, 1992.
11. J. L. Peterson : *Computation Sequence Sets.* JCSS **13**, pp. 1-24, 1976.
12. K. Rüdiger Reischuk : *Komplexitätstheorie, Band I: Grundlagen,* B. G. Teubner, 1999.

Received: June 28, 2009

Revised: April 25, 2010

FINITELY EXPANDABLE DEEP PDAs

PETER LEUPOLD*

Fachbereich Elektrotechnik/Informatik,
Fachgebiet Theoretische Informatik,
Universität Kassel, Kassel, Germany
eMail: Peter.Leupold@web.de

ALEXANDER MEDUNA

Brno University of Technology,
Faculty of Information Technology, Department of Information Systems,
Božetěchova 2, Brno 61266, Czech Republic
eMail: meduna@fit.vutbr.cz

As special cases of deep pushdown automata, we discuss *finitely expandable deep pushdown automata* that always contain a bounded number of non-input symbols in their pushdown stores. Based on these automata, it establishes an infinite hierarchy of language families that coincides with the hierarchy resulting from matrix grammars of finite index. Thus finitely expandable deep pushdown automata can are an automaton counterpart to these grammars. It also follows that deleting transitions do not add more power to finitely expandable deep pushdown automata. In a final section, we suggest some open problems.

Keywords: Pushdown Automata, Matrix Grammars.

1. Introduction

Consider the standard conversion of context-free grammars to equivalent pushdown automata that act as general top-down parsers, for example as presented by Autebert et al.,[1] page 176. Recall that during every move, the resulting pushdown automata either pop or expand their pushdown stores depending on the symbol occurring on the pushdown top. If an input symbol occurs there, the automata make a pop. If a non-input symbol occurs on the pushdown top, these automata expand their pushdown stores so they replace the top nonterminal with a string. Inspired by the way

*This work was done while Peter Leupold was funded as a post-doctoral fellow by the Japanese Society for the Promotion of Science under grant number P07810.

general top-down parsers work, deep pushdown automata, introduced by Meduna,[2] represent their slight generalization. Indeed, they work exactly as these parsers except that they can make expansions deeper in the pushdown store. More precisely, they can make an expansion of depth $4m$ so they change the m-th non-input symbol occurring in the pushdown store to a string, where $m \geq 1$. In the present paper, we discuss n-expandable deep pushdown automata that always contain n or fewer non-input symbols occurring in their pushdown stores, where $n \geq 0$. We demonstrate that n-expandable deep pushdown automata are equivalent to matrix grammars of index n (see the book by Dassow and Păun[3] for the definition of these grammars). Based on this equivalence, we demonstrate that the language family accepted by j-expandable deep pushdown automata represents a proper language subfamily of the language family accepted by $j+1$-expandable deep pushdown automata language, for all $j \geq 0$. As this infinite hierarchy of language families coinciding with the hierarchy resulting from matrix grammars of finite index, finitely expandable deep pushdown automata can be seen as the automaton-based counterpart to matrix grammars of finite index. In its conclusion, this paper formulates some open problems.

2. Definitions

Terms and notation from general formal language theory, logic and set theory are assumed to be known to the reader and can be looked up in standard textbooks like the ones by Harrison[4] or Meduna[5] . We will denote the empty word by λ. $|u|$ shall be the length of the word u. $|u|_\Sigma$ denotes the number of occurences of symbols from the set Σ in the word u.

With this, we already come to the central definition in our context, the one of deep push-down automata.

Definition 2.1. A *deep pushdown automaton (Deep PDA)* is a 7-tuple $M = (Q, \Sigma, \Gamma, R, s, S, F)$, where Q is a finite set of states, Σ is the input alphabet, and Γ is the pushdown alphabet; Σ is a subset of Γ, there is a bottom symbol $\#$ in $\Gamma \setminus \Sigma$. $s \in Q$ is the start state, $S \in \Gamma$ is the start pushdown symbol, $F \subset Q$ is the set of final states. The transition relation R is a finite set that contains elements of $(I \times Q \times (\Gamma \setminus (\Sigma \cup \{\#\}))) \times (Q \times (\Gamma \setminus \{\#\})^+)$ and of $(I \times Q \times \{\#\}) \times (Q \times (\Gamma \setminus \{\#\})^* \{\#\})$; here I is a a set of integers of the form $\{i : 1 \leq i \leq n\}$ for some n. We will write $mqA \rightarrow pv \in R$ instead of $(m, q, A, p, v) \in R$.

The operation of such a deep PDA M is as follows. M *pops* its pushdown from x to y, symbolically written as $x \;_p\!\Rightarrow y$, if $x = (q, au, az), y = (q, u, z)$, where $a \in \Sigma, u \in \Sigma^*, z \in \Gamma^*$. M *expands* its pushdown from x to y, symbolically written as $x \,{}_e\!\Rightarrow y$, if $x = (q, w, uAz), y = (p, w, uvz), mqA \to pv \in R$, where $q, p \in Q, w \in \Sigma^*, A \in \Gamma, u, v, z \in \Gamma^*$, and $|u|_{\Gamma \setminus \Sigma} = m - 1$. We call a sequence of pops, expensions and moves a *computation* of M.

To express that M makes $x \,{}_e\!\Rightarrow y$ according to $mqA \to pv$, we write $x \,{}_e\!\Rightarrow y\ [mqA \to pv]$. We say that $mqA \to pv$ is a *rule of depth* m; accordingly, $x \,{}_e\!\Rightarrow y\ [mqA \to pv]$ is an *expansion of depth* m. M makes a *move* from x to y, symbolically written as $x \Rightarrow y$, if M makes either $x \,{}_e\!\Rightarrow y$ or $x \,{}_p\!\Rightarrow y$.

If $n \in I$ is the minimal positive integer such that each of M's rules is of depth n or less, we say that M *is of depth* n, symbolically written as ${}_nM$. In the standard manner, extend ${}_p\!\Rightarrow, {}_e\!\Rightarrow$, and $\Rightarrow$ to ${}_p\!\Rightarrow^m, {}_e\!\Rightarrow^m$, and $\Rightarrow^m$, respectively, for $m \geq 0$; then, based on ${}_p\!\Rightarrow^m$, ${}_e\!\Rightarrow^m$, and $\Rightarrow^m$, define ${}_p\!\Rightarrow^+, {}_p\!\Rightarrow^*, {}_e\!\Rightarrow^+, {}_e\!\Rightarrow^*, \Rightarrow^+$, and $\Rightarrow^*$. Let M be of depth n, for some $n \in I$. We define the *language accepted by* ${}_nM$, $L({}_nM)$, as $L({}_nM) = \{w \in \Sigma^* | (s, w, S\#) \Rightarrow^* (f, \lambda, \#)$ in ${}_nM$ with $f \in F\}$.

This is the original type of deep pushdown automaton. As explained in the introduction, here we want to investigate automata with a constant bound on the number of nonterminals that can be present on the stack at any given time. Such an *n-expandable deep pushdown automaton* is an 8-tuple $M = (Q, \Sigma, \Gamma, R, s, S, F, n)$, where all components are as for the original deep PDA, and n is a positive integer. Expanding transitions can only be applied in a step $x \,{}_e\!\Rightarrow y$ if y will not contain more than n nonterminals. Otherwise either some other transition must be applied or the computation stops without accepting. A deep PDA that is n-expandable for some n is called *finitely expandable*.

We do not use automata of a specified depth here. Depths greater than n would not have any effect anyway. For a depth k smaller than n it might be an interesting question, if the corresponding deep PDAs are equivalent to k-expandable ones, but we do not treat this problem here. Therefore our n-limited deep PDAs are able to expand any of their up to n nonterminals at any given time,

Finally, we recall matrix grammars from regulated rewriting, since they will play a central role in the next section.

Definition 2.2. A *matrix grammar* is a quadruple (N, T, M, S), where N and T are two disjoint, finite alphabets and $S \in N$. M is a finite set of sequences of the form $(r_1, r_2, \ldots, r_k)$, where the r_i are all rewrite rules from the set $N \times (N \cup T)^+$.

The sequences of the form $(r_1, r_2, \ldots, r_k)$ are called *matrices*. In a derivation of a matrix grammar, only complete matrices can be applied. That is, all the rewrite rules in the matrix must be applied in the order given. Then the next matrix can be applied. For example, the matrix $(A \to ab, B \to bb, A \to a)$ cannot be applied to the string $aACCBbB$, because the first rule will rewrite the only A in the string; then the third rule cannot be applied any more.

This definition is narrower than the usual one, where more general rewrite rules are admitted; but here we will only use context-free rewrite rules and therefore include only these in our definition. A matrix grammar of index k for a positive integer is one whose sentential form must not contain more than k nonterminals in any derivation step, otherwise the derivation is stopped and not successful. A matrix grammar that is of index k for some positive integer k is called of finite index.

3. A Hierarchy of Finitely Expandable Deep PDAs

With the original definition of deep PDA as given in the preceding section, computations are in principle free to alternate between steps that pop terminals and ones that expand non-terminals. However, if there exists an accepting computation for a word, there is always a computation that first does all the expansions and only then starts to pop. Already the fact that expanding transitions do not read any input suggest that this is possible, because in this way there is no dependence between the two types of operations.

Lemma 3.1. *If a deep PDA M accepts a word w, then there is an accepting computation of M on w, which performs only pops after the first pop operation.*

Proof. Let γ be a computation of a deep PDA M. If γ already fulfills our lemma's conditions there is nothing to show. Otherwise there exists a sequence of two configurations in γ such that the first one is a pop operation, the following one an expansion of a non-terminal. Let the former pop the symbol a, and let the latter be the transition $nqA \to pU$. This means that the configuration just before these two transitions must have the symbol a on the top of the push-down. Since the pop does not change the state or introduce or change non-terminals, q is already the current state and A is the n-th non-terminal in the push-down.

We will write this configuration $aw; q; auAv$, where the first component is the contents of the input tape, the second the current state, and the third one the stack contents with the top on the left hand side. u contains $n-1$ nonterminals. The two computation steps described above are

$$aw; q; auAv \;_p\!\Rightarrow w; q; uAv \;_e\!\Rightarrow w; p; uUv \; [nqA \rightarrow pU].$$

However, $nqA \rightarrow pU$ can also be applied before the pop, because the depth of A is already n and the current state is q. This expansion, on the other hand, does not change the fact that a is the top-most symbol on the push-down, and thus it can be popped now, because the pop operation is defined for all combinations of terminals and states. The resulting configuration is the same as for popping first and then applying $nqA \rightarrow pU$:

$$aw; q; auAv \;_e\!\Rightarrow aw; p; auUv \; [nqA \rightarrow pU] \;_p\!\Rightarrow w; p; uUv.$$

Summarizing, we can obtain another computation γ' where the pop and the transition are exchanged, but the result is the same as for γ. Since this process can be iterated, all the pop operations can be moved to the end of the computation step by step, and this can be done for any computation. □

So during a first phase in any deep PDA computation it is possible to do only expansions on the stack and basically work like a grammar deriving a word of only terminal symbols; though, of course, the state of the automaton controls which rules are applicable where in every step. Then the word generated in this way is matched against the input. The computation accepts if and only if the two words on the input tape and the stack are equal and if the last state reached in the expansion phase was final, since the pops will not change the state anymore. The same proof works for finitely expandable deep PDAs, and we will now use this property in proving that these automata are equivalent to matrix grammars of finite index.

Theorem 3.1. *For all integers $n > 0$, matrix grammars of index n generate the same class of languages that is accepted by n-expandable deep PDAs.*

Proof. It is almost straight-forward to see that finitely expandable deep PDAs can simulate matrix grammars of finite index. They can simulate exactly the grammar's derivation in their stack. All n or less non-terminals can be expanded at any time. and in its state the automaton can store information about what matrix it is currently simulating. We will now define such a deep PDA for a given matrix grammar (N, T, M, S). All the transitions we define below will be defined for all possible depths, i.e. for all positions from one to n. To increase readability, we will leave away the depth in the transitions and will take it for understood that this means defining them for all depths from one to n.

So let $G = (N, T, M, S)$ be a matrix grammar of index n. Our deep PDA is $\mathcal{M} = (\{s\} \cup (M \times \{1, 2, \ldots, \ell\}), T, \{S\} \cup N \cup T, R, s, S, \{s\}, n)$, where ℓ is the maximal number of rules in a matrix of G. The only component left to define is the transition function R. For every matrix $m : A_1 \to v_1, A_2 \to v_2, \ldots, A_k \to v_k$ from M it contains the following transitions where m_i denotes the element (m, i) from the set of states:

$$sA_1 \to m_1v_1, m_1A_2 \to m_2v_2, m_2A_3 \to m_3v_3, \ldots$$

$$\ldots, m_{k-2}A_{k-2} \to m_{k-1}v_{k-2}, m_{k-1}A_k \to sv_k.$$

We omit the depths of the transitions. All of them should be defined for all possible depths 1 to n such that every grammar rule can be simulated in every nonterminal position. The states m_i are used only in these transitions simulating m, thus there is only one possible transition for each of these states. This means that the automaton either executes the entire sequence of transitions corresponding to the matrix m, or it will stop without accepting the word.

To see that G and $\mathcal{M}$ are equivalent, let us look at the sentential forms of a derivation and the corresponding stack contents. Both start with S. As stated above, rules of the grammar and expanding transitions of the deep PDA enact exactly the same changes. Thus as long as we do not use any pops, the sentential forms and the corresponding stack contents are identical, and it is obvious that every terminal word that can be derived by the grammar can also be generated on the stack. Lemma 3.1 shows that all computations of M can be normalized in this manner. Finally, after finishing the simulation of a matrix, M is always in its final state s and can thus accept the word on the stack if it consists only of terminals and matches the input word. Therefore the deep PDA accepts the same language that the grammar generates.

For showing the inverse inclusion, we need a more sophisticated argumentation. We will construct a matrix grammar for a given deep PDA, and again the sentential forms and the stack contents will be in close correspondence. However, since the grammar does not distinguish between different positions of non-terminals while the automaton does, we need to store information about their position in the non-terminals themselves. Whenever new ones are introduced or when one is rewritten to a string of only terminals, the following ones need to be updated, because their positions change. This will be ensured by putting all the rules involved in one matrix.

We now proceed to define a matrix grammar simulating an n-expandable deep PDA $(Q, \Sigma, \Gamma, R, s, S, F, n)$. The set of non-terminals will be

$$(N \times \{2, \ldots, n\}) \cup (N \times Q) \cup (\{\varepsilon\} \times \{2, \ldots, n\}) \cup \{S\},$$

where $N = \Gamma \setminus \Sigma$, and the second component will contain the position of a given stack symbol, if it is not the first one, or it will contain the state, if the symbol is the first one. The start symbol will be S, and the matrix $(S \to (S, s)(\varepsilon, 2)(\varepsilon, 3) \cdots (\varepsilon, n))$ is the only one containing a rule rewriting S. The symbols with ε are placeholders for further non-terminals.

For the matrices simulating the automaton's transitions, the simplest case is when exactly one non-terminal is on the right-hand side of the rule. Then the positions of non-terminals in the other parts of the sentential form are not affected. So for every transition $kpA \to quBv$ with $uv \in T^*$ we add the matrices $((X, p) \to (X, q), (A, k) \to u(B, k)v)$ for all non-terminals X and all k between two and n. The first rule changes the state that is stored in the sentential form's first nonterminal, the second rule does the expansion at position k. If $k = 1$, then the matrix is singleton and contains the rule $(A, p) \to u(B, q)v$, which rewrites the non-terminal and changes the state in a single step.

If a rule of the grammar produces more than one non-terminal, then the depth in the simulated stack of all the non-terminals to the right of that application side is changed. Therefore these positions need to be updated accordingly. The problem here is that we do not know, how many non-terminals there are. This is where the function of the symbols containing ε becomes clear. With their presence, we always have exactly n non-terminals during the simulation of the deep PDA, and thus the matrices can be designed with this condition.

So for a transition $kpA \to qu_0B_1u_1B_2 \ldots u_{\ell-1}B_\ell u_\ell$, first note that it can only be executed, if there are at most $n - \ell + 1$ nonterminals on the stack; otherwise the condition of being finitely expandable is violated. This, on the other hand, means that the current sentential form must contain at least $\ell - 1$ placeholders containing ε. We test for this by starting the matrices with sequences $(X, i) \to (X, i), (\varepsilon, i + 1) \to (\varepsilon, i + 1)$, which do not change anything but establish that the first placeholder is at position $i + 1$. As a second step, all the nonterminals from positions $k + 1$ to i are moved $\ell - 1$ positions to the right and replaced by the nonterminal put there by the deep PDA's expansion. This is done by sequences $(Y, k + j) \to (B_{j+1}, k + j)u_{j+1}, (Z, k+j+\ell-1) \to (Y, k+j+\ell-1)$ for the first $\ell - 1$ nonterminals, which are replaced by the new ones from the expanding transition. The following ones are copied by sequences $(Y, m) \to (Y, m), (Z, m + \ell - 1) \to (Y, m+\ell-1)$; here the first rule checks, which nonterminal is at the original position, the second rule copies it $\ell - 1$ positions to the right. This process must start from the right side in order not to delete any nonterminal before it has been copied. Finally in position k we must apply $(A, k) \to u_0(B_1, k)u_1$ and in position one $(X, p) \to (X, q)$ to update the state.

Summarizing, the matrices for simulating $kpA \to qu_0B_1u_1B_2 \ldots u_{\ell-1}B_\ell u_\ell$ are the following, where i is such that $n \geq i > k$:

$$
\begin{aligned}
&((X, i) \to (X, i), (\varepsilon, i+1) \to (\varepsilon, i+1),\\
&\left.\begin{aligned}&(Y_m, m) \to (Y_m, m),\\ &(Z_m, m+\ell-1) \to (Y_m, m+\ell-1),\end{aligned}\right\}\text{for } m \text{ from } i \text{ down to } k+\ell\\
&\left.\begin{aligned}&(Y_m, m) \to (B_{m-k+1}, m)u_{m-k+1},\\ &(Z_m, m+\ell-1) \to (Y_m, m+\ell-1),\end{aligned}\right\}\text{for } m \text{ from } k+\ell-1 \text{ down to } k+1\\
&(A, k) \to u_0(B_1, k)u_1, (X', p) \to (X', q))
\end{aligned}
$$

One matrix is defined for each possible combination of $X, X', Y_m \in N$, $Z \in N \cup \{\varepsilon\}$. If $k = 1$, then the last line is condensed into one single rule $(A, p) \to u_0(B_1, q)u_1$. Notice that the number of nonterminals in the sentential form remains constant during application of these matrices and thus the condition of being of finite index is complied with.

again we need matrices that change state and do the expansion in one, so the first two rules are contracted into $(A, p) \to u_1(B_1, q)u_2(B_2, 2)u_3 \ldots u_\ell(B_\ell, \ell)u_{\ell+1}$.

Finally, there remains the case where a non-terminal is expanded to a string of terminals only. In this case one new placeholder must be introduced, and the position of the nonterminals right of the rule application must be decreased by one. Let the transition be $kpA \to qu$ with $u \in T^*$. Again, we start by testing for the border between nonterminals and placeholders by $(X, i) \to (X, i), (\varepsilon, i+1) \to (\varepsilon, i+1)$. Then we can go from left to right, first update the state, then simulate the rule application, and finally simulate the movement of the other nonterminals to the left by adjusting their position numbers and insert a new ε. The matrices are:

$$
\begin{aligned}
&((X, i) \to (X, i), (\varepsilon, i+1) \to (\varepsilon, i+1),\\
&(X', p) \to (X', q), (A, k) \to u,\\
&(Y_m, m) \to (Y_m, m-1) \text{ for } m \text{ from } k+1 \text{ to } i\\
&(X, i) \to (X, i-1)(\varepsilon, i))
\end{aligned}
$$

In this case, there is also the possibility that we have n nonterminals present and no ε. In that case the first line simply contains $(X, n) \to (X, n)$, which establishes that the last position is occupied by a nonterminal. For $k = 1$ there is a little modification necessary, because deleting the first nonterminal would also delete the state of the deep PDA. Therefore the second line is $(A, p) \to u, (Y, 2) \to (Y, q)$, and the third line is only for m from 3 to i. Notice that the rule $(X, i) \to (X, i-1)(\varepsilon, i)$, which increases the number of nonterminals, is applied only after the nonterminal at position k has been deleted. In this way the condition of being of finite index is complied with also here.

When removing the left-most non-terminal, however, there is also the possibility that it is the only one. In this case, this would be the last expansion of a non-terminal, and the simulation should stop. The only matter left to resolve, when the last non-terminal is deleted, is the fact that in the end of the simulation, in addition to the word of terminals, we have the remaining placeholders. They should be removed after the last expansion of a non-terminal in an accepting computation. Such an expansion can only take place in the first position, otherwise there are other non-terminals left that would be expanded further. Therefore we add for all transitions removing the first non-terminal matrices that do this expansion and in addition remove placeholders from positions 2 to n; this way these matrices are only applicable if the non-terminal that is expanded is really the last one. Formally, for every transition $1pA \to qu$ with $u \in T^*$ we add the matrix $((A, p) \to u, (\varepsilon, 2) \to \lambda, \ldots, (\varepsilon, n) \to \lambda)$, if q is a final state. This way, also the state of the deep PDA disappears, if it was final, and the grammar's derivation terminates with the same string of terminals that are on the stack of the simulated deep PDA in a computation according to the pattern of Lemma 3.1. Here it must be recalled that context-free matrix grammars of finite index have the same generative power whether there are deleting rules or not, see Lemma 3.1.2 in the book on regulated rewriting.[3]

Here the derivation graphs do not correspond to each other as directly as above. However, the correspondence is still rather evident. Let us define the following mapping that projects the compound nonterminals to their first component and deletes the placeholders:

$$\rho(x) := \begin{cases} X & \text{if } x = (X, i), X \in N, i \in \{1, 2, \ldots, n\} \cup Q \\ \lambda & \text{if } x = (\varepsilon, i), i \in \{1, 2, \ldots, n\} \\ x & \text{if } x \in T. \end{cases}$$

Obviously, the deep PDA's stack contents at the start of a computation are the same as the image under ρ of the sentential form obtained after applying the only matrix applicable in the beginning, namely $(S \rightarrow (S, s)(\varepsilon, 2)(\varepsilon, 3) \cdots (\varepsilon, n))$. Also the state stored in the sentential form is equal to the one of the deep PDA. As the explanations throughout the definition of our matrix grammar illustrate, these two features –equality of stack contents and the sentential form's image under ρ plus the equality of states– are preserved by simultaneous application of an expanding transition in the deep PDA and the corresponding matrix on the sentential form. Due to the one-to-one correspondence between expanding transitions and matrices (that always have to apply all of their rules) it is clear that any computation according to Lemma 3.1, before popping is started, leads to a terminal string on the stack that can also be generated by the grammar, if th e current state is final. The inverse is equally obvious and thus the two devices are equivalent. □

For matrix grammars of finite index it is known that they give rise to an infinite hierarchy of classes of languages. For every positive integer n, the class of languages generated by matrix grammars of index n is properly contained in the class of languages generated by matrix grammars of index $n + 1$, see Theorem 3.1.7 the book on regulated rewriting.[3] Since these classes are equal to the ones accepted by finitely expandable deep PDAs, also these devices induce an infinite hierarchy.

Corollary 3.1. *For all integers $n > 0$, the class of languages accepted by n-expandable deep PDAs is properly contained in the class of languages accepted by $n + 1$-expandable deep PDAs.*

For deep PDAs in general it is an open question whether admitting deleting transition of the form $kpA \rightarrow q\lambda$ changes their computational power. In the case of finitely expandable deep PDAs we can give a negative answer to the corresponding question. Looking back at the proof of Theorem 3.1, we can see that the introduction of deleting transitions in the deep PDA simulated will still let us allow to construct an equivalent matrix grammar in the same way. Only some more of the rules in the matrix grammar will now be deleting, but as mentioned in the proof this does not change their generative power.

Corollary 3.2. *For all integers $n > 0$, n-expandable deep PDAs with deleting transitions accept the same class of languages as n-expandable deep PDAs without deleting transitions.*

4. Open Problems

In the preceding investigations, only finitely expandable deep PDAs have been considered that can access all of the non-terminals on their pushdown store. On the other hand, in preceding work the so-called depth has been the focus of investigations. This is the number of non-terminals counted from the top that the device can rewrite, while more non-terminals might be present deeper down in the pushdown storage. When looking at the interplay between these two types of restrictions, it seems most interesting to determine whether a finite number of nonterminals beyond a deep PDA's depth can increase its power. More formally: are k-expandable deep PDAs equivalent to $k+1$-expandable deep PDAs od depth k? And for which integers i are k-expandable deep PDAs of depth $k-i$ equivalent to k-expandable deep PDAs of depth $k-i-1$?

A related question is the relation between finitely expandable and general deep PDAs. It is rather obvious that k-expandable deep PDAs can be simulated by deep PDAs of depth k. The converse is probably not true. However, the languages that were used to separate the classes of languages accepted by deep PDAs of depths k and $k+1$ (originally used by Kasai to separate classes of state grammars[6]) can also be accepted by k-expandable deep PDAs, respectively $k+1$-expandable deep PDAs. This indicates that the restriction of expandability might not be much weaker than that of depth.

References

1. J. Autebert, J. Berstel and L. Boasson, *Handbook of Formal Languages, Volume 1* (Springer, Berlin, 1997), Berlin, ch. Context-Free Languages and Pushdown Automata.
2. A. Meduna, Deep Pushdown Automata, *Acta Informatica* **42**, 541 (2006).
3. J. Dassow and G. Păun, *Regulated Rewriting in Formal Language Theory* (Springer-Verlag, Berlin, 1989).
4. M. A. Harrison, *Introduction to Formal Language Theory* (Addison-Wesley, Reading, Massachusetts, 1978).
5. A. Meduna, *Automata and Languages: Theory and Applications* (Springer, London, 2000).
6. T. Kasai, An Hierarchy Between Context-Free and Context-Sensitive Languages, *Journal of Computer and Systems Sciences* **4**, 492 (1970).

Received: June 21, 2009

Revised: April 14, 2010

THE PRIMITIVITY DISTANCE OF WORDS

GERHARD LISCHKE*

Institute of Informatics, Faculty of Mathematics and Informatics, Friedrich Schiller University Jena, Ernst-Abbe-Platz 1-4, D-07743 Jena, Germany
e-mail: gerhard.lischke@uni-jena.de

The Hamming distance between two words of equal length is the number of positions where the two words differ. This distance is extended to a distance between words and languages and to a maximal distance of words of given length and a language. We investigate these distances between words and various sets of primitive words and various sets of periodic words which have been introduced in the paper ITO/LISCHKE: *Generalized periodicity and primitivity for words*, Math. Log. Quart. 53, 2007. The distance from an arbitrary word to one of the sets of primitive words is not greater than one. In the opposite direction, from primitive words to nonprimitive words this distance may be greater, and we determine it exactly to the periodic words and to the semi-periodic words, depending from the lengths of the words and from the cardinality of the alphabet, as well as to the remaining sets of nonprimitive words if the alphabet has at least three letters.

Keywords: Periodicity of words, primitivity of words, Hamming distance.

1. Introduction

A Hamming distance-based measure h from coding theory [2] was used in [4,5] to study similarity of languages and to create some uncountable hierarchies of languages. In [3], some special kinds of periodicity and primitivity for words have been considered, and the set theoretical relationship between the sets of these words was given. It was the idea of Sándor Horváth in Budapest to ask for the distance in the sense of h between arbitrary words and these languages. Studying such distances between words and languages, as well as the analogously defined edit distances, is not new, see, for instance,

*Most of this work was done while visiting the Faculty of Science of Kyoto Sangyo University in Spring 2008, and a preliminary version of it was presented at the workshop AFLAS 2008 in Kyoto.

[7], and the references there. But, whereas Manthey and Reischuk consider computing the Hamming distance and computing the edit distance from a complexity theory point of view, we are interested in the exact distance between a given word and the sets of words which are primitive in a special sense. This should also be interesting and important to know. We shall show that this distance is not greater than 1 whatever special kind of primitivity we use. On the other hand, the distance between words and sets of words which are periodic in some special sense is more complicated and not yet clear in all cases.

In Section 2 we recall our basic definitions and relationships from [3,4,5] which are needed for our investigations and we give the new definitions. In Section 3, first we show that for each non primitive resp. non strongly primitive word there exists a primitive resp. strongly primitive word with the distance 1. Then we generalize this result to the other kinds of primitivity. In Section 4 we consider the distances between words and the complements of the sets of primitive words which are the periodic words and we determine them exactly for the periodic words and for the semi-periodic words, depending from the cardinality of the alphabet, as well as to the remaining sets of periodic words if the alphabet has at least three letters. In Section 5 we make some concluding remarks and propose the use of so-called vectors of primitivity distances to classify primitive words.

2. Basic definitions and relationships

For our whole paper, let X be a fixed nonempty, finite alphabet. Furthermore, we assume that X is a nontrivial alphabet in the sense that it has at least two symbols a and b (otherwise, all of our results become trivial or meaningless), and that we have a fixed ordering of X. $\mathbb{N} = \{0, 1, 2, 3, \ldots\}$ denotes the set of all natural numbers. X^* is the free monoid generated by X or the set of all words over X. The empty word is denoted by e, and $X^+ =_{Df} X^* \setminus \{e\}$.

For a word $p \in X^*$, $|p|$ denotes the length of p. For $k \in \mathbb{N}$, p^k denotes the concatenation of k copies of the word p. For $1 \leq i \leq |p|$, $p[i]$ is the letter at the i-th position of p. For $p \in X^+$ with $|p| \geq 2$ define $p^\circ =_{Df} p[1]p[2] \cdots p[|p| - 1]x$ where x is the first letter from $X \setminus \{p[|p|]\}$ according to the fixed ordering of X (in fact it doesn't matter which letter $x \neq p[|p|]$ it is).

For words $p, q \in X^*$, we say p is a *prefix of* q, in symbols $p \sqsubseteq q$, if there exists $r \in X^*$ such that $q = pr$. p is a *strict prefix of* q, in symbols $p \sqsubset q$, if $p \sqsubseteq q$ and $p \neq q$. $Pr(q) =_{Df} \{p : p \sqsubset q\}$ is the *set of all strict prefixes* *of* q (including e if $q \neq e$). p is a *subword of* q, in symbols $p \top q$, if there exist $r, s \in X^*$ such that $q = rps$. $p \not\top q$ means that p does not occur as a subword of q.

We consider the following folding operation:
For $p, q \in X^*$, $\quad p \otimes q =_{Df} \{w_1 w_2 w_3 : w_1 w_3 \neq e \wedge w_1 w_2 = p \wedge w_2 w_3 = q\}$,
$p^{\otimes 0} =_{Df} \{e\}$, $\;p^{\otimes k+1} =_{Df} \bigcup\{w \otimes p : w \in p^{\otimes k}\}$ for $k \in \mathbb{N}$.
For sets $A, B \subseteq X^*$, $\quad A \otimes B =_{Df} \bigcup\{p \otimes q : p \in A \wedge q \in B\}$.

The following example illustrates this operation.
Let $p = aabaa$. Then $p \otimes p = p^{\otimes 2} = \{aabaaaabaa, aabaaabaa, aabaabaa\}$.

Now we define the following sets of words which are periodic or primitive in different senses:

$Per \;=_{Df}\; \{u : \exists v \exists n (v \sqsubset u \wedge n \geq 2 \wedge u = v^n)\}$
is the set of *periodic* words.

$Q \;=_{Df}\; X^+ \setminus Per$ is the set of *primitive* words.

$SPer \;=_{Df}\; \{u : \exists v \exists n (v \sqsubset u \wedge n \geq 2 \wedge u \in v^n \cdot Pr(v))\}$
is the set of *semi-periodic* words.

$SQ \;=_{Df}\; X^+ \setminus SPer$ is the set of *strongly primitive* words.

$QPer \;=_{Df}\; \{u : \exists v \exists n (v \sqsubset u \wedge n \geq 2 \wedge u \in v^{\otimes n})\}$
is the set of *quasi-periodic* words.

$HQ \;=_{Df}\; X^+ \setminus QPer$ is the set of *hyper primitive* words.

$PSPer \;=_{Df}\; \{u : \exists v \exists n (v \sqsubset u \wedge n \geq 2 \wedge u \in \{v^n\} \otimes Pr(v))\}$
is the set of *pre-periodic* words.

$SSQ \;=_{Df}\; X^+ \setminus PSPer$ is the set of *super strongly primitive* words.

$SQPer \;=_{Df}\; \{u : \exists v \exists n (v \sqsubset u \wedge n \geq 2 \wedge u \in v^{\otimes n} \cdot Pr(v))\}$
is the set of *semi-quasi-periodic* words.

$SHQ \;=_{Df}\; X^+ \setminus SQPer$ is the set of *strongly hyper primitive* words.

$QQPer \;=_{Df}\; \{u : \exists v \exists n (v \sqsubset u \wedge n \geq 2 \wedge u \in v^{\otimes n} \otimes Pr(v))\}$
is the set of *quasi-quasi-periodic* words.

$HHQ \;=_{Df}\; X^+ \setminus QQPer$ is the set of *hyper hyper primitive* words.

The inclusion structure between these sets is given in the following figure, where the lines denote strict inclusion from bottom to top. Sets which are not connected by such a line, are incomparable under inclusion.

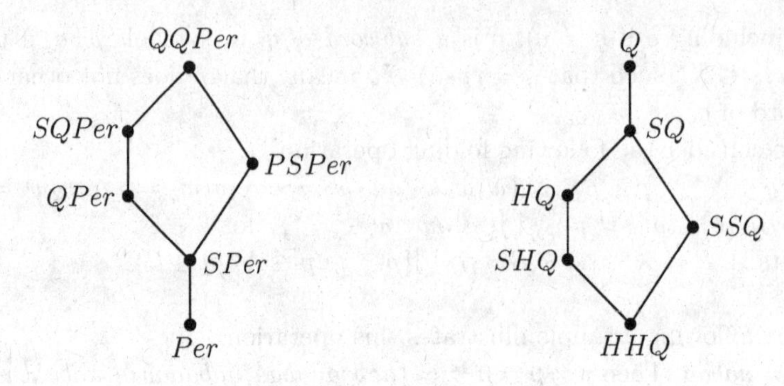

Figure 1

Recall, that the sets Q and Per have received special interest and play an important role in the algebraic theory of codes and formal languages (see [6,8,10]), but also the remaining sets deserve some attention [3].

Let $u \in X^+$. The shortest word v such that there exists a natural number n with $u = v^n$ is called the *root* of u, denoted by $root(u)$. The shortest word v such that there exists a natural number n with $u \in v^n \cdot Pr(v)$ is called the *strong root* of u, denoted by $sroot(u)$.
Four further kinds of roots can be defined in ways corresponding to the definitions of the sets above, but they are not explicitly used in this paper. Let us remark that all these roots are primitive words and that their prefix relationship is investigated in [3].

For two words u and v of the same length, the *Hamming distance* is $h(u,v) =_{Df} |\{i : 1 \leq i \leq |u| \wedge u[i] \neq v[i]\}|$, where $|M|$ for a set M denotes its cardinality.

For $k \in \mathbb{N}$, two words u and v are called *k-similar*, denoted by $u \underset{k}{\sim} v$, if $|u| = |v|$ and $h(u,v) \leq k$.
For a nonnegative real number $\delta < 1$, words u and v are called *δ-similar*, denoted by $u \underset{\delta}{\sim} v$, if $|u| = |v|$ and $h(u,v) \leq \delta \cdot |u|$.
Two languages $L_1, L_2 \subseteq X^*$ are called *k-similar*, denoted by $L_1 \underset{k}{\sim} L_2$, if $\forall u \exists v (u \in L_1 \rightarrow v \in L_2 \wedge u \underset{k}{\sim} v) \wedge \forall v \exists u (v \in L_2 \rightarrow u \in L_1 \wedge u \underset{k}{\sim} v)$.
$L_1 \underset{\delta}{\sim} L_2$ is defined accordingly for $0 \leq \delta < 1$.

Now we define the *distance* between words and languages.

For $p \in X^*$ and $L \subseteq X^*$, assuming L contains words of length $|p|$, $d(p,L) =_{Df} \min\{h(p,q) : |q| = |p| \wedge q \in L\}$.

Of course we have $0 \leq d(p, L) \leq |p|$ and $d(p, L) = 0 \leftrightarrow p \in L$.

Because in the following we are interested in the distance between words and sets of primitive words and there are no nonprimitive words of length 1 (and no primitive words of length 0), in considering $d(p, L)$ we assume in principle that $|p| \geq 2$ and $\{|q| : q \in L\} \supseteq \mathbb{N} \setminus \{0, 1\}$.

The *maximal distance of words of length* n *and a language* $L \subseteq X^*$ is $md(n, L) =_{Df} \max\{d(p, L) : p \in X^* \wedge |p| = n\}$ for $n \in \mathbb{N}$.

Because this distance may depend from the cardinality of the alphabet X, we shall write $md_k(n, L)$ instead of $md(n, L)$ in such cases, where $k = |X|$ is the cardinality of X.

Lemma 2.1. *If for some language* L *there exists a word* p *with* $d(p, L) = m$, *then for each natural number* $l \leq m$ *there exist words* p_l *of the same length as* p *such that* $d(p_l, L) = l$.

Proof. Let $q \in L$ with $h(p, q) = m$. Then for $l \in \{0, 1, \ldots, m\}$, change $m - l$ of those letters $p[i]$ where $p[i] \neq q[i]$ into $q[i]$ to get p_l. Assume $d(p_l, L) = r < l$. Then there must exist $q' \in L$ with $h(p_l, q') = r$, and $h(p, q') \leq h(p, p_l) + h(p_l, q') = m - l + r < m$ contradicting $d(p, L) = m$. □

Therefore for our languages L and natural numbers $n \geq 2$, in the following it is only interesting to determine the values $md_k(n, L)$. Then it is clear that for each natural number $l \leq md_k(n, L)$ there exist words p of length n such that $d(p, L) = l$ but no word p of length n with $d(p, L) > md_k(n, L)$.

For a finite sequence $x_1, x_2, \ldots, x_s$ of letters from X, let $\text{MAJ}(x_1, x_2, \ldots, x_s)$ be that letter which is in the majority of $x_1, x_2, \ldots x_s$, and if two or more letters have the same majority then it should be the first one of them according to the fixed ordering of X (in fact it doesn't matter which of them we take in such case).

Finally, let us remark that for a real number r, $\lfloor r \rfloor$ denotes the greatest integer which is smaller or equal to r, and $\lceil r \rceil$ denotes the smallest integer which is greater or equal to r. By a strict divisor of a natural number n we mean any divisor $s > 1$ of n (including n itself).

3. The distance to primitive words

It is easy to see that for each nonprimitive or periodic word p there exists a primitive word with the distance 1, namely p°. The same is true for semi-periodic words. This is a consequence of the famous Theorem of Fine and Wilf:

Lemma 3.1. ([1, 6]) *Let u and v be nonempty words and $i, j \in \mathbb{N}$. If u^i and v^j contain a common prefix of length $|u|+|v|-gcd(|u|,|v|)$ where $gcd(|u|,|v|)$ denotes the greatest common divisor of $|u|$ and $|v|$, then $root(u) = root(v)$.*

Theorem 3.1. *If $p \in Per$ then $p^\circ \in Q$. If $p \in SPer$ then $p^\circ \in SQ$.*

Proof. First assume $p \in SPer$ and $sroot(p) = v$ where $p = v^n v_i$ with $n \geq 2$, $|v_i| = i \neq 0$, and $v_i \sqsubset v$. Then $p^\circ = v^n v_i^\circ$. Assume $p^\circ \notin SQ$ and $sroot(p^\circ) = u$ which means that $p^\circ = u^m u'$ for some $m \geq 2$ and $u' \sqsubset u$. Then v^{n+1} and u^{m+1} have the common prefix $p[1] \cdots p[|p|-1]$ of length $n \cdot |v| + i - 1$. We have $m \cdot |u| - 1 \leq n \cdot |v| + i - 1$ and therefore $|u|+|v|-gcd(|u|,|v|) \leq |u|+|v|-1 \leq \frac{n \cdot |v|+i}{m} + |v| - 1 \leq \frac{n}{2}|v| + |v| + \frac{i}{2} - 1 \leq n \cdot |v| + i - 1$. It follows by Lemma 3.1 that $root(u) = u = root(v) = v$. On the other side, because of $|p| = |p^\circ|$ and $p \neq p^\circ$, $u = v$ is not possible.

If $i = 0$ which means $p \in Per$ we can argue in exactly the same way and get $p^\circ \in SQ \subseteq Q$. □

Corollary 3.1. *$d(p, Q) \in \{0, 1\}$ and $d(p, SQ) \in \{0, 1\}$ for arbitrary $p \in X^+$, and $md(n, Q) = md(n, SQ) = 1$ for $n \geq 2$.*

Instead of proving analogoues results for each of the sets HQ, SSQ, SHQ, and HHQ we prove a more general result and use the inclusion structure from Figure 1.

We use the following definition.

For $x \in X$ and $p \in X^*$, an *x-block in p* is a longest subword of consecutive letters x in p.

Theorem 3.2. *If $p \in QQPer$ then there exists $q \in HHQ$ such that $h(p, q) = 1$.*

Proof. Assume $p \in QQPer$. Let a be the first letter of p.
Case 1). There is no other letter in p, i.e. $p = a^i$, $i \geq 2$. Then $q =_{Df} p^\circ \in HHQ$.

Case 2). There is still another letter in p, let's say b, and i should be the greatest length of a b-block in p. This means $p = p_1ab^ip_2$ where $p_1 = e$ or $a \sqsubseteq p_1$, $b^i \not\top p_1$ and $p_2 = e$ or $a \sqsubseteq p_2$ with $b^{i+1} \not\top p_2$. Then define $q =_{Df} p_1bb^ip_2$. Assume $q \in QQPer$. This means that $q \in u^{\otimes m} \otimes Pr(u)$ for some $u \sqsubset q$ and $m \geq 2$. Then $p_1b^{i+1} \sqsubseteq u$ must follow and therefore $b^{i+1} \top p_2$ which is a contradiction.
In both cases we found $q \in HHQ$ with $h(p,q) = 1$ which proves the theorem. □

The following are consequences of the inclusion structure from Figure 1:

Corollary 3.2. *$d(p, HQ), d(p, SSQ), d(p, SHQ), d(p, HHQ) \in \{0, 1\}$ for arbitrary $p \in X^+$, and $md(n, HQ) = md(n, SSQ) = md(n, SHQ) = md(n, HHQ) = 1$ for each $n \geq 2$.*

Also Corollary 3.1 is a consequence of Theorem 3.2 and it is interesting to remark that we proved it without using the Theorem of Fine and Wilf. But, on the other hand, the construction of $q = p^\circ$ in Theorem 3.1 is much easier than that of Theorem 3.2.

4. The distance to nonprimitive words

We have seen that it is not far from an arbitrary word to some word which is primitive in any sense. More exactly, it is enough to change one digit in an arbitrary word to convert it into a primitive one of any sense. But in the opposite direction, from some primtive word to a nonprimitive one the distance may be greater.

Theorem 4.1. *For natural numbers $n, k \geq 2$ holds that*

$$md_k(n, Per) = \begin{cases} n - \frac{n}{s}(\lfloor \frac{s}{k} \rfloor + 1) & \text{if there is a strict divisor of } n \\ & \text{which is not dividable by } k, \text{ and} \\ & s \text{ is the smallest such divisor} \\ n - \frac{n}{k} & \text{if } k \text{ is prime and } n \text{ is a power of } k. \end{cases}$$

Proof. First remark, that there is no strict divisor of n which is not dividable by k if and only if k is prime and n is a power of k. If $2 \leq n \leq k$ and s divides n, then let p be a word of length n which has all of its letters different each other. Then $q =_{Df} (p[1] \cdots p[\frac{n}{s}])^s \in Per$ with $h(p,q) = n - \frac{n}{s}$, and this is the smallest distance to a periodic word if s is the smallest strict divisor of n. Also, in this case, $s = k$ if $n = k$ and k is prime, or $s < k$ and therefore $\lfloor \frac{s}{k} \rfloor = 0$, and this corresponds to the theorem. Now let $n > k$ and $p \in X^*$ be an arbitrary word of length n. Let $n = r \cdot s$ with $s \geq 2$ (If n is prime then $s = n$ and $r = 1$). We define a word $q \in Per$ in the following way: Let $q =_{Df} u^s$ where $|u| = r$ and $u[i] =_{Df}$ $\text{MAJ}(p[i], p[i+r], p[i+2r], \ldots, p[i+(s-1)r])$, for $i = 1, ..., r$. Obviously, p can be transformed into q by changing of at most $r \cdot \lfloor s - \frac{s}{k} \rfloor$ letters. If k divides s then $r \cdot \lfloor s - \frac{s}{k} \rfloor = r \cdot s - \frac{rs}{k} = n - \frac{n}{k}$. If k doesn't divide s then $r \cdot \lfloor s - \frac{s}{k} \rfloor = r \cdot (s - \lfloor \frac{s}{k} \rfloor - 1) = n - \frac{n}{s}(\lfloor \frac{s}{k} \rfloor + 1)$, which is smaller than $n - \frac{n}{k}$ and has its smallest value if s is the smallest strict divisor of n which is not dividable by k.

We get the given value for $d(p, Per)$ if we choose p such that for $i = 1, \ldots, r$, each letter from X occurs with frequency $\lfloor \frac{s}{k} \rfloor$ or $\lfloor \frac{s}{k} \rfloor + 1$ under $p[i], \ldots, p[i+(s-1)r]$, where $n = r \cdot s$ and s is the smallest strict divisor of n which is not dividable by k, or $s = k$ if k is prime, n is a power of k and $r = \frac{n}{k}$. □

Corollary 4.1. *Some special values are*

$$md_k(n, Per) = \begin{cases} \frac{n}{2} & \text{if } k > 2 \text{ and } n \text{ is even,} \\ & \text{or } k = 2 \text{ and } n \text{ is a power of } 2 \\ (n - \frac{n}{s})/2 & \text{if } k = 2 \text{ and } s \geq 3 \text{ is the smallest} \\ & \text{odd and strict divisor of } n \\ n - 1 & \text{if } k = n \text{ and } n \text{ is prime} \\ n - (\lfloor \frac{n}{k} \rfloor + 1) & \text{if } k \neq n \text{ and } n \text{ is prime} \\ \frac{n-1}{2} & \text{if } k = 2 \text{ and } n > 2 \text{ is prime.} \end{cases}$$

Remark 4.1. If we define $f_k(n) =_{Df} md_k(n, Per)$ for fixed $k \geq 2$, then f_k is not monotone for each $k \geq 2$. This is illustrated by the following examples. For $k = 2$ let n be such that both $n - 1$ and $n + 1$ are prime and $n \geq 12$ is not a power of 2. Then $f_2(n-1) = \frac{n-2}{2} > f_2(n) = \frac{n - \frac{n}{s}}{2}$ with $s \geq 3$, and $f_2(n) < f_2(n+1) = \frac{n}{2}$. For $k > 2$ let n be prime with $n > \frac{3k}{k-2}$. Then $f_k(n-1) = \frac{n-1}{2} < f_k(n) = n - (\lfloor \frac{n}{k} \rfloor + 1)$, and $f_k(n) > f_k(n+1) = \frac{n+1}{2}$.

Now we interpret our results in the language of similarity of languages. For a two-letter alphabet X we know from Corollary 3.1 and Corollary 4.1 that for each $p \in Per$ there is a $q \in Q$ with $p \underset{1}{\sim} q$ and for each $p \in Q$ of length $n \geq 2$ there is a $q \in Per$ with $p \underset{\frac{n}{2}}{\sim} q$ but not $p \underset{\frac{n}{2}-1}{\sim} q$ if n is prime. This means:

Corollary 4.2. *Let Q' be the set of all primitive words of length at least 2 over a two-letter alphabet. Then Q' and Per are not c-similar for any natural constant c, but* $Q' \underset{\frac{1}{2}}{\sim} Per$.

Corollary 4.3. *For an alphabet X with k symbols,* $Q' \underset{1-\frac{1}{k}}{\sim} Per$ *holds.*

Remark, that such $(1-\frac{1}{k})$-similarities have been established already for other language classes in [5].

It is trivial that $L_1 \subseteq L_2$ implies $d(p, L_2) \leq d(p, L_1)$ and $md(n, L_2) \leq md(n, L_1)$. Because of the inclusion structure given in Figure 1 it is therefore interesting to investigate whether the distances to $SPer$, $QPer$, $PSPer$, $SQPer$, and $QQPer$ diminish the values from Theorem 4.1.

Theorem 4.2. *For natural numbers $n, k \geq 2$ holds that*

$$md_k(n, SPer) = \begin{cases} \lceil \frac{n}{3} \rceil & \text{if } k = 2 \\ \lceil \frac{n}{2} \rceil & \text{if } k > 2. \end{cases}$$

Proof. For $n = 2$ everything is clear. Let $p \in X^*$ with $n = |p| > 2$ and $k \geq 2$ be the cardinality of the alphabet X. We want to convert p into a semi-periodic word q with $sroot(q) = u$, this means $q = u^s u'$ for some $s \geq 2$ and $u' \sqsubset u$. Let $r =_{Df} |u|$ and $t =_{Df} |u'|$. Then

$$n = r \cdot s + t \text{ and } 0 \leq t < r. \tag{1}$$

For $i = 1, \ldots, t$, let $u[i] =_{Df} \text{MAJ}(p[i], p[i+r], p[i+2r], \ldots, p[i+sr])$, and for $i = t+1, \ldots, r$, let $u[i] =_{Df} \text{MAJ}(p[i], p[i+r], \ldots, p[i+(s-1)r])$. Then

$$h(p,q) \leq t \cdot \lfloor s + 1 - \frac{s+1}{k} \rfloor + (r-t) \cdot \lfloor s - \frac{s}{k} \rfloor,$$

where the equality may hold. Let us denote

$$m_k(n, r, s, t) =_{Df} t \cdot \lfloor s + 1 - \frac{s+1}{k} \rfloor + (r-t) \cdot \lfloor s - \frac{s}{k} \rfloor,$$

and let be $s = \alpha k + \gamma$ where $0 \leq \gamma < k$.
If $\gamma = 0$, then $\alpha = \frac{s}{k}$ and

$$\begin{aligned} m_k(n, r, s, t) &= t \cdot \lfloor s+1-\alpha-\tfrac{1}{k} \rfloor + (r-t)\lfloor s-\alpha \rfloor = t(s+1-\alpha-1)+(r-t)(s-\alpha) \\ &= r(s-\alpha) = rs(1-\tfrac{1}{k}). \end{aligned}$$

This has its smallest value (under the conditions $\gamma = 0$ and $s \geq 2$) if $s = k$,

$\alpha = 1$, and r is the smallest integer with $r(s+1) > n$. This is $r = \lfloor \frac{n}{s+1} \rfloor + 1$, and we get

$$m_k(n, r, s, t) = (\lfloor \frac{n}{k+1} \rfloor + 1)(k-1). \tag{2}$$

If $\gamma \neq 0$, then $\alpha = \frac{s-\gamma}{k}$ and $m_k(n,r,s,t) = t \cdot \lfloor s+1-\alpha-\frac{\gamma+1}{k} \rfloor + (r-t)\lfloor s - \alpha - \frac{\gamma}{k} \rfloor = t(s-\alpha) + (r-t)(s-\alpha-1) = r(s-\alpha-1) + t = n - r(1+\alpha) = n - r(1 + \frac{s-\gamma}{k})$. This has its smallest value for $s = 3$ if $k = 2$, and for $s = 2$ if $k \neq 2$, and $r = \lfloor \frac{n}{s} \rfloor$. In the first case, we have $\alpha = 1$ and

$$m_k(n, r, s, t) = n - \lfloor \frac{n}{3} \rfloor \cdot 2 = \begin{cases} \frac{n}{3} & \text{if } n \equiv 0 \mod 3 \\ \frac{n+2}{3} & \text{if } n \equiv 1 \mod 3 \\ \frac{n+4}{3} & \text{if } n \equiv 2 \mod 3 \end{cases}. \tag{3}$$

In the second case, we have $\alpha = 0$ and

$$m_k(n, r, s, t) = n - \lfloor \frac{n}{2} \rfloor = \lceil \frac{n}{2} \rceil. \tag{4}$$

$md_k(n, SPer)$ is now the smallest value of (2), (3), (4) under the condition (1). For $k = 2$ we get from (2) the value $\lfloor \frac{n}{3} \rfloor + 1 = \begin{cases} \frac{n}{3} + 1 & \text{if } n \equiv 0 \mod 3 \\ \frac{n+2}{3} & \text{if } n \equiv 1 \mod 3 \\ \frac{n+1}{3} & \text{if } n \equiv 2 \mod 3 \end{cases}$.

It follows with (3) that $md_2(n, SPer) = \begin{cases} \frac{n}{3} & \text{if } n \equiv 0 \mod 3 \\ \frac{n+2}{3} & \text{if } n \equiv 1 \mod 3 \\ \frac{n+1}{3} & \text{if } n \equiv 2 \mod 3 \end{cases}$.

This means that $md_2(n, SPer) = \lceil \frac{n}{3} \rceil$. For $k > 2$, the smaller value of (2) and (4) is given by (4), which completes the proof. □

Corollary 4.4. $SQ' \underset{\frac{2}{3}}{\sim} SPer$ *for a two-letter alphabet and* $SQ' \underset{\frac{3}{4}}{\sim} SPer$ *otherwise.*
(SQ' is the set of all strongly primitive words with length at least 2.)

Here we see that it would be convenient to define *f-similarity* for a function f. Then we would have $SQ' \underset{f}{\sim} SPer$ where

$$f(n) = \begin{cases} \frac{1}{3} + \frac{2}{3n} & \text{if } |X| = 2 \\ \frac{1}{2} + \frac{1}{2n} & \text{otherwise.} \end{cases}$$

Investigating the distances to $QPer$, $PSPer$, $SQPer$ and $QQPer$ becomes more complicated because of the use of the folding operation. For a k-letter alphabet with $k > 2$ the maximal distance of words of length n to one of these languages is the same as to the set $SPer$. We presume that this is also true for a two letter alphabet, but we have not yet a complete proof.

Theorem 4.3. *If $\mathcal{L}$ is one of the sets $QPer$, $PSPer$, $SQPer$, and $QQPer$ then for natural numbers $n \geq 2$ and $k \geq 3$ holds that*
$md_k(n, \mathcal{L}) = md_3(n, SPer) = \lceil \frac{n}{2} \rceil$.

Proof. Let $\mathcal{L} \in \{QPer, PSPer, SQPer, QQPer\}$. It is clear that $md_k(n, \mathcal{L}) \leq md_k(n, SPer)$ for all $n, k \geq 2$. To show the equality we have to find for each $n \geq 2$ a word p of length n over a k-letter alphabet with $d(p, \mathcal{L}) = md_k(n, SPer)$. The case $k = 2$ will be discussed after the proof. Let $k \geq 3$ and a, b, c be three pairwise different letters from the alphabet X, and $p =_{Df} a^{\lceil \frac{n}{3} \rceil} b^{\lfloor \frac{n+1}{3} \rfloor} c^{\lfloor \frac{n}{3} \rfloor}$. Then $|p| = n$, and it is easy to see that a word $q \in QQPer$ with the smallest distance to p has a hyper-hyper-root u with no overlappings in q. This means, q is of the form $u^s u' \in u^{\otimes s} \otimes Pr(u)$ where $s \geq 2$ and $u' \in Pr(u)$. Indeed, it is $q = u^2 u' \in SPer$ where u' is the first letter of u or it is empty (depending whether n is odd or not) and u is the prefix of length $\lfloor \frac{n}{2} \rfloor$ of p. Therefore $d(p, \mathcal{L}) = md_3(n, SPer) = \lceil \frac{n}{2} \rceil$. □

Conjecture. If $\mathcal{L}$ is one of the sets $QPer$, $PSPer$, $SQPer$, and $QQPer$ then for natural numbers $n \geq 2$ holds that $md_2(n, \mathcal{L}) = md_2(n, SPer) = \lceil \frac{n}{3} \rceil$.

To prove this conjecture we have to find for each $n \geq 2$ a word p of length n over $\{a, b\}$ with $d(p, \mathcal{L}) = \lceil \frac{n}{3} \rceil$. For $n \in \{2, 3\}$ this is done by $p = ab$ resp. $p = abb$. Now let $n > 3$ and therefore $\lceil \frac{n}{3} \rceil > 1$, and let $p =_{Df} a^{\lceil \frac{n}{3} \rceil} b^{\lfloor \frac{2n}{3} \rfloor}$. Then $|p| = n$ and $h(p, q') = \lceil \frac{n}{3} \rceil$ with $q' = b^n \in \mathcal{L}$. To show that there is no word $q \in QPer$ with $|q| = n$ and $h(p, q) < \lceil \frac{n}{3} \rceil$ we assume the opposite. This means, $q \in u^{\otimes s}$ for some $s \geq 2$. We could show that $|u| < \lceil \frac{n}{3} \rceil - 1$ and $s > 3$ must follow, and that there must be overlappings of u in q. But a final proof was not yet successful. By some computer experiments [9] this conjecture was reinforced for all words with length up to 15.

5. Concluding remarks

Comparing Theorem 4.1 and Theorem 4.2 we see that for a k-letter alphabet and words of length n, the maximal primitivity distance — this should mean the distance to the sets Per, $SPer$, and so on — is the same for $SPer$ as for Per if $k > 2$ and n is even or $k = 2$ and n is dividable by 3, but in the most other cases it is smaller for $SPer$. Of course, for a fixed primitive word p of length $n > 2$ we have $0 < d(p, SPer) \leq d(p, Per) \leq md(n, Per)$, where $d(p, SPer) < d(p, Per)$ is possible.

If, for instance, $p = (ab)^m b$, $m > 1$, $n = 2m + 1$, then $d(p, SPer) = 1$, and $d(p, Per) = m = \frac{n-1}{2}$.

For a more instructive description we propose the use of the following so-called *vector of primitivity distances*:

$$D(p) =_{Df} [d(p, Per), d(p, SPer)],$$

or more generally

$$Dv(p) =_{Df} [d(p, Per), d(p, SPer), d(p, QPer), d(p, PSPer), d(p, SQPer), d(p, QQPer)].$$

To be independent from the lengths of the words consider the *relative distance vector*, which is

$$Dv'(p) =_{Df} \frac{1}{|p|} \cdot [d(p, Per), d(p, SPer), d(p, QPer), d(p, PSPer), d(p, SQPer), d(p, QQPer)].$$

Even though by Theorem 4.3 and our conjecture the maximal distances to the sets $SPer$, $QPer$, ... are the same, for single words these distances may be different each other.

Because of Figure 1 and Theorem 4.1, if $Dv(p) = [z_1, z_2, \ldots, z_6]$ then $0 \leq z_6 \leq z_5 \leq z_3 \leq z_2 \leq z_1 \leq md(|p|, Per)$ and $z_6 \leq z_4 \leq z_2$. It is still open whether there exist words p such that all components in $Dv(p)$ are different from each other. Also it should be interesting to study more general relationships between the components of $Dv(p)$ and to classify the primitive words according to such vectors.

Acknowledgement

I am very grateful to Masami Ito for our cooperation and for his hospitality and generosity during my stay in Kyoto in Spring 2008, and to Péter Burcsi in Budapest for his interest and comments which exposed some mistakes in my preliminary version and led to a complete revision of it.

References

1. N.J.Fine, H.S.Wilf, *Uniqueness theorems for periodic functions*, Proc. AMS 16 (1965), 109–114.
2. R.W.Hamming, *Error detecting and error correcting codes*, Bell System Techn. Journ., 29 (1950), 147–160.

3. M.ITO, G.LISCHKE, *Generalized periodicity and primitivity for words*, Math. Log. Quart. 53 (2007), 91–106, Corrigendum in Math. Log. Quart. 53 (2007), 642–643.
4. G.LISCHKE, *Restorations of punctured languages and similarity of languages*, Math. Log. Quart. 52 (2006), 20–28.
5. G.LISCHKE, *Some uncountable hierarchies of formal languages*, Annales Univ. Sci. Budapest., Sect. Comp. 26 (2006), 171–179
6. M.LOTHAIRE, *Combinatorics on Words*, Addison-Wesley, Reading (Mass.), 1983.
7. B.MANTHEY, R.REISCHUK, *The intractability of computing the Hamming distance*, Theoret. Comput. Sci. 337 (2005), 331–346.
8. H.J.SHYR, *Free Monoids and Languages*, Hon Min Book Company, Taichung, 1991.
9. L.WOLFF, Projektarbeit, Gymnasium Bergschule, Apolda, Germany, 2008.
10. S.S.YU, *Languages and Codes*, Tsang Hai Book Publishing Co., Taichung, 2005.

Received: June 21, 2009

Revised: March 31, 2010

FINE CONVERGENCE OF FUNCTIONS AND ITS EFFECTIVIZATION

TAKAKAZU MORI
Faculty of Science, Kyoto Sangyo University,
E-mail: morita@cc.kyoto-su.ac.jp

YOSHIKI TSUJII
Faculty of Science, Kyoto Sangyo University,
E-mail: tsujiiy@cc.kyoto-su.ac.jp

MARIKO YASUGI*
Kyoto Sangyo University,
E-mail: yasugi@cc.kyoto-su.ac.jp

In this article, we first discuss the Fine continuity and the Fine convergence in relation to the continuous convergence on $[0, 1)$. Subsequently, we treat computability and the effective Fine convergence for a sequence of functions with respect to the Fine topology. We prove that the Fine computability does not depend on the choice of an effective separating set and that the limit of a Fine computable sequence of functions under the effective Fine convergence is Fine computable. Finally, we generalize the result of Brattka, which asserts the existence of a Fine computable but not locally uniformly Fine continuous function.

Keywords: Fine topology, dyadic interval, Fine continuous function, Fine computable function, effective Fine convergence, continuous convergence.

1. Introduction

The main objective of this article is to effectivize the notions of convergence of a sequence of continuous functions and the limit of such a sequence in the Fine space.

Fine introduced the *Fine metric* on the unit interval and initiated the theory of Walsh-Fourier analysis by proving some fundamental

*This work has been supported in part by Kayamori Foundations of Informational Sciences Advancement K15VIII and Science Foundations of JSPS No.16340028.

theorems.[4,10] He defined the Fine metric between two real numbers as the weighted sum of differences of corresponding bits in their binary expansions with infinitely many 0's. Many topological properties concerning the Fine metric are derived from the property that a dyadic interval, that is, an interval $[a, b)$ with dyadic rationals a and b, is open and closed (clopen) with respect to the Fine metric. The topology generated by the set of all dyadic intervals is equivalent to the one induced by the Fine metric. We call this topology the Fine topology and $[0, 1)$ with this topology *Fine space.*

In this article, we use the term "function" as a mapping from some space to the real line $\mathbb{R}$ with the ordinary Euclidean topology. In order to specify the topological properties with respect to the Fine topology, we prefix "Fine" to the relevant terms. For example, the convergence of a sequence in $[0, 1)$ with respect to the Fine topology is called *Fine convergence.* Topological notions with no prefix or with the prefix "$\mathbb{E}$-" will mean the notions with respect to the Euclidean metric.

Let $\mathcal{C}_F$ be the set of all Fine continuous functions. It is well known that a function belongs to $\mathcal{C}_F$ if and only if it is $\mathbb{E}$-continuous at every dyadic irrational and right $\mathbb{E}$-continuous at every dyadic rational.[4,10] Moreover, a function in $\mathcal{C}_F$ is uniformly Fine continuous if and only if it has a left limit at every dyadic rational. Fine continuous function may diverge. For example, $f(x) = \frac{1}{1-2x}\chi_{[0,\frac{1}{2})}(x)$ is locally uniformly Fine continuous and diverges at $\frac{1}{2}$, where χ_A denotes the indicator (characteristic) function of the set A. Brattka proved the existence of a Fine computable function, hence Fine continuous, which is not locally uniformly Fine continuous.

In Section 2, we consider various notions of continuity and the corresponding notions of convergence for functions on the Fine space. We define *t-Fine Fine convergence* (Definition 2.5) and "Fine convergence" (Definition 2.6) and prove their equivalence (Proposition 2.2). Both are stronger than pointwise convergence and weaker than locally uniform Fine convergence. They preserve Fine continuity (Proposition 2.4). Their relation to continuous convergence is also discussed. In Section 4, we prove that the notion of effective Fine continuity of functions does not depend on the choice of an effective separating set (Theorem 4.1). In Section 5, we introduce the notion of effective Fine convergence of functions (Definition 5.1). We prove that the limit of an effectively Fine continuous sequence of functions under this effective Fine convergence is also effectively Fine continuous (Theorem 5.2) and that the effective Fine limit of a Fine computable sequence of functions is Fine computable (Theorem 5.3). We also define the notion of a computable sequence of dyadic step functions and prove that a function

f is Fine computable if and only if there exists a computable sequence of dyadic step functions which Fine converges effectively to f (Theorem 5.4). In Section 6, we work on Brattka's example, which is Fine computable but not locally uniformly Fine continuous. We prove that this example satisfies a recursive functional equation, which is related to self-similarity. We modify this equation and obtain other examples (Theorems 6.1 and 6.2).

Let us remark that, some of the notions and propositions in this article could be generalized to any metric space with a computability structure and an effective separating set. The advantage of the Fine space that there is a universal algorithm to determine, for any computable element x, which fundamental neighborhood x belongs to.

Let us note that we employ Pour-El and Richards' approach[8] in treating effectivity and computability, and will not mention other approaches in order to avoid complication.

2. Fine convergence and continuous Fine convergence

In this section, we define various classical notions of continuity and convergence of real functions with respect to the Fine topology. We will then effectivize them in Sections 3-5. The domain of discurse will be the real functions on $[0,1)$. Those who are interested only effective results can skip this section.

The *Fine metric* d_F on $[0,1)$ is defined[4] to be $d_F(x,y) = \sum_{k=1}^{\infty} |\sigma_k - \tau_k|2^{-k}$ for $x, y \in [0,1)$, where $0.\sigma_1\sigma_2\cdots$ and $0.\tau_1\tau_2\cdots$ are the binary expansions of x and y with infinitely many zeros respectively.

A left-closed right-open interval with dyadic rational end points is called a *dyadic interval.* It is easy to see that a dyadic interval is open and closed with respect to the Fine metric. This property corresponds to prohibition of left convergence to dyadic rationals and makes some $\mathbb{E}$-discontinuous functions Fine continuous.

We use the following notations for special dyadic intervals.

$I(n,k) = [k\,2^{-n}, (k+1)2^{-n}),\ 0 \leqslant k \leqslant 2^n - 1,$

$J(x,n) =$ such $I(n,k)$ that includes x.

Since the intervals $\{I(n,k)\}_k$ are disjoint and $\cup_{k=0}^{2^n-1} I(n,k) = [0,1)$, $J(x,n)$ is uniquely determined for each n and x. We call $I(n,k)$ a *fundamental dyadic interval* (*of level* n) and $J(x,n)$ a *dyadic neighborhood of* x (*of level* n). It is obvious that $I(n,k) = \{x \mid d_F(x, k\,2^{-n}) < 2^{-n}\}$ holds.

The Fine space is totally bounded. However, it is not complete, since, for any dyadic rational r, the sequence $\{r - 2^{-n}\}$ is a Fine Cauchy sequence but does not Fine converge.

We state a simple property for later use.

Lemma 2.1. *The following three are equivalent for any $x, y \in [0,1)$ and any positive integer n.*
(i) $y \in J(x,n)$. (ii) $x \in J(y,n)$. (iii) $J(x,n) = J(y,n)$.

It is obvious that the sequence $\{J(x,n)\}_n$ forms a fundamental system of neighborhoods of x and the set of all fundamental dyadic intervals becomes an open base for the topology introduced by the Fine metric. If we define $V_n(x) = J(x,n)$, then it is easy to show that $\{V_n\}$ satisfies the axioms of an effective uniform topology.[12] It holds that $J(x, n+1) \subset \{y \mid d_F(x,y) < 2^{-n}\}$. So the topology induced from $\{V_n\}$ is equivalent to the one induced from the Fine metric.

The notion of continuity on the Fine space is formulated as follows.

Definition 2.1. (t-Fine continuity) A function f is said to be *topologically Fine continuous (t-Fine continuous)* if, for each k and x, there exists a positive integer $N(k,x)$ such that $y \in J(x, N(k,x))$ implies $|f(y) - f(x)| < 2^{-k}$.

We define the following Fine continuity using an enumeration of all dyadic rationals $\{e_i\}$. Notice that, we can select a sequence of dyadic rationals which Fine converges to x for each $x \in [0,1)$, and that we can select an e_i such that $x \in J(e_i, n)$, or $e_i \in J(x,n)$, for each x and n.

Definition 2.2. (Fine continuity) A function f is said to be *Fine continuous* if, for each k and i, there exists an integer $N(k,i)$ such that
(a) $x \in J(e_i, N(k,i))$ implies $|f(x) - f(e_i)| < 2^{-k}$.
(b) $\bigcup_i J(e_i, N(k,i)) = [0,1)$.

Proposition 2.1. *The t-Fine continuity and the Fine continuity are equivalent.*

Proof. Suppose first that f is Fine continuous with respect to $N_2(k,i)$. Then, by (b), for each k and x, there exists an i such that $x \in J(e_i, N_2(k+1,i))$. Define, for such an i, $N_1(k,x) = N_2(k+1,i)$. Recall that $J(e_i, N_2(k+1,i)) = J(x, N_2(k+1,i))$ holds. So, if $y \in J(x, N_1(k,x)) = J(x, N_2(k+1,i))$, then $y \in J(e_i, N_2(k+1,i))$. Since $x \in J(e_i, N_2(k+1,i))$, by (a) for x and y, it follows
$|f(y) - f(x)| \leqslant |f(y) - f(e_i)| + |f(x) - f(e_i)| < 2^{-(k+1)} + 2^{-(k+1)} = 2^{-k}$.
So, f is t-Fine continuous with respect to $N_1(k,x)$.

Conversely, assume that f is t-Fine continuous with respect to $N_1(k,x)$. Define $N_2(k,i) = \min\{N_1(k+1,x) \mid e_i \in J(x, N_1(k+1,x)), x \in [0,1)\}$. Notice that the minimum is attained by some $z \in [0,1)$. For such a z, $N_2(k,i) = N_1(k+1,z)$. Now suppose $x \in J(e_i, N_2(k,i)) = J(e_i, N_1(k+1,z))$. Notice that $J(e_i, N_1(k+1,z)) = J(z, N_1(k+1,z))$. Then $x \in J(z, N_1(k+1,z))$ and $e_i \in J(z, N_1(k+1,z))$. So, using the t-Fine continuity, we have,
$|f(x) - f(e_i)| \leqslant |f(x) - f(z)| + |f(z) - f(e_i)| < 2^{-(k+1)} + 2^{-(k+1)} = 2^{-k}$.
This proves (a).

Notice next that, for each k and x, there is an e_i such that $e_i \in J(x, N_1(k+1,x))$. Take a z as above. Then, since $N_1(k+1,z) \leqslant N_1(k+1,x)$,

$$x \in J(e_i, N_1(k+1,x)) \subset J(e_i, N_1(k+1,z)) \text{ and hence}$$
$$x \in J(x, N_1(k+1,x)) \subset J(z, N_1(k+1,z)) = J(e_i, N_2(k,i)).$$

Hence, we obtain $x \in J(e_i, N_2(k,i))$. This proves (b). □

Let us remark that t-Fine continuity of a function is the usual topological definition of continuity. We have introduced *Fine continuity* since it is suitable for effectivization.

Uniform convergence and locally uniform convergence are fundamental concepts of the calculus.

Definition 2.3. (Uniform Fine continuity) A function f is said to be uniformly Fine continuous if, for any k, there exists a positive integer $N(k)$ such that $y \in J(x, N(k))$ implies $|f(x) - f(y)| < 2^{-k}$.

Definition 2.4. (Locally uniform Fine continuity) A function f is said to be locally uniformly Fine continuous if, for k and i, there exist positive integers $N(i)$ and $M(k,i)$ such that $x, y \in J(e_i, N(i))$ and $y \in J(x, M(k,i))$ imply $|f(x) - f(y)| < 2^{-k}$.

Definition 2.5. (t-Fine convergence) A sequence of functions $\{f_n\}$ is said to *t-Fine converge* to f if, for each k and x, there exist positive integers $N(k,x)$ and $M(k,x)$ such that $y \in J(x, N(k,x))$ and $n \geqslant M(k,x)$ imply $|f_n(y) - f(y)| < 2^{-k}$.

Let us note that we obtain locally uniform Fine convergence from Definition 2.5 if $N(k,x)$ does not depend on k. We also define the following convergence, for the sake of effectivization, similarly to Definition 2.2.

Definition 2.6. (Fine convergence) A sequence of functions $\{f_n\}$ is said to *Fine converge* to f if, for each k and i, there exist positive integers $N(k,i)$ and $M(k,i)$ such that

(a) $x \in J(e_i, N(k,i))$ and $n \geqslant M(k,i)$ imply $|f_n(x) - f(x)| < 2^{-k}$,
(b) $\bigcup_{i=1}^{\infty} J(e_i, N(k,i)) = [0,1)$ for each k.

Similarly to Proposition 2.1, we can prove the following equivalence.

Proposition 2.2. *t-Fine convergence and Fine convergence are equivalent.*

For a sequence of Fine continuous functions, we obtain the following proposition.

Proposition 2.3. *If a sequence of t-Fine continuous functions t-Fine converges to f, then f is also t-Fine continuous.*

Proof. Let $\{f_n\}$ be a sequence of t-Fine continuous functions with respect to $N_1(n,k,x)$ and suppose that it t-Fine converges to f with respect to $N_2(k,x)$ and $M(k,x)$. Define $N(k,x) = \max\{N_2(k+2,x), N_1(M(k+2,x), k+2, x)\}$. Then $y \in J(x, N(k,x))$ implies

$$\begin{aligned} &|f(y) - f(x)| \\ &\leqslant |f(y) - f_{M(k+2,x)}(y)| + |f_{M(k+2,x)}(y) - f_{M(k+2,x)}(x)| \\ &\quad + |f_{M(k+2,x)}(x) - f(x)| < 3 \cdot 2^{-(k+2)} < 2^{-k}. \end{aligned}$$

□

From Propositions 2.1, 2.2 and 2.3, we obtain the following proposition.

Proposition 2.4. *If a sequence of Fine continuous functions Fine converges to f, then f is also Fine continuous.*

t-Fine convergence reminds us of continuous convergence. According to Binz,[1] continuous convergence and locally uniform convergence do not coincide on the Fine space. Schröder[9] investigated the notion of continuous convergence of a function sequence in relation to the admissible representation of the space of all continuous functions

Definition 2.7. (Continuous Fine Convergence[9]) $\{f_n\}$ is said to *Fine converge continuously* to f if $\{f_n(x_n)\}$ $\mathbb{E}$-converges to $f(x)$ for every sequence $\{x_n\}$ which Fine converges to x.

Notice that the continuous Fine convergence is equivalent to the following; for each k and x there exist integers $N(k,x)$ and $M(k,x)$ which satisfy that $y \in J(x, N(k,x))$ and $n \geqslant M(k,x)$ imply $|f_n(y) - f(x)| < 2^{-k}$.

Proposition 2.5. *If a sequence of Fine continuous functions $\{f_n\}$ Fine converges continuously to f, then f is Fine continuous.*

Proof. Let us assume that a sequence $\{x_m\}$ Fine converges to x. For each m, $\{f_n(x_m)\}_n$ $\mathbb{E}$-converges to $f(x_m)$ by virtue of continuous convergence. So, we can choose a strictly increasing sequence of positive integers $\{n_m\}$ such that $|f_{n_m}(x_m) - f(x_m)| < 2^{-m}$. If we define $y_n = x_m$ if $n = n_m$ for some m and $= x$ otherwise, then $\{y_n\}$ Fine converges to x. From the continuous convergence, $\{f_n(y_n)\}$ $\mathbb{E}$-converges to $f(x)$. Hence, the subsequence $\{f_{n_m}(x_m)\}$ $\mathbb{E}$-converges to $f(x)$. From the above inequality, $\{f(x_m)\}$ also $\mathbb{E}$-converges to $f(x)$. □

For a sequence of Fine continuous, we obtain the following proposition.

Proposition 2.6. *For a sequence of Fine continuous functions, t-Fine convergence and continuous Fine convergence are equivalent.*

From Propositions 2.2 and 2.6, we obtain the following theorem.

Theorem 2.1. *For a sequence of Fine continuous functions, Fine convergence and continuous Fine convergence are equivalent.*

3. Fine computability

A sequence of rationals $\{r_n\}$ is called *recursive* if there exist recursive functions $\alpha(n)$, $\beta(n)$ and $\gamma(n)$ which satisfy $r_n = (-1)^{\gamma(n)}\frac{\beta(n)}{\alpha(n)}$. We will subsequently treat the computability of sequences from the Fine space and the computability of functions on the Fine space. So, we assume that a number x or a sequence $\{x_n\}$ is in $[0,1)$ unless otherwise stated.

A double sequence $\{x_{n,m}\}$ is said to *Fine converge effectively* to a sequence $\{x_n\}$ if there exists a recursive function $\alpha(n,k)$ such that $x_{n,m} \in J(x_n,k)$ for all n,k and all $m \geqslant \alpha(n,k)$.

A sequence $\{x_n\}$ is said to be *Fine computable* if there exists a recursive double sequence of rationals $\{r_{n,m}\}$ which Fine converges effectively to $\{x_n\}$. For this definition, it is sufficient to take a recursive sequence of dyadic rationals instead of a recursive sequence of rationals in general. A single element x is called Fine computable if the sequence $\{x,x,x,\ldots\}$ is Fine computable. The definition of Fine computability can be extended to multiple sequences in an obvious manner.

If we use the Euclidean metric instead of the Fine metric in the above, then we obtain the usual notion of computability on the real line. We call this computability $\mathbb{E}$-*computability*. Notice that a single real number is $\mathbb{E}$-computable if and only if it is Fine computable, and that a Fine computable sequence of real numbers is also an $\mathbb{E}$-computable sequence,[2,6,13] but the converse of the latter fact does not hold. It also holds that a recursive sequence of rationals is Fine computable, while an $\mathbb{E}$-computable sequence of rationals is not necessarily Fine computable.[2,6]

There have been several approaches to weaker notions of computable functions in order to make some simple $\mathbb{E}$-discontinuous functions computable. We quote only some recent works,[2,6,7,12–14] which are closely related to this article. In the last two, the computability on the range is weakened by replacing the recursive modulus of convergence with the limiting recursive one in the definition of computable sequences of reals. Another method is that the topology on the domain of definition is replaced by the Fine metric, which is stronger than the Euclidean metric.[2,6,7] The latter approach is generalized to the computability with respect to an effective uniformity.[12,13]

The uniform Fine computability of a function is introduced by Mori.[6] The locally uniform Fine computability is also treated together with effective locally uniform Fine convergence. A similar but slightly different notion of computability is also introduced[12] for functions on a space with an effective uniform topology.

In the rest of this section, we review the above two definitions of computability for functions on the Fine space, together with the corresponding effective convergence. Another will be introduced in the next section. We take a recursive enumeration of all dyadic rationals in $[0,1)$ $\{e_i\}$, as an effective separating set, and use it throughout this article. (An effective separating set is a computable sequence which is dense in $[0,1)$ and which effectively approximates every computable sequence.)

Definition 3.1. (Uniformly Fine computable sequence of functions[6]) A sequence of functions $\{f_n\}$ is said to be *uniformly Fine computable* if (i) and (ii) below hold.

(i) (*Sequential Fine computability*) The double sequence $\{f_n(x_m)\}$ is $\mathbb{E}$-computable for any Fine computable sequence $\{x_m\}$.

(ii) (*Effectively uniform Fine continuity*) There exists a recursive function $\alpha(n,k)$ such that, for all n,k and all $x,y \in [0,1)$, $y \in J(x,\alpha(n,k))$ implies $|f_n(x) - f_n(y)| < 2^{-k}$.

The uniform Fine computability of a single function f is defined by that of the sequence $\{f,f,\ldots\}$. Notice that the computability of the sequence $\{f_n(x_m)\}$ in (i) is $\mathbb{E}$-computability.

Definition 3.2. (Effectively uniform convergence of functions[6]). A sequence of functions $\{f_n\}$ is said to *converge effectively uniformly* to a function f if there exists a recursive function $\alpha(k)$ such that, for all n and k, $n \geqslant \alpha(k)$ implies $|f_n(x) - f(x)| < 2^{-k}$ for all x.

Theorem 3.1. *If a uniformly Fine computable sequence of functions $\{f_n\}$ converges effectively uniformly to a function f, then f is also uniformly Fine computable.*

The proof is similar to that of the corresponding theorem.[8]

Definition 3.3. (Locally uniformly Fine computable sequence of functions[5]) A sequence of functions $\{f_n\}$ is said to be *locally uniformly Fine computable* if the following (i) and (ii) hold.

(i) $\{f_n\}$ is sequentially Fine computable.

(ii) *(Effectively locally uniform Fine continuity)* There exist recursive functions $\alpha(n,k,i)$ and $\beta(n,i)$ which satisfy the following (ii-a) and (ii-b).

(ii-a) For all i, n and k, $|f_n(x) - f_n(y)| < 2^{-k}$ if $x, y \in J(e_i, \beta(n,i))$ and $y \in J(x, \alpha(n,k,i))$.

(ii-b) $\bigcup_{i=1}^{\infty} J(e_i, \beta(n,i)) = [0,1)$ for each n.

It is proved in Example 4.1[5] that the function f defined by $f(x) = \frac{1}{1-2x}$ if $x < \frac{1}{2}$ and $= 0$ if $x \geq \frac{1}{2}$ is locally uniformly Fine computable but not uniformly Fine continuous, since it diverges at $\frac{1}{2}$.

Definition 3.4. (Effectively locally uniform Fine convergence[5]). A sequence of functions $\{f_n\}$ is said to *Fine converge effectively locally uniformly* to a function f if there exist recursive functions $\gamma(i)$ and $\delta(k,i)$ such that

(a) $|f_n(x) - f(x)| < 2^{-k}$ for $x \in J(e_i, \gamma(i))$ and $n \geq \delta(k,i)$,

(b) $\cup_{i=1}^{\infty} J(e_i, \gamma(i)) = [0,1)$.

Theorem 3.2.[5] *If a locally uniformly Fine computable sequence of functions $\{f_n\}$ Fine converges effectively locally uniformly to f, then f is locally uniformly Fine computable.*

Theorem 3.2 can be proved similarly to the proof of Theorem 5.3 in Section 4. The above two definitions of computable functions can be carried over to an effectively separable metric space with a computability structure or to a space with effective uniformity.

The notion of the Fine computable functions is introduced by Brattka[2] for a single function. We extend it to that of a function sequence. and prove a theorem similar to Theorem 3.2.

Recall that $\{e_i\}$ is a recursive enumeration of all dyadic rationals in $[0,1)$ and that it is an effective separating set.

Definition 3.5. (Fine computable sequence of functions) A sequence of functions $\{f_n\}$ is said to be *Fine computable* if it satisfies the following.

(i) $\{f_n\}$ is sequentially Fine computable.

(ii) (Effective Fine Continuity) There exists a recursive function $\alpha(n,k,i)$ such that

(ii-a) $x \in J(e_i, \alpha(n,k,i))$ implies $|f_n(x) - f_n(e_i)| < 2^{-k}$,

(ii-b) $\bigcup_{i=1}^{\infty} J(e_i, \alpha(n,k,i)) = [0,1)$ for each n, k.

Fine computability of a single function f is defined by replacing $\alpha(n,k,i)$ with $\alpha(k,i)$. It is equivalent to saying that the sequence $\{f, f, \ldots\}$ is computable.

Definition 3.6. (Effective Fine continuity with respect to $\{r_i\}$) If the requirement (ii) in Definition 3.5 holds for a Fine computable sequence $\{r_i\}$ instead of $\{e_i\}$, we say that f is *effectively Fine continuous with respect to* $\{r_i\}$.

We proposed[12] a slightly different notion of computability of functions on an effective uniform topological space, that is, we required the sequential computability, the effective continuity with respect to some effective separating set and the relative computability.

4. Fine computable functions

On the Fine space, we can prove that the effective Fine continuity of a function sequence does not depend on the choice of an effective separating set. We will state and prove this fact for a single function f as below.

Theorem 4.1. *If f is effectively Fine continuous with respect to an effective separating set $\{r_i\}$, then f is effectively Fine continuous with respect to any effective separating set $\{t_j\}$.*

For the proof of this theorem, we prepare some elementary properties concerning dyadic intervals. Classically, they are self-evident.

We say that a sequence of dyadic intervals $I_j = [a_j, b_j)$ $(a_j < b_j)$ *recursive* if $\{a_j\}$ and $\{b_j\}$ are recursive sequences of dyadic rationals. A recursive dyadic intervals $\{I_j\}$ is called *recursive dyadic covering of a dyadic interval* I if it is recursive and satisfies $\cup_j I_j = I$.

Lemma 4.1. *The following hold.*

(i) *Let $\{I_j\}$ be a recursive sequence of dyadic intervals. For any Fine computable sequence of numbers $\{x_l\}$, $x_l \in I_j$ or $x_l \notin I_j$ can be determined effectively.*

(ii) *Let* $\{s_i\}$ *be an effective separating set. Then we can find effectively an* i *such that* $s_i \in I(n,k)$ *holds for any* n *and* k, *that is , there is a recursive function of* n *and* k *which computes* i. *In this case,* $I(n,k) = J(s_i, n)$.

(iii) *Let* $\{I_j\}$ *be a recursive dyadic covering of* $[0,1)$ *and let* $\{x_n\}$ *be Fine computable. Then there is a recursive* $j = j(n)$ *such that* $x_n \in I_{j(n)}$.

(iv) *Let* I *and* J *denote dyadic intervals. Then, we can decide effectively whether* $I \cap J = \phi$ *or not, and whether* $I \subseteq J$ *or not.*

(v) *If a dyadic interval* $[a,b)$ *is not a fundamental dyadic interval, then we can decompose it effectively into finitely many disjoint fundamental dyadic intervals.*

From the condition (ii-b) in Definition 3.5, it follows that the set of dyadic neighborhoods $\{J(e_i, \alpha(n,k,i))\}_i$ is a recursive dyadic covering of $[0,1)$ for each n,k.

For a covering consisting of dyadic intervals, the following lemma holds.

Lemma 4.2. *Let* $\{J_p\}$ *be a recursive dyadic covering of a dyadic interval* I. *Then, we can construct a recursive dyadic covering* $\{I_q\}$ *of* I, *which satisfies the following conditions.*

(i) *Each* I_q *is a fundamental dyadic interval.*
(ii) I_q *is a subinterval of* J_p *for some* p.
(iii) I_q*'s are disjoint.*

Proof. Let us first note that we can claim the following (a) and (b) by using Lemma 4.1: (a) The complement of a dyadic interval, say $[a,b)^C$, is a disjoint union of intervals $[0,a) \cup [b,1)$. (b) The complement of a finite union of dyadic intervals $(\cup_{i=1}^n [a_i,b_i))^C = \cap_{i=1}^n [a_i,b_i)^C$ can be represented by a finite disjoint union of fundamental dyadic intervals.

We only outline the construction of $\{I_q\}$ according to a routine procedure in measure theory. The construction itself will explain that the whole procedure can be done effectively.

First, J_1 is a dyadic interval by definition. So, we can decompose it into finitely many disjoint fundamental dyadic intervals, say, $I_1, \ldots, I_{\tau_1}$ by Lemma 4.1 (v).

Second, consider $(J_2 \cap (J_1)^C) = (J_2 \cap (\cup_{q=1}^{\tau_1} I_q)^C)$. It is a finite disjoint union of dyadic intervals by (b) just above. So, we decompose them and obtain a finite sequence of disjoint fundamental dyadic intervals $I_{\tau_1+1}, \ldots, I_{\tau_1+\tau_2}$, the union of which is $(J_2 \cap (J_1)^C)$.

Next, try the same for $(J_3 \cap (J_1 \cup J_2)^C) = (J_3 \cap (\cup_{q=1}^{\tau_1+\tau_2} I_q)^C)$, and so on. If we continue the above process, we obtain $\{I_q\}$, which is the desired sequence.

The construction above suggests the following: if $J_p = J(r_p, \alpha(p))$ for some recursive function $\alpha(p)$ and a recursive sequence of dyadic rationals $\{r_p\}$, then we can obtain recursive functions $\beta(q)$ and $\gamma(q)$ $(0 \leqslant \gamma(q) \leqslant 2^{\beta(q)} - 1)$ so that $I_q = I(\beta(q), \gamma(q))$. □

Proposition 4.1. *Let $\{r_i\}$ be an effective separating set and let f be a function on $[0, 1)$. Then, f is effectively Fine continuous with respect to $\{r_i\}$ if and only if there exist a Fine computable double sequence $\{s_{k,q}\}$ and a recursive function $\delta(k, q)$ which satisfy the following.*

(a) *$\{s_{k,q}\}_q$ is a subset of $\{r_i\}$ for each k.*

(b) *$\{J(s_{k,q}, \delta(k, q))\}_q$ is disjoint for each k.*

(c) *$x \in J(s_{k,q}, \delta(k, q))$ implies $|f(x) - f(s_{k,q})| < 2^{-k}$.*

(d) *$\bigcup_{q=1}^{\infty} J(s_{k,q}, \delta(k, q)) = [0, 1)$ for each k.*

Proof. First, we prove the "if" part. For each k and i, we can find effectively such $q = q(k, i)$ that $r_i \in J(s_{k+1,q}, \delta(k+1, q))$. It is sufficient to take $\alpha(k, i) = \delta(k+1, q)$ (cf. Definitions 3.5 and 3.6), since

$$|f(x) - f(r_i)| \leqslant |f(x) - f(s_{k+1,q})| + |f(s_{k+1,q}) - f(r_i)| < 2^{-k}$$

for $x \in J(r_i, \alpha(k, i)) = J(s_{k+1,q}, \delta(k+1, q))$.

To prove the "only if" part, let $\alpha(k, i)$ be a recursive modulus of continuity of f and let us consider $\{J(r_p, \alpha(k+1, p))\}_p$ for each k. If we apply Lemma 4.2 to this sequence with $I = [0, 1)$, then we obtain recursive functions $\beta(k, q)$ and $\gamma(k, q)$ so that the sequence $\{I_{k,q}\} = \{I(\beta(k, q), \gamma(k, q))\}$ is a recursive dyadic covering of $[0, 1)$ and satisfies (i) through (iii) of Lemma 4.2 for each k. We define $\delta(k, q) = \beta(k, q)$. For each q, we can select $p = p(k, q)$ and $i = i(k, q)$ so that $r_i \in I_{k,q} \subseteq J(r_p, \alpha(k+1, p))$. If we put $s_{k,q} = r_i$, then it holds that

$$|f(x) - f(s_{k,q})| \leqslant |f(x) - f(r_p)| + |f(r_p) - f(r_i)| < 2^{-k},$$

for $x \in J(s_{k,q}, \delta(k, q)) = I_{k,q}$. □

Proof of Theorem 4.1. Assume that f is effectively Fine continuous with respect to an effective separating set $\{r_i\}$ and that $\{t_j\}$ is an effective separating set. Let $\{s_{k,q}\}$ and $\delta(k, q)$ satisfy the requirements (a) through (d) in Proposition 4.1. For each k, q, choose some $t_j \in J(s_{k+1,q}, \delta(k+1, q))$ and denote it by $u_{k,q}$. (We can do this effectively, hence $\{u_{k,q}\}$ is Fine computable). It holds that $J(s_{k+1,q}, \delta(k+1, q)) = J(u_{k,q}, \delta(k+1, q))$ and

$$|f(y) - f(u_{k,q})| \leqslant |f(y) - f(s_{k+1,q})| + |f(s_{k+1,q}) - f(u_{k,q})| < 2^{-k}$$

for $y \in J(u_{k,q}, \delta(k+1, q))$. If we define $\tilde{\delta}(k, q) = \delta(k+1, q)$, then $\tilde{\delta}(k, q)$

is recursive and the conditions (a) through (d) of Proposition 4.1 hold for $\{u_{k,q}\}$ and $\tilde{\delta}(k,q)$ with respect to $\{t_j\}$. If we apply Proposition 4.1 again, we obtain that f is effectively Fine continuous with respect to $\{t_j\}$. □

Let us consider the maximum of a Fine computable function. It is proved that a uniformly Fine computable function has the computable supremum.[7] But the corresponding property does not hold for locally uniformly Fine computable functions. To see this, let us define

$$\chi_c(x) = \chi_{[0,c)}(x),\ \tilde{\chi}_n(x) = \chi_{[1-2^{-(n-1)},1-2^{-n})}(x). \tag{1}$$

Proposition 4.2. *There exists a bounded locally uniformly Fine computable function, the supremum of which is not computable.*

Proof. Let a be a one-to-one recursive function from positive integers to positive integers whose range is not recursive. Define $c_n = \sum_{k=1}^{n} 2^{-a(k)}$. Then $\{c_n\}$ is an $\mathbb{E}$-computable sequence of real numbers, which is monotonically increasing and converges to a non-computable limit c.[8] Define also $f(x) = \sum_{n=1}^{\infty} c_n \tilde{\chi}_n(x)$. $I_n = [1-2^{-(n-1)}, 1-2^{-n}) = [\frac{2^n-2}{2^n}, \frac{2^n-1}{2^n})$ is a fundamental dyadic interval and $\{I_n\}$ is a partition of $[0,1)$. Let us define $n = \beta(i) = \alpha(k,i)$ if $e_i \in I_n$. Then α and β are recursive, and f is locally uniformly Fine computable with respect to β and α. On the other hand, $\sup_{0 \leqslant x<1} f(x) = c$ is not $\mathbb{E}$-computable. □

At the end of this section, we give a simple example of a function which satisfies neither the sequential computability nor the effective Fine continuity (cf. Definition 3.5). In the following proposition, $\frac{1}{3}$ is not essential, and the proposition remains valid if we replace $\frac{1}{3}$ with any dyadic irrational.

Proposition 4.3. $\chi_{\frac{1}{3}}$ *satisfies the following:*
(i) $\chi_{\frac{1}{3}}$ *is not "Fine continuous."*
(ii) *It is not sequentially Fine computable.*

5. Effective Fine convergence

In this section, we define effective Fine convergence of a sequence of functions, and prove that the space of effectively Fine continuous functions is closed with respect to this convergence.

Definition 5.1. (Effective Fine convergence of functions) We say that a sequence of functions $\{f_n\}$ *Fine converges effectively* to a function f if there exist recursive functions $\beta(k,i)$ and $\gamma(k,i)$ which satisfy

(a) $x \in J(e_i, \beta(k,i))$ and $n \geqslant \gamma(k,i)$ imply $|f_n(x) - f(x)| < 2^{-k}$,
(b) $\bigcup_{i=1}^{\infty} J(e_i, \beta(k,i)) = [0,1)$ for each k.

Definition 5.2. (Computable sequence of dyadic step functions[6]) A sequence of functions $\{\varphi_n\}$ is called a *computable sequence of dyadic step functions* if there exist a monotonically increasing recursive function $\delta(n)$ and an $\mathbb{E}$-computable sequence of reals $\{c_{n,j}\}$ $(0 \leqslant j < 2^{\delta(n)}, n = 1, 2, \ldots)$ such that

$$\varphi_n(x) = \sum_{j=0}^{2^{\delta(n)}-1} c_{n,j} \chi_{I(\delta(n),j)}(x). \tag{2}$$

A computable sequence of dyadic step functions is uniformly Fine continuous, since $\varphi_n(x) = \varphi_n(y)$ if $x, y \in I(\delta(n), j)$ for some j. Typical examples of computable sequences of dyadic step functions are the system of Walsh functions, that of Haar functions and that of Rademacher functions.

Theorem 5.1. *Let f be a Fine computable function. Define a "computable sequence of dyadic step functions" $\{\varphi_n\}$ by*

$$\varphi_n(x) = \sum_{j=0}^{2^n-1} f(j2^{-n}) \chi_{I(n,j)}(x). \tag{3}$$

Then $\{\varphi_n\}$ Fine converges effectively to f.

Proof. Let f be a Fine computable function with respect to $\alpha(k,i)$.

If $n \geqslant \alpha(k+1,i)$, then $J(e_i, \alpha(k+1,i)) = \bigcup_{j2^{-n} \in J(e_i, \alpha(k+1,i))} I(n,j)$. Assume further that $x \in J(e_i, \alpha(k+1,i))$. Then, $x \in I(n,j)$ for some j which satisfies $j2^{-n} \in J(e_i, \alpha(k+1,i))$ and $\varphi_n(x) = f(j2^{-n})$. So we obtain

$$\begin{aligned} |\varphi_n(x) - f(x)| &= |f(j2^{-n}) - f(x)| \leqslant |f(j2^{-n}) - f(e_i)| + |f(e_i) - f(x)| \\ &< 2^{-(k+1)} + 2^{-(k+1)} = 2^{-k}. \end{aligned}$$

Therefore, $\{\varphi_n\}$ Fine converges effectively to f with respect to $\gamma(k,i) = \beta(k,i) = \alpha(k+1,i)$. □

Similarly to Proposition 4.1, we can obtain the following proposition.

Proposition 5.1. *A sequence of functions $\{f_n\}$ Fine converges effectively to f if and only if there exist a recursive sequence of dyadic rationals $\{s_{k,i}\}$ and recursive functions $\beta(k,i)$ and $\gamma(k,i)$ which satisfy the following:*
(a) *$x \in J(s_{k,i}, \beta(k,i))$ and $n \geqslant \gamma(k,i)$ imply $|f_n(x) - f(x)| < 2^{-k}$.*
(b) *$\bigcup_{i=1}^{\infty} J(s_{k,i}, \beta(k,i)) = [0,1)$ for each k.*
(c) *$\{J(s_{k,i}, \beta(k,i))\}_i$ are disjoint for each k.*

We can also define the notion of effective Fine convergence with respect to any effective separating set, and prove that the notion of effective Fine convergence does not depend on the choice of an effective separating set.

Now, we prove the closedness of the space of Fine computable functions under effective Fine convergence.

Theorem 5.2. *If an effectively Fine continuous sequence of functions* $\{f_n\}$ *Fine converges effectively to* f*, then* f *is effectively Fine continuous.*

Proof. Let $\{f_n\}$ be effectively Fine continuous with respect to $\alpha(n,k,p)$, that is, $x \in J(e_p, \alpha(n,k,p))$ implies $|f_n(x) - f_n(e_p)| < 2^{-k}$ and $\bigcup_{p=1}^{\infty} J(e_p, \alpha(n,k,p)) = [0,1)$ for each n, k. From effective Fine convergence, we obtain $\{s_{k,i}\}$, $\beta(k,i)$ and $\gamma(k,i)$ satisfying the conditions (a), (b) and (c) in Proposition 5.1. In particular, $\{J(s_{k,i}, \beta(k,i))\}_i$ are are disjoint for each k.

From the requirement (ii-b) of Definition 3.5 for $\alpha(n,k,p)$, we have

$$J(s_{k+2,i}, \beta(k+2,i)) \subseteq [0,1) = \bigcup_{p=1}^{\infty} J(e_p, \alpha(\gamma(k+2,i), k+2, p)).$$

If we set $I = J(s_{k+2,i}, \beta(k+2,i))$ and $\{J_{k,i,p}\}_p = \{J(e_p, \alpha(\gamma(k+2,i), k+2,p)) \cap I\}_p$, and apply Lemma 4.2, we obtain a recursive dyadic covering of $J(s_{k+2,i}, \beta(k+2,i))$, say $\{I_{k,i,q}\} = \{I(\xi(k,i,q), \eta(k,i,q))\}$, which satisfies (i)$\sim$(iii) of Lemma 4.2 for each pair k, i . Let us remark that $I_{k,i,q}$ is a subinterval of $J_{k,i,p}$ for some p, and that $\xi(k,i,q)$ and $\eta(k,i,q)$ are recursive functions.

We can find effectively some $p = p(k,i,q)$ such that $e_p \in I_{k,i,q}$. Define $r_{k,i,q} = e_p$ and $\delta(k,i,q) = \xi(k,i,q)$, and assume $x \in J(r_{k,i,q}, \delta(k,i,q)) = I_{k,i,q}$. Since $J(r_{k,i,q}, \delta(k,i,q)) \subseteq J(s_{k+2,i}, \beta(k+2,i))$, $|f(x) - f_{\gamma(k+2,i)}(x)| < 2^{-(k+2)}$ and $|f(r_{k,i,q}) - f_{\gamma(k+2,i)}(r_{k,i,q})| < 2^{-(k+2)}$ hold. So

$$\begin{aligned} &|f(x) - f(r_{k,i,q})| \\ &\leqslant |f(x) - f_{\gamma(k+2,i)}(x)| + |f_{\gamma(k+2,i)}(x) - f_{\gamma(k+2,i)}(r_{k,i,q})| \\ &\quad + |f_{\gamma(k+2,i)}(r_{k,i,q}) - f(r_{k,i,q})| \\ &< |f_{\gamma(k+2,i)}(x) - f_{\gamma(k+2,i)}(r_{k,i,q})| + 2^{-(k+1)}. \end{aligned}$$

On the other hand, $I_{k,i,q} = J(r_{k,i,q}, \delta(k,i,q)) \subseteq J(e_p, \alpha(\gamma(k+2,i), k+2,p)) = J_{k,i,p}$ and $r_{k,i,q} = e_p$ imply that

$$|f_{\gamma(k+2,i)}(x) - f_{\gamma(k+2,i)}(r_{k,i,q})| = |f_{\gamma(k+2,i)}(x) - f_{\gamma(k+2,i)}(e_p)| < 2^{-(k+2)}.$$

Therefore, $x \in J(r_{k,i,q}, \delta(k,i,q))$ implies $|f(x) - f(r_{k,i,q})| < 2^{-k}$.

Furthermore, $\cup_i \cup_q J(r_{k,i,q}, \delta(k,i,q)) = \cup_i J(s_{k+2,i}, \beta(k+2,i)) = [0,1)$ due to the assumption for $\{s_{k,i}\}$, β, γ and δ.

We can perform the above procedure effectively in i. So, taking some recursive pairing function, $\langle i,q \rangle = i + \frac{1}{2}(i+q)(i+q+1)$ for example, define $r_{k,\ell} = r_{k,i,q}$ and $\delta(k,\ell) = \delta(k,i,q)$ for $\ell = \langle i,q \rangle$. Then, the necessary condition of Proposition 4.1 (with respect to k and ℓ) holds for f, $\{r_{k,\ell}\}$ and $\delta(k,\ell)$ for each ℓ. We can thus conclude that f is effectively continuous.

□

Theorem 5.3. *If a Fine computable sequence of functions $\{f_n\}$ Fine converges effectively to f, then f is Fine computable.*

Proof. Effective Fine continuity is guaranteed by Theorem 5.2.

Let us assume that $\{f_n\}$ Fine converges effectively to f with respect to $\beta(k,i)$ and $\gamma(k,i)$. To prove the sequential computability, let $\{x_m\}$ be Fine computable. We can find effectively an $i = i(k,m)$ so that $x_m \in J(e_i, \beta(k,i))$. If $n \geqslant \gamma(k,i)$, then $|f_n(x_m) - f(x_m)| < 2^{-k}$. So the $\mathbb{E}$-computable sequence $\{f_n(x_m)\}_n$ converges effectively to $\{f(x_m)\}$ effectively in m, and hence $\{f(x_m)\}_m$ is an $\mathbb{E}$-computable sequence. □

Combining Theorem 5.3 with Theorem 5.1, we obtain the following.

Theorem 5.4. (Equivalent condition for Fine computable function) *A function f is Fine computable if and only if there exists a computable sequence of dyadic step functions which Fine converges effectively to f.*

We can extend Theorem 5.3 to the case where a computable double sequence $\{f_{m,n}\}$ Fine converges effectively to a sequence $\{f_m\}$, by suitably extending the notions of the Fine computable sequence, the effective Fine convergence and the computable sequence of dyadic step functions.

Theorem 5.5. *If a Fine computable double sequence of functions $\{f_{m,n}\}$ Fine converges effectively to a sequence $\{f_m\}$, then $\{f_m\}$ is Fine computable.*

Theorem 5.6. *A sequence of functions $\{f_m\}$ is Fine computable if and only if there exists a computable double sequence of dyadic step functions $\{\varphi_{m,n}\}$, which Fine converges effectively to $\{f_m\}$.*

Example 5.1. (Counter Example) Let us consider $\chi_{\frac{1}{3}}$ in Proposition 4.3. Define x_n to be $\frac{1}{3}(1-4^{-n})$. Then $\{x_n\}$ is Fine computable and Fine converges to $\frac{1}{3}$. Hence, χ_{x_n} converges pointwise to $\chi_{\frac{1}{3}}$. Moreover, $\{\chi_{x_n}\}$ is a computable sequence of dyadic step functions (Definition 5.2). However, the convergence is neither Fine nor continuous due to Propositions 2.4 and 2.5.

6. Recursive functional equations and Fine computable functions

In this section, we provide several examples concerning Fine computability of functions. Some of them are represented as linear combinations of $\chi_c(x)$'s and $\tilde{\chi}_n(x)$'s, which have been introduced in Section 4 (Equation (1)).

Example 6.1. Define $f_n = \sum_{i=1}^{n} 2^{-i}\chi_{e_i}$ and $f = \sum_{i=1}^{\infty} 2^{-i}\chi_{e_i}$. Then, $\{f_n\}$ is uniformly Fine computable , $|f_n(x) - f_m(x)| \leqslant \sum_{i=n+1}^{m} 2^{-i} < 2^{-n}$ holds for $n < m$, and $\{f_n\}$ converges effectively uniformly to f. So, f is uniformly Fine computable by Theorem 3.1. On the other hand, f is $\mathbb{E}$-discontinuous at every dyadic rational, since $f(x) - f(e_i) \geqslant 2^{-i}$ for any $x < e_i$.

Example 6.2. Let a be a one-to-one recursive function from positive integers to positive integers, whose range is not recursive, and let us define $f_n(x) = \sum_{k=1}^{n} \tilde{\chi}_{a(k)}(x)$ and $f(x) = \sum_{k=1}^{\infty} \tilde{\chi}_{a(k)}(x)$. Then, $\{f_n\}$ is a computable sequence of dyadic step functions. Classically, $\{f_n\}$ converges to f and f is Fine continuous. However, f does not satisfy the sequential computability, since $f(1-2^{-m}) = 1$ if $m = a(k)$ for some k and $= 0$ otherwise. So, the convergence is not effectively Fine by Theorem 5.3. On the other hand, f is effectively locally uniformly Fine continuous.

The existence of an example which is Fine computable but not locally uniformly Fine computable has been proved by Brattka.[2]

Example 6.3. (Brattka[2]) The example of Brattka is the following:

$$v(x) = \tag{4}$$
$$\begin{cases} \sum_{i=0}^{\infty}(\ell_i \bmod 2)2^{-n_i-\sum_{j=0}^{i-1}(n_j+\ell_j)} & \text{if} \quad \mu(x) = 0^{n_0}1^{\ell_0}0^{n_1}1^{\ell_1}0^{n_2}\cdots \\ \sum_{i=0}^{m}(\ell_i \bmod 2)2^{-n_i-\sum_{j=0}^{i-1}(n_j+\ell_j)} & \text{if} \quad \mu(x) = 0^{n_0}1^{\ell_0}0^{n_1}1^{\ell_1}0^{n_2}\cdots 1^{\ell_m}0^{\omega} \end{cases},$$

where $n_0 \geqslant 0$, $n_i > 0$ for $i > 0$ and $\ell_i > 0$ for all $i \geqslant 0$. ($\mu(x)$ expresses the binary expansion of x with infinitely many zero's.)

For investigation of this example and its generalizations, we introduce the following fundamental dyadic intervals and mappings.

$A_\ell = [1-2^{-(\ell-1)}, 1-2^{-\ell})$, $S_\ell(t) = 1-2^{-(\ell-1)} + 2^{-\ell}t : [0,1) \to A_\ell$

$B_\ell = [1-2^{-\ell}, 1) = \bigcup_{j=\ell+1}^{\infty} A_j$, $R_\ell(t) = 1-2^{-\ell} + 2^{-\ell}t : [0,1) \to B_\ell$

Obviously, $\{A_\ell\}_{\ell=1}^{\infty}$ is an infinite partition of $[0,1)$ and $\{A_1, \ldots, A_j, B_j\}$ is a finite partition of $[0,1)$ for each j. Furthermore, S_ℓ is a bijection from $[0,1)$ onto A_ℓ and $S_\ell^{-1}(x) = 2^\ell(x-(1-2^{-(\ell-1)}))$. R_ℓ is a bijection from $[0,1)$ onto B_ℓ and $R_\ell^{-1}(x) = 2^\ell(x-(1-2^{-\ell}))$.

We note that $x \in A_\ell$ if and only if $\mu(x)$ is expressed as $1^{\ell-1}0 * * \cdots$.

First, we treat the approximating sequence of dyadic step functions $\{v_n\}$, which is obtained from v by Equation (3) in Theorem 5.1. Since v is known to be Fine computable, $\{v_n\}$ Fine converges effectively to v by virtue of Theorem 5.1.

The fact that the v is not locally uniformly Fine continuous[2] assures that it is not locally uniformly Fine computable. From this and Theorem 3.2, it follows that the convergence of $\{v_n\}$ to v cannot be locally uniformly Fine, let alone effectively locally uniformly Fine.

It is easy to see that the sequence $\{v_n\}$ satisfies the following recurrence equation.

$$v_1(x) = \begin{cases} 0 \text{ if } \ x \in A_1 = [0, \frac{1}{2}) \\ 1 \text{ if } \ x \in B_1 = [\frac{1}{2}, 1) \end{cases},$$

$$v_n(x) = \begin{cases} \frac{1+(-1)^i}{2} + 2^{-i} v_{n-i}(S_i^{-1}(x)) \ \text{ if } \ x \in A_i \ (1 \leqslant i \leqslant n-1) \\ \frac{1+(-1)^n}{2} \ \text{ if } \ x \in A_n \\ \frac{1+(-1)^{n+1}}{2} \ \text{ if } \ x \in B_n \end{cases}. \quad (5)$$

We illustrate the first four of $\{v_n\}$ in Figure 1. Let us examine the graph of v_4. The restriction of v_4 to $A_1 = [0, \frac{1}{2})$ is the contraction of the graph of v_3 with scale $\frac{1}{2}$. The restriction of v_4 to $A_2 = [\frac{1}{2}, \frac{3}{4})$ is the vertical translation of the contraction of the graph of v_2 with scale 2^{-2}. The restriction of v_4 to $A_3 = [\frac{3}{4}, \frac{7}{8})$ is the contraction of the graph of v_1 with scale 2^{-3}. $v_4(x) = 1$ if $x \in A_4 = [\frac{7}{8}, \frac{15}{16})$ and $v_4(x) = 0$ if $x \in B_4 = [\frac{15}{16}, 1)$.

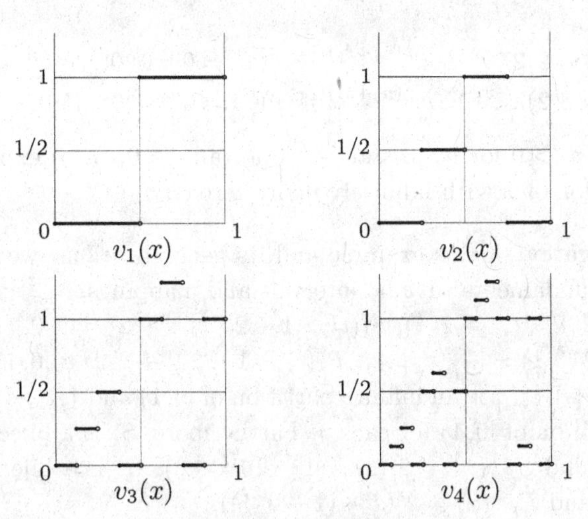

Fig. 1. $v_n(x)$ for $n = 1, 2, 3, 4$

By definition, it holds that $v_\ell(k2^{-n}) = v_n(k2^{-n})$ for $\ell \geqslant n$, and hence they are equal to $v(k2^{-n})$ for any $k < 2^n$. This shows that the value $v(x)$ is determined by $v_n(x)$ if x is a dyadic rational of level n.

In Figure 2, we present an approximating graph of v by drawing a line from $(k2^{-6}, v(k2^{-6}))$ to $((k+1)2^{-6}, v(k2^{-6}))$ for $0 \leqslant k \leqslant 2^6 - 1$.

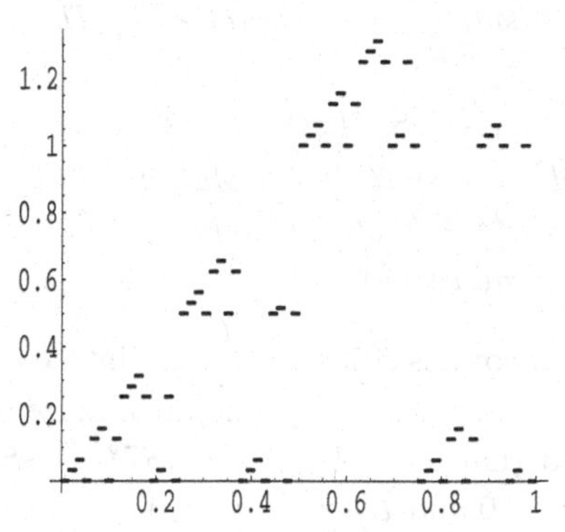

Fig. 2. $v(x)$ for $x = k2^{-6}, 0 \leqslant k \leqslant 2^6 - 1$

To prove some properties of the function v, we derive a simple recurrence equation. It is easily proved from Equation (4) that $v(x)$ satisfies $v(0) = 0$ and the following functional equation

$$v(x) = \tfrac{1+(-1)^\ell}{2} + 2^{-\ell} v(S_\ell^{-1}(x)) \ \text{ if } \ x \in A_\ell \ (\ell = 1, 2, \ldots). \tag{6}$$

Equation (6) suggests that the graph of v has a certain properties of the fractal. This property is called invariance for an infinite systems of contractions.[11] If we replace the first term in the right hand side of Equation (6) with a computable sequence from $[0, 1)$, we can obtain other examples of Fine computable functions.

Theorem 6.1. *Assume that $\{h(\ell)\}_\ell$ is an $\mathbb{E}$-computable sequence from $[0, 1]$ and that $h(1) = 0$.*

(i) *The equation*

$$f(x) = h(\ell) + 2^{-\ell} f(S_\ell^{-1}(x)) \ \ \textit{if} \ \ x \in A_\ell \ (\ell = 1, 2, \ldots) \tag{7}$$

has a unique bounded Fine computable solution.

(ii) *If $\liminf_{\ell\to\infty} h(\ell) \neq \limsup_{\ell\to\infty} h(\ell)$, then the bounded solution of Equation (7) is not locally uniformly Fine continuous.*

(iii) *If $\liminf_{\ell\to\infty} h(\ell) = \limsup_{\ell\to\infty} h(\ell) = \lim_{\ell\to\infty} h(\ell) = a$ and this convergence is effective, then the bounded solution of Equation (7) is uniformly Fine computable.*

If h is given by $h(\ell) = 0$ for an odd ℓ and $= 1$ for an even ℓ, then we obtain the example of Brattka.

We can also get Fine computable functions by the following equation, which is similar to Equation (7) but slightly different.

Theorem 6.2. *Let h satisfy the assumption of Theorem 6.1.*

(i) *The equation*

$$f(x) = h(\ell) + \tfrac{1}{2} f(S_\ell^{-1}(x)) \quad \text{if } x \in A_\ell \ (\ell = 1, 2, \ldots) \tag{8}$$

has a unique bounded Fine computable solution.

(ii) *If h is not constant, then the bounded solution of Equation (8) is not locally uniformly Fine continuous.*

For the proof of Theorems 6.1 and 6.2, we introduce the following notations: For each $x \in [0,1)$, we can define an infinite sequence of positive integers $\{\ell_i(x)\}_{i=1}^{\infty}$ so that $x \in A_{\ell_1(x)}$ and $S_{\ell_i(x)}^{-1} \cdots S_{\ell_1(x)}^{-1}(x) \in A_{\ell_{i+1}(x)}$, and then define $L_0(x) = 0$ and $L_j(x) = \ell_1(x) + \ell_2(x) + \cdots + \ell_j(x)$ for $j > 0$.

For a dyadic rational r, we define its level by

$$lev(r) = \min\{n \in \mathbb{N} \mid \exists j. r = j2^{-n}\}. \tag{9}$$

We have mentioned the level of a fundamental dyadic interval I in Section 3. We denote this with $lev(I)$. If $\{r_n\}$ is a recursive sequence of dyadic rationals, then $\{lev(r_n)\}_n$ is recursive.

We list up some properties concerning $\{S_\ell\}$ and $\{\ell_i(x)\}$.

Fact 1: $\ell_j(x) \geqslant 1$ *and* $L_j(x) \geqslant j$.

Fact 2: *For any positive integers $\ell_1, \ell_2, \ldots, \ell_k$, we define $L_k = \ell_1 + \ldots + \ell_k$. Then $S_{\ell_1} S_{\ell_2} \cdots S_{\ell_k}([0,1))$
$= [1 - 2^{-L_1} - 2^{-L_2} - \cdots - 2^{-(L_k - 1)}, 1 - 2^{-L_1} - 2^{-L_2} - \cdots - 2^{-L_k})$ is a fundamental dyadic interval of level L_k.*

Fact 3: *For any positive integers $\ell_1, \ell_2, \ldots, \ell_k$,
if $x \in S_{\ell_1} S_{\ell_2} \cdots S_{\ell_k}([0,1))$, then $\ell_i(x) = \ell_i$, $i = 1, 2, \ldots, k$.*

Fact 4: *If a dyadic rational r is of level n and lies in A_ℓ, then the level of $S_\ell^{-1}(r)$ is equal to or less than $n - \ell$. Hence, if $L_j(r) > lev(r)$, then $S_{\ell_j(r)}^{-1} \cdots S_{\ell_2(r)}^{-1} S_{\ell_1(r)}^{-1}(r) = 0$.*

Fact 5: *If $\{x_n\}$ is a Fine computable sequence of reals, then the double sequence of integers $\{\ell_i(x_n)\}$ is recursive in i and n.*

Fact 6: *Let f be a solution of Equation (7).
Put $t = S_{\ell_j(x)}^{-1} \cdots S_{\ell_2(x)}^{-1} S_{\ell_1(x)}^{-1}(x)$ for $x \in [0,1)$. Then we obtain $f(x) =$*

$$h(\ell_1(x)) + 2^{-L_1(x)} h(\ell_2(x)) + \cdots + 2^{-L_{j-1}(x)} h(\ell_j(x)) + 2^{-L_j(x)} f(t). \tag{10}$$

Moreover, if r is dyadic rational and $L_j(r) > lev(r)$, then it holds by Fact 4 that

$$f(r) = h(\ell_1(r)) + 2^{-L_1(r)} h(\ell_2(r)) + \cdots + 2^{-L_{j-1}(r)} h(\ell_j(r)). \tag{11}$$

Fact 7: *Let f satisfy Equation (8). Put* $t = S^{-1}_{\ell_j(x)} \cdots S^{-1}_{\ell_2(x)} S^{-1}_{\ell_1(x)}(x)$ *for* $x \in [0,1)$. *Then, we obtain*

$$f(x) = h(\ell_1(x)) + 2^{-1}h(\ell_2(x)) + \cdots + 2^{-(j-1)}h(\ell_j(x)) + 2^{-j}f(t), \quad (12)$$

$$f(r) = h(\ell_1(r)) + 2^{-1}h(\ell_2(r)) + \cdots + 2^{-(j-1)}h(\ell_j(r)). \quad (13)$$

for dyadic rational r with $L_j(r) > lev(r)$.

Subsequently, $||f||$ will denote the supremum of a function f (if it exists).

Proof of Theorem 6.2 (i). Let f be a bounded solution of Equation (8) (or Equation (7)). Since, $0 \in A_1$ and $S_1^{-1}(0) = 0$, we obtain $f(0) = \frac{1}{2}f(0)$ and hence $f(0) = 0$. From Equation (12) (or Equation (10)) and the assumption that $h(\ell) \in [0,1]$, we obtain $|f(x)| \leqslant 1 + 2^{-1} + \cdots + 2^{-(j-1)} + 2^{-j}||f||$. Letting j tend to infinity, we obtain $|f(x)| \leqslant \sum_{j=0}^{\infty} 2^{-j} = 2$. Hence $||f|| \leqslant 2$.

Existence: Since $||h|| \leqslant 1$, $\sum_{j=1}^{\infty} 2^{-(j-1)}h(\ell_j(x))$ converges absolutely and uniformly in x. If we denote this limit function by f, then it is easy to prove that f satisfies Equation (8).

Uniqueness: Suppose that f and g are bounded solutions of Equation (8) or of Equation (7). Then, from Equation (12) (or from Equation (10)),

$$|f(x) - g(x)| \leqslant 2^{-j}(||f|| + ||g||)$$

holds for all j. Since the right-hand side tends to zero as j tends to infinity, we obtain $f = g$.

We remark that the unique bounded solution of Equation (8) is given by $f(x) = \sum_{j=1}^{\infty} 2^{-(j-1)}h(\ell_j(x))$ by *Existence* and *Uniqueness*, The convergence in the right-hand side is effectively uniform.

Effective Fine Continuity: We temporarily fix an arbitrary k. From the definition of $\{S_\ell\}$ and Fact 2, the set of intervals $\{S_{\ell_1}S_{\ell_2}\cdots S_{\ell_{k+3}}([0,1))\}_{\ell_1,\ell_2,\ldots,\ell_{k+3}}$ is a partition of $[0,1)$ consisting of fundamental dyadic intervals. ($\ell_1, \ell_2, \ldots, \ell_{k+3}$ range over all positive integers.) Therefore, each e_i is contained in some $I = S_{\ell_1}S_{\ell_2}\cdots S_{\ell_{k+3}}([0,1))$. Note that we can find such I effectively in k and i. If we define $\gamma(k,i)$ to be the level of I, then $J(e_i, \gamma(k,i)) = I$ and γ is recursive.

Assume that $x \in J(e_i, \gamma(k,i))$. Then, $\ell_j(x) = \ell_j(e_i) = \ell_j$ for $1 \leqslant j \leqslant k+3$ by Fact 3, and we obtain by Equation (12)

$$f(x) = h(\ell_1) + 2^{-1}h(\ell_2) + \cdots + 2^{-(k+2)}h(\ell_{k+3}) + 2^{-(k+3)}f(t),$$

$$f(e_i) = h(\ell_1) + 2^{-1}h(\ell_2) + \cdots + 2^{-(k+2)}h(\ell_{k+3}) + 2^{-(k+3)}f(s),$$

where, $t = S^{-1}_{\ell_{k+3}} \cdots S^{-1}_{\ell_2} S^{-1}_{\ell_1}(x)$ and $s = S^{-1}_{\ell_{k+3}} \cdots S^{-1}_{\ell_2} S^{-1}_{\ell_1}(e_i)$. Therefore,

$$|f(x) - f(e_i)| \leqslant 2\, 2^{-(k+3)}||f|| \leqslant 4\, 2^{-(k+3)} < 2^{-k}.$$

This proves the effective Fine continuity of f.

Sequential Computability: Let $\{x_n\}$ be a Fine computable sequence in $[0, 1)$. Define $y_{n,m} = h(\ell_1(x_n)) + 2^{-1}h(\ell_2(x_n)) + \cdots + 2^{-(m-1)}h(\ell_m(x_n))$.

Then, the double sequence $\{y_{n,m}\}$ is $\mathbb{E}$-computable by Fact 5 and $\mathbb{E}$-converges effectively to $\{f(x_n)\}$ by the remark in the proof of *uniqueness*. Therefore, $\{f(x_n)\}$ is an $\mathbb{E}$-computable sequence of reals. □

Theorem 6.1 (i) can be proved similarly by replacing (8) with (7) and (12) with (11).

Proof of Theorem 6.1 (ii). Let us assume that $\liminf_{m\to\infty} h(m) \neq \limsup_{m\to\infty} h(m)$, and suppose that f were locally uniformly Fine continuous with respect to functions $\alpha(k, i)$ and $\beta(i)$, that is, for all k, $|f(x) - f(y)| < 2^{-k}$ if $x, y \in J(e_i, \beta(i))$ and $y \in J(x, \alpha(k, i))$, and $\bigcup_{i=1}^{\infty} J(e_i, \beta(i)) = [0, 1)$.

Put $\delta = \limsup_{\ell\to\infty} h(\ell) - \liminf_{\ell\to\infty} h(\ell)$ and consider any fixed i and the corresponding $J(e_i, \beta(i))$.

Now, take k so large that $2^{-k} < \delta\, 2^{-(\beta(i)+1)}$. From the definition of δ, there exist $m_1 > \alpha(k, i)$ and $m_2 > \alpha(k, i)$ such that $h(m_2) - h(m_1) > \frac{\delta}{2}$.

Let z be the left end point of $J(e_i, \beta(i))$. Then it holds that $lev(z) \leqslant \beta(i)$. Put further $x = z + 2^{-(\beta(i)+1)}(1 - 2^{-(m_1-1)})$ and $y = z + 2^{-(\beta(i)+1)}(1 - 2^{-(m_2-1)})$. Then z, x and y are dyadic rationals and z can be expressed as $j2^{-\beta(i)}$ for some integer j. From the last property above, there exists an integer n such that $L_n(z) = \beta(i) + 1$. $\ell_j(z) = \ell_j(x) = \ell_j(y)$ if $j \leqslant n$, $\ell_j(z) = 1$ if $j > n$, $\ell_{n+1}(x) = m_1$, $\ell_{n+1}(y) = m_2$ and $\ell_j(x) = \ell_j(y) = 1$ if $j > n + 1$.

By Equation (11) and Fact 1, we obtain

$$f(y) - f(x) = 2^{-L_n(z)}(h(m_2) - h(m_1)) > 2^{-(\beta(i)+2)}\delta. \tag{14}$$

From Equation (14) and the choice of k, $f(y) - f(x) > 2^{-k}$ holds.

On the other hand, $x, y \in J(e_i, \beta(i))$ and $y \in J(x, \alpha(k, i))$ hold. This implies, from the assumption, $|f(x) - f(y)| < 2^{-k}$, contradicting Equation (14). f is thus not locally uniformly Fine continuous. □

Proof of Theorem 6.2 (ii). Assume that $h(m_1) < h(m_2)$. For any i, there exists an integer n such that $L_n(e_i) = lev(e_i) + 1$. Put, for any m,

$$x = e_i + 2^{-(lev(e_i)+1)}(1 - 2^{-(m-1)}) + 2^{-(m+lev(e_i)+1)}(1 - 2^{-(m_1-1)})$$

and

$$y = e_i + 2^{-(lev(e_i)+1)}(1 - 2^{-(m-1)}) + 2^{-(m+lev(e_i)+1)}(1 - 2^{-(m_2-1)}).$$

Then, x and y are dyadic rationals and satisfy

$$\ell_{n+1}(x) = \ell_{n+1}(y) = m,\ \ell_{n+2}(x) = m_1,\ \ell_{n+2}(y) = m_2,$$
$$\ell_{n+3}(x) = \ell_{n+3}(y) = 1,\ \ell_{n+4}(x) = \ell_{n+4}(y) = 1,\ \cdots.$$

So we obtain

$$f(x) = f(e_i) + 2^{-(lev(e_i)+1)}h(m) + 2^{-(lev(e_i)+2)}h(m_1)$$
$$f(y) = f(e_i) + 2^{-(lev(e_i)+1)}h(m) + 2^{-(lev(e_i)+2)}h(m_2)$$

by Equation (13). Hence, $f(y) - f(x) = 2^{-(lev(e_i)+2)}(h(m_2) - h(m_1)) > 0$.

On the other hand, it holds that $x, y \in J(z, lev(e_i) + m)$ and $y \in J(e_i, lev(e_i) + m)$. If f were locally uniformly Fine continuous, then $f(y) - f(x)$ would be arbitrarily small for sufficiently large m, contradicting the last inequality. □

Proof of Theorem 6.1 (iii). For any $\ell_1, \ell_2, \ldots, \ell_j$ and $x \in [0, 1)$, define $t = S_{\ell_j}^{-1} \cdots S_{\ell_1}^{-1}(x)$. Then it holds that $\ell_i(x) = \ell_i$ for $1 \leqslant i \leqslant j$ and we obtain by Fact 6 (10),

$$f(x) = h(\ell_1) + 2^{-L_1}h(\ell_2) + \cdots + 2^{-L_{j-1}}h(\ell_j) + 2^{-L_j}f(t). \qquad (15)$$

Let $\alpha(k)$ be a modulus of convergence of h to some number a, that is, α is a recursive function which satisfies that $\ell \geqslant \alpha(k)$ implies $|h(\ell) - a| < 2^{-k}$. We can assume that $\alpha(k) \geqslant k$.

Let us consider the finite partition of $[0, 1)$ consisting of all sets of the form $U_1U_2 \cdots U_{k+3}[0, 1)$, where U_i is chosen from the family of sets $\{S_1, S_2, \ldots, S_{\alpha(k+3)}, R_{\alpha(k+3)}\}$. By Fact 2, each $U_1U_2 \cdots U_{k+3}[0, 1)$ is a fundamental dyadic interval. So we can define $\beta(k)$ to be the maximum of their levels.

Suppose $y \in J(x, \beta(k))$. Then x and y are contained in some $U_1U_2 \cdots U_{k+3}[0, 1)$.

If $R_{\alpha(k+3)}$ does not appear in $U_1, U_2, \ldots, U_{k+3}$, then it holds that $\ell_i(x) = \ell_i(y)$ for $1 \leqslant i \leqslant k + 3$ from Fact 3. So we obtain by Equation (15)

$$|f(x) - f(y)| \leqslant 2\,2^{-L_{k+3}}||f|| \leqslant 4\,2^{-(k+3)} < 2^{-k}.$$

Otherwise, there exists at least one $R_{\alpha(k+3)}$ in $U_1, U_2, \ldots, U_{k+3}$. Let U_j be the first occurrence of $R_{\alpha(k+3)}$ in $U_1, U_2, \ldots, U_{k+3}$. If $j \geqslant 2$, then $\ell_i(x) = \ell_i(y)$ for $1 \leqslant i \leqslant j - 1$. Since $R_{\alpha(k+3)}([0, 1)) = B_{\alpha(k+3)} = \bigcup_{i=\alpha(k+3)+1}^{\infty} A_i$, we obtain from (15), for some $t, s \in [0, 1)$,

$$f(x) = h(\ell_1(x)) + 2^{-L_1(x)}h(\ell_2(x)) + \cdots + 2^{-L_{j-2}(x)}h(\ell_{j-1}(x))$$
$$+2^{-L_{j-1}(x)}h(\ell_j(x)) + 2^{-L_j(x)}f(t),$$
$$f(y) = h(\ell_1(y)) + 2^{-L_1(y)}h(\ell_2(y)) + \cdots + 2^{-L_{j-2}(y)}h(\ell_{j-1}(y))$$
$$+2^{-L_{j-1}(y)}h(\ell_j(y)) + 2^{-L_j(y)}f(s).$$

It holds that $\ell_j(x), \ell_j(y) \geqslant \alpha(k + 3) \geqslant k + 3$. So we obtain

$$
\begin{aligned}
&|f(x) - f(y)| \\
&\quad \leqslant 2^{-L_{j-1}(x)}|h(\ell_j(x)) - h(\ell_j(y))| + 2^{-L_j(x)}|f(t)| + 2^{-L_j(y)}|f(s)| \\
&\quad \leqslant 2^{-(k+3)} + 4\,2^{-\alpha(k+3)} \leqslant 5\,2^{-(k+3)} < 2^{-k}.
\end{aligned}
$$

Therefore, $y \in J(x, \beta(k))$ implies $|f(x) - f(y)| < 2^{-k}$, and the effectively uniform Fine continuity holds. □

References

1. E. Binz. *Continuous Convergence on C(X)*. Lecture Notes in Mathematics 469. Springer, 1975.
2. V. Brattka. Some Notes on Fine Computability. *Journal of Universal Computer Science*, 8:382-395, 2002.
3. H.-P. Butzmann and M. Schroder. Spaces making continuous convergence and locally uniformly convergence coincide, their very weak P-property, and their topological behavior. *Math. Scand.*, 67:227-254, 1990.
4. N. J. Fine. On the Walsh Functions. *Trans. Amer. Math. Soc.*, 65:373-414, 1949.
5. T. Mori. Computabilities of Fine-Continuous Functions. *Computability and Complexity in Analysis, (4th International Workshop, CCA2000. Swansea)*, ed. by Blanck, J. *et al.*, 200-221. Springer, 2001.
6. T. Mori. On the computability of Walsh functions. *Theoretical Computer Science*, 284:419-436, 2002.
7. T. Mori. Computabilities of Fine continuous functions. *Acta Humanistica et Scientifica Universitatis Sangio Kyotiensis, Natural Science Series I,* 31:163-220, 2002. (in Japanese)
8. M.B. Pour-El and J. I. Richards. *Computability in Analysis and Physics.* Springer-Verlag, 1989.
9. M. Schröder. Extended admissibility *Theoretical Computer Science*, 284:519-538, 2002.
10. F. Schipp, W.R. Wade and P. Simon. *Walsh Series.* Adam Hilger, 1990.
11. Y. Tsujii, M. Mori, M. Yasugi and H. Tsuiki . Fine-Continuous Functions and Fractals Defined by Infinite Systems of Contractions. *Lecture Notes in Computer Science*, Vol. 5489, 109-125. Springer, 2009.
12. Y. Tsujii, M. Yasugi and T. Mori. Some Properties of the Effective Uniform Topological Space. *Computability and Complexity in Analysis, (Lecture Notes in Computer Science 2064)*, ed. by Blanck, J. *et al.*, 336-356. Springer, 2001.
13. M. Yasugi, T. Mori and Y. Tsujii. Effective sequence of uniformities and its effective limit, CCA2005 Proceedings (Informatik Berichte 326-7/2005 Fern Universität in Hagen),301-318,2005.
14. M. Yasugi and M. Washihara. A note on Rademacher functions and computability. *Words, Languages and Combinatorics III*, ed. by Masami Ito, Teruo Imaoka, 466-475. World Scientific, 2003.

Received: August 17, 2009

Revised: December 14, 2009

ON A HIERARCHY OF PERMUTATION LANGUAGES

BENEDEK NAGY
Department of Computer Science, Faculty of Informatics, University of Debrecen,
4010 Debrecen, PO Box. 12, Hungary,
E-mail: nbenedek@inf.unideb.hu
www.inf.unideb.hu/~nbenedek

The family of context-free grammars and languages are frequently used. Unfortunately several important languages are not context-free. In this paper a possible family of extensions is investigated. In our derivations branch-interchanging steps are allowed: language families obtained by context-free and permutation rules are analysed. In permutation rules both sides of the rule contain the same symbols (with the same multiplicities). The simplest permutation rules are of the form $AB \to BA$. Various families of permutation languages are defined based on the length of non-context-free productions. Only semi-linear languages can be generated in this way, therefore these language families are between the context-free and context-sensitive families. Interchange lemmas are proven for various families. It is shown that the generative power is increasing by allowing permutation rules with length three instead of only two. Closure properties and other properties are also detailed.

Keywords: Chomsky hierarchy; Permutation languages; Interchange (permutation) rule; Semi-linear languages; Mildly context-sensitivity.

1. Introduction

The Chomsky type grammars and the generated language families belong to the most basic and most important fields of theoretical computer science. The field is fairly old, the basic concepts and results are from the middle of the last century (see, for instance, Refs. 1–4). The context-free grammars (and languages) are widely used due to their generating power and due to the simple way of derivation. The derivation trees represent the context-free derivations. In these derivations the direction left-to-right is preserved. Moreover the branches of the tree are independent of each-other. Well known that there is a big gap between the efficiency of context-free and context-sensitive grammars. There are very simple non-context-free languages as well, for instance $\{a^{2^n} | n \in \mathbb{N}\}$, $\{a^n b^n c^n | n \in \mathbb{N}\}$, etc. So, context-free grammars are not enough to describe several phenomena of the world,[5] but the context-sensitive family is too large, the context-sensitive grammars are too powerful and they have some inconvenient properties. Therefore several branches of extensions of context-free grammars were introduced by controlling the derivations in another way.[5] It was known in the early 70's that every context-sensitive grammar has an equivalent one using rules of the following types $AB \to AC, AB \to BA, A \to BC, A \to B$ and $A \to a$ (where A, B, C are non-terminals and a is a terminal symbol). In 1974 Penttonen showed that one-side context-sensitivity is enough to obtain the whole context-sensitive language class,[6] so grammars with only rules of type $AB \to AC, A \to BC, A \to B, A \to a$ are enough. In Turing-machine simulations the rules of type $AB \to BA$ are frequently used to represent the movement of the head of the machine. We use the term *permutation rule* (or interchange rule) for those rules which have the same multiset of symbols in both sides. They allow to permute some consecutive letters in the sentential form. The grammars having non-context-free rules only in the form $AB \to BA$ was characterized in Refs. 7,8. Now we are continuing the research by allowing longer permutation rules, e.g., rules of length 3 as rules type $ABC \to CBA$. We note here that in Ref. 9 long permutation rules are allowed without fixed points.

These rules are monotone rules having exactly the same letters in both sides. We will show that the context-free rules and interchange rules are more efficient than the context-free derivations, moreover the generative capacity is increasing by allowing permutation rules of length 3 instead of 2, but they are not enough to get all context-sensitive languages.

Our investigation has an interest for concurrency and parallelisation theory as well, where the order of some processes can be interchanged.

The work has some linguistic motivations as well: in some morphologically rich languages (as, for instance, Japanese, Finnish and Hungarian) the word order is not strict in a sentence. Example: 'A kutya hangosan ugat.' 'Hangosan ugat a kutya.' 'A kutya ugat hangosan.' 'Hangosan a kutya ugat.' 'Ugat a kutya hangosan.' 'Ugat hangosan a kutya.' are all correct sentences about the same meaning: The dog (a kutya) barks (ugat) loudly (hangosan). So, usually some of the parts of the sentences can freely be interchanged. Some linguistical applications of permutation languages are shown in Ref. 10.

The structure of the paper is as follows. In the next section we recall some basic definitions and facts that we need later on. After this, Sec. 3 is devoted to introduce and analyse new families of languages. Some of our results are extensions of the results presented in Ref. 8 to wider classes of languages. Moreover our main result is the proof of the strict inclusion between the language classes generated by context-free and permutation rules of length 2 and 3, respectively. We will prove that these families contain only semi-linear context-sensitive languages and all the context-free languages. Examples and several properties, such as, closure properties are detailed.

2. Basic Definitions and Preliminaries

First some definitions about Chomsky-type grammars and generated languages are recalled and our notations are fixed.

A grammar is a construct $G = (N, T, S, H)$, where N, T are the nonterminal and terminal alphabets, with $N \cap T = \emptyset$; they are finite sets. $S \in N$ is a special symbol, called initial letter or start symbol. H is a finite set of pairs, where a pair (v, w) is usually denoted by $v \to w$ with $v \in (N \cup T)^*N(N \cup T)^*$ and $w \in (N \cup T)^*$, (where we used the well-known notation of Kleene-star). H is the set of derivation rules; $v \Rightarrow w$ ($v, w \in (N \cup T)^*$) is a direct derivation if there exist $v_1, v_2, v', w' \in (N \cup T)^*$ such that $v = v_1 v' v_2$, $w = v_1 w' v_2$ and $v' \to w' \in H$. The transitive and reflexive closure of the direct derivation is the derivation denoted by $v \Rightarrow^* u$. We say that $v \in (N \cup T)^*$ is a sentential form if $S \Rightarrow^* v$ holds. The language generated by a grammar G is the set of terminal words which can be derived from the initial letter: $L(G) = \{w | S \Rightarrow^* w, w \in T^*\}$.

We use λ to denote the empty word. For any word and sentential form u we will use $|u|$ to sign its length, i.e., the number of letters it contains. Note that $|\lambda| = 0$. Two grammars are equivalent if they generate the same language up to the empty word.

Depending on the possible structures of the derivation rules various classes of grammars are defined. We recall the most important classes.

- monotone grammars: each rule $v \to u$ satisfies the condition $|v| \leq |u|$ but the possible rule $S \to \lambda$, in which case S does not occur on any right hand side of a rule.
- context-free grammars: for every rule the next scheme holds: $A \to v$ with $A \in N$ and $v \in (N \cup T)^*$.
- regular grammars: each derivation rule is one of the following forms: $A \to w$, $A \to wB$; where $A, B \in N$ and $w \in T^*$.

A language is regular/ context-free/ context-sensitive if it can be generated by a regular/ context-free/ monotone grammar, respectively. For these families the notations $\mathbf{L}_{reg}$, $\mathbf{L}_{CF}$ and $\mathbf{L}_{CS}$ are used. The language families generated by generative grammars form the Chomsky hierarchy: $\mathbf{L}_{reg} \subsetneq \mathbf{L}_{CF} \subsetneq \mathbf{L}_{CS}$.

Let u be a word, then u^T denotes its mirror word (i.e., its reading from the end to the beginning).

Let v and u be two words over the alphabet T. The shuffle of v and u is defined as $u \diamond v = \{u_1 v_1 ... u_n v_n | u = u_1 ... u_n, v = v_1 ... v_n, u_i \in T^*, v_i \in T^*, 1 \leq i \leq n, n \in \mathbb{N}, n \geq 1\}$. Consequently, the shuffle of languages L_1 and L_2 is: $L_1 \diamond L_2 = \{w | w = u \diamond v, u \in L_1, v \in L_2\}$.

Let T and T' be two alphabets. A mapping $h : T^* \rightarrow (T')^*$ is called homomorphism if $h(\lambda) = \lambda$ and $h(uv) = h(u)h(v)$ for all $u, v \in T^*$.

Let the terminal alphabet T be ordered. For each word its Parikh-vector is assigned (Parikh-mapping). The elements of this vector are the occurrences of the letters of the alphabet in the word. Formally, using alphabet $T = (a_1, a_2, ..., a_n)$: $\Psi : T^* \rightarrow \mathbb{N}^n$, $\Psi(w) = (|w|_{a_1}, |w|_{a_2}, ..., |w|_{a_n})$, where $w \in T^*$ and $|w|_{a_i}$ is the number of occurrences of the letter a_i in w. The set of Parikh-vectors of the words of a language is called the Parikh-set of the language. Formally: $\Psi(L) = \{\Psi(w) | w \in L\}$. Two languages are letter-equivalent if their Parikh-sets are identical. A language is linear (in Parikh-sense) if its Parikh set can be written in the form of a linear set: $\left\{ \underline{v}_0 + \sum_{i=1}^{m} x_i \underline{v}_i | x_i \in \mathbb{N} \right\}$, for some $\underline{v}_j \in \mathbb{N}^n, 0 \leq j \leq m$. A language is semi-linear (in Parikh-sense) if its Parikh set can be written as a finite union of linear sets. Due to Parikh's theorem[11] it is known that every context-free language is semi-linear. For every semi-linear set there is a regular language such that its Parikh set equals to the given semi-linear set. Non semi-linear context-sensitive languages are known, for instance $L_\square = \{a^{n^2} | a \in T\}$.

The commutative closure of a language L is the set of all words having Parikh-vectors included in the Parikh-map of the language, i.e., $\{w | \Psi(w) \in \Psi(L)\}$. A language is called commutative if it is the commutative closure of itself.

The context-free grammars are very popular ones because the concept of derivation trees fits very well in these derivations. It is an important property of the (context-free) derivations that the direction left-to-right is preserved. The letters in the beginning of the sentential form refer for the beginning of the derived word, and have no influence to the end-part.

Now we refine the Chomsky hierarchy. We will obtain language classes between the context-free and the context-sensitive ones. We do this by allowing some permutations of the branches of the derivation trees, i.e., the direction left-to-right is not necessarily preserved in derivations.

3. Context-Free Grammars Extended with Permutation Rules

First we are defining the grammar and language class we are dealing with in a formal way.

A rule is a permutation rule if the left hand side contains exactly the same non-terminals as its right hand side (with multiplicities). One can characterize these rules by the length of its sides. There is only one type of permutation rules with length two: $AB \to BA$, where $A, B \in N$. For rules having larger length there are several possibilities.

A grammar $G = (N, T, S, H)$ is a permutation grammar if besides the context-free rules H contains only special type of non-context-free rules, namely: permutation rules. We denote the language family generated in this way by $\mathbf{L}_{perm}$. A permutation grammar is said to be of order n if it has permutation rules only of length n. The languages that can be generated by permutation grammars having permutation rules only of length n are denoted by $\mathbf{L}_{perm_n}$.

We note here that in some papers[7,8] only permutation rules of order 2 are used, while in Ref. 9 longer rules are also allowed but without fixed points. In this paper we are using the most general form as we defined above.

In the derivation these new rules allow to permute some branches of the derivation tree, so the left-to-right property of the context-free case is violated. Now, let us see an example for a grammar that generates a language in $\mathbf{L}_{perm_2}$.

Example 3.1. Let $G = (\{S, A, B, C\}, \{a, b, c\}, S, H)$ be a permutation grammar with $H = \{S \to ABC, S \to SABC, AB \to BA, BA \to AB, AC \to CA, CA \to AC, BC \to CB, CB \to BC, A \to a, B \to b, C \to c\}$. Figure 1 shows the "derivation-tree" of the word *aaccbb* in this system.

The language containing all words with the same number of a, b and c is generated in the previous example. This language is non-context-free, since intersected by the regular language described by the expression $a^*b^*c^*$ the language $\{a^n b^n c^n | n \in \mathbb{N}\}$ is obtained which does not satisfies the usual pumping lemma for context-free languages (Bar-Hillel[12]). So, we can state, that the generating power of the grammars is increasing if we allow interchange rules. Obviously without any (efficiently applicable) interchange rule one can generate any context-free language. Looking at the "derivation tree" shown in Fig. 1 one can observe that some branches are interchanged violating the left-to-right property of context-free derivations.

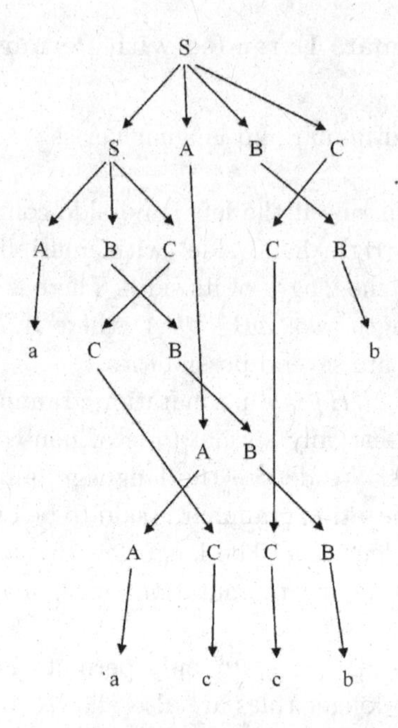

Fig. 1. Derivation in a permutation grammar.

One may ask how many interchange rules are needed in the grammar to generate a non-context-free language. The next example answers the question: one interchange rule is enough to increase the generating power.

Example 3.2. Let $G = (\{S, A, B, C\}, \{a, b, c\}, S, H)$ be a grammar with rules: $H = \{S \to ABC, S \to ABSC, BA \to AB, A \to a, B \to b, C \to c\}$. The generated language intersected by the regular language $a^*b^*c^*$ is $a^nb^nc^n$.

Definition 3.1. Let $G = (N, T, S, H)$ be a permutation grammar that generates L. The grammar $G_b = (N, T, S, H')$ obtained from G by deleting the non-context-free rules from H is the basis grammar of G. The generated context-free language L_b is a basis language of L.

A basis language is letter equivalent to the original one. Since the permutation rules do not modify the multiset of the symbols of a sentential form, the Parikh-vector/set of the generated word/language is the same as the Parikh-vector/set of the word/language generated in a context-free way without the permutation rules. This fact can be formalized by the following results.

Lemma 3.1. *Each basis language of a permutation language L is a subset of L.*

Theorem 3.1. *All languages which can be generated with permutation grammars are semi-linear in Parikh-sense.*

Proof. Consider a permutation grammar that generates the language L. Each generated word w of L has a letter-equivalent w' in the basis context-free language of L. Therefore L and L_b are letter equivalent. Since all context-free languages are semi-linear the classes $\mathbf{L}_{perm}$ and $\mathbf{L}_{perm_n}$ ($n \in \mathbb{N}, n \geq 2$) are semi-linear as well. □

Corollary 3.1. *There are context-sensitive languages that cannot be generated by permutation grammars.*

The proof of this corollary may go by the following example. The context-sensitive language $L_\square$ is not semi-linear, therefore it cannot be generated using only permutation rules as non-context-free rules.

We would like to know something on the generative powers of the permutation grammars having various order.

Lemma 3.2. *Let $G = (N, T, S, H)$ be a permutation grammar with a permutation rule $u \to v$ of length n. There is a permutation grammar $G' = (N', T, S, H')$ that is equivalent to G and the set H' is obtained from H by deleting the rule $u \to v$, adding some context-free rules, and adding a permutation rule of length $n + 1$.*

Proof. Let A be the first nonterminal of u, i.e., $u = Au'$. Let B_1 be a newly introduced nonterminal (that is in $N' \setminus N$). Let H' contain the rules $A \to B_1A$, $B_1Au' \to B_1v$, and $B_1 \to \lambda$ (instead of the original rule $u \to v$). One can see that these new rules by the help of the new nonterminal symbol simulate exactly the original derivation rule $u \to v$, i.e., the sequence of these new rules can and must be applied in a terminating derivation in G' if and only if the original rule is applied at that place in a derivation in G. □

Let n be the length of the longest permutation rule(s) of the grammar G. By the previous lemma one can replace every shorter permutation rule by a set of context-free rules and permutation rules of length n. Therefore the generated language belongs to the class $\mathbf{L}_{perm_n}$ for every permutation grammar having permutation rules of length at most n. As consequences of the previous sequence of ideas we can state the following results about the hierarchy of these language classes.

Theorem 3.2. *The language classes of permutation languages are in the following relation:*

$$\boldsymbol{L}_{perm_n} \subseteq \boldsymbol{L}_{perm_{n+1}}, \quad \text{for } n \in \mathbb{N}, n > 1, \quad \text{and } \boldsymbol{L}_{perm} = \bigcup_{n=1}^{\infty} \boldsymbol{L}_{perm_n}.$$

Proof. The statements follow directly from Lemma 3.2. □

Further, we can use the term 'order' of a permutation grammar in the sense of its longest permutation rule. Now, we are detailing some further results about $\mathbf{L}_{perm}$.

Based on Lemma 3.1 we can state the following.

Proposition 3.1. *For every permutation language L there is a number $n \in \mathbb{N}$ such that for any word $w \in L$ with $|w| > n$ there is a word $w' \in L$ with the following properties:*

- *w' and w have identical Parikh-vectors,*
- *every context-free pumping lemma works on w', i.e., there are infinitely many words in L that can be obtained from w' by pumping.*

So context-free pumping lemmas (Bar-Hillel,[12] Ogden,[13] Bader-Moura,[14] Dömösi-Ito-Katsura-Nehaniv,[15] etc.) can be applied in this way to permutation languages. Later in this section some interchange lemmas will be shown for permutation languages.

Now we show an example for a language in $\mathbf{L}_{perm_3}$. The idea is to compose the words of three different languages (two linear context-free and a permutation language of order 2 are used). So every word of the language can be divided to three scattered subword in the following way: the letters having the same position modulo 3 form the desired subwords.

Example 3.3. Let $G = (\{S, S', S'', A, B, C, D, E, E', F, F', G, G', H, H', I, J, K, M, O, P, Q, R\}, \{a, b, c, d, e, e', f, f', g, g', h, h'\}, S, \{S \to E'AIBS'CODG, S \to E'AJBS''CQDG, S \to F'AKBS'CPDH, S \to F'AMBS''CRDH, S' \to IAIBS'CODO, S' \to IAJBS''CQDO, S' \to JAKBS'CPDQ, S' \to JAMBS''CRDQ, S'' \to KAIBS'CODP, S'' \to KAJBS''CQDP, S'' \to MAKBS'CPDR, S'' \to MAMBS''CRDR, AIB \to BIA, BIA \to AIB, AJB \to BJA, BJA \to AJB, AKB \to BKA, BKA \to AKB, AMB \to BMA,$

$BMA \to AMB$, $AOB \to BOA$, $BOA \to AOB$, $APB \to BPA$,
$BPA \to APB$, $AQB \to BQA$, $BQA \to AQB$, $ARB \to BRA$,
$BRA \to ARB$, $AS'B \to BS'A$, $BS'A \to AS'B$, $AS''B \to BS''A$,
$BS''A \to AS''B$, $AIC \to CIA, CIA \to AIC$, $AJC \to CJA, CJA \to AJC$,
$AKC \to CKA, CKA \to AKC, AMC \to CMA$, $CMA \to AMC$,
$AOC \to COA, COA \to AOC, APC \to CPA, CPA \to APC$,
$AQC \to CQA$, $CQA \to AQC, ARC \to CRA, CRA \to ARC$,
$AS'C \to CS'A$, $CS'A \to AS'C$, $AS''C \to CS''A$, $CS''A \to AS''C$,
$AID \to DIA$, $DIA \to AID$, $AJD \to DJA, DJA \to AJD$,
$AKD \to DKA, DKA \to AKD, AMD \to DMA, DMA \to AMD$,
$AOD \to DOA, DOA \to AOD, APD \to DPA, DPA \to APD$,
$AQD \to DQA, DQA \to AQD, ARD \to DRA, DRA \to ARD$,
$AS'D \to DS'A$, $DS'A \to AS'D, AS''D \to DS''A, DS''A \to AS''D$,
$BIC \to CIB, CIB \to BIC$, $BJC \to CJB, CJB \to BJC$, $BKC \to CKB$,
$CKB \to BKC$, $BMC \to CMB$, $CMB \to BMC$, $BOC \to COB$,
$COB \to BOC$, $BPC \to CPB$, $CPB \to BPC$, $BQC \to CQB$,
$CQB \to BQC$, $BRC \to CRB$, $CRB \to BRC$, $BS'C \to CS'B$,
$CS'B \to BS'C$, $BS''C \to CS''B$, $CS''B \to BS''C$, $BID \to DIB$,
$DIB \to BID$, $BJD \to DJB$, $DJB \to BJD$, $BKD \to DKB$,
$DKB \to BKD$, $BMD \to DMB$, $DMB \to BMD$, $BOD \to DOB$,
$DOB \to BOD$, $BPD \to DPB$, $DPB \to BPD$, $BQD \to DQB$,
$DQB \to BQD$, $BRD \to DRB$, $DRB \to BRD$, $BS'D \to DS'B$,
$DS'B \to BS'D$, $BS''D \to DS''B$, $DS''B \to BS''D$, $CID \to DIC$,
$DIC \to CID$, $CJD \to DJC$, $DJC \to CJD$, $CKD \to DKC$,
$DKC \to CKD$, $CMD \to DMC$, $DMC \to CMD$, $COD \to DOC$,
$DOC \to COD$, $CPD \to DPC$, $DPC \to CPD$, $CQD \to DQC$,
$DQC \to CQD$, $CRD \to DRC$, $DRC \to CRD$, $CS'D \to DS'C$,
$DS'C \to CS'D$, $CS''D \to DS''C$, $DS''C \to CS''D$, $I \to EE'$, $J \to EF'$,
$K \to FE'$, $M \to FF'$, $O \to GG'$, $P \to GH'$, $Q \to HG'$, $R \to HH'$,
$S' \to EG'$, $S'' \to FH'$, $A \to a$, $B \to b$, $C \to c$, $D \to d$, $E \to e$, $E' \to e'$,
$F \to f, F' \to f'$, $G \to g$, $G' \to g'$, $H \to h$, $H' \to h'\})$

be a grammar that generates L_d. Then clearly $L_d \in \mathbf{L}_{perm_3}$. The language consists of the words which can be described in the following way: every word has length divisible by 12. Regarding the letters of positions 2 modulo 3: they form words having the same number of a, b, c and d. The letters of positions divisible by 3 form words $\{ww' | w \in \{e, f\}^*, w' \in \{g, h\}^*$, and w'

equals to the mirror of w (i.e., w^T) with respect to the mapping $M : M(e) = g, M(f) = h\}$. For every word of L_d the words of letters having positions 1 modulo 3 form the similar word as the word of every third positions over the alphabet $\{e', f', g', h'\}$.

Now we present a result that can be used to decide whether a language is not a permutation language. The next theorem is an interchange lemma for the family $\mathbf{L}_{perm_2}$.

Theorem 3.3. *Let $L \in \mathbf{L}_{perm_2}$ and let L_b be any of its basis languages. For any word $w \in L$, $w \notin L_b$ there exists a word $w' \in L$ such that there exist $u, v, x, y \in T^*$: $w = uxyv, w' = uyxv$ and $w \neq w'$.*

Proof. Let G be a permutation grammar of order 2 that generates L such that it generates L_b without the permutation rules. Since $w \notin L_b$ there exists a/last rule $AB \to BA$ that is applied in a derivation of w for some $A, B \in N$. Then there is a derivation, such that $S \Rightarrow^* uABv \Rightarrow uBAv \Rightarrow^* uxyv = w$ in which w can be generated in context-free way from $uBAv$. Obviously $w' = uyxv \in L$ generated without the last application of the interchange rule $AB \to BA$. □

Now we show that Example 3.3 does not satisfy the condition of Theorem 3.3.

Language L_d is clearly not context-free.

Let us consider the words w'_n of the following shape: let u_n be the word of the letters of 2 modulo 3 positions (that can be obtained by homomorphism $M_1 : M_1(a) = a, M_1(b) = b, M_1(c) = c, M_1(d) = d, M_1(e) = \lambda, M_1(e') = \lambda, M_1(f) = \lambda, M_1(f') = \lambda, M_1(g) = \lambda, M_1(g') = \lambda, M_1(h) = \lambda, M_1(h') = \lambda$), and v_n be the word formed by letters of 0 modulo 3 positions (that can be obtained by homomorphism $M_2 : M_2(a) = \lambda, M_2(b) = \lambda, M_2(c) = \lambda, M_2(d) = \lambda, M_2(e) = e, M_2(e') = \lambda, M_2(f) = f, M_2(f') = \lambda, M_2(g) = g, M_2(g') = \lambda, M_2(h) = h, M_2(h') = \lambda$). Since the letters of positions 1 modulo 3 have to be exactly the signed (′) versions of the next letters of position 0 modulo 3, u_n and v_n identify the word of the language. Let $u_n = a^n b^n c^n d^n$ and $v_n = e^n f^n h^n g^n$.

Let us assume that $L_d \in \mathbf{L}_{perm_2}$: then let G_i be any permutation grammar of order 2 that generates L_d, and let L_{bi} be the correspondent basis language.

L_d is not context-free. Moreover there is no context-free language L_c such that $L_c \subseteq L_d$ and L_c contains all the words w'_n for all $n \in \mathbb{N}$ (because the pumping[12] of any long enough words w'_n yields words that are not in L_d). Therefore there is a value of $m \in \mathbb{N}$ such that w'_n cannot be generated in context-free way if $n > m$. Then there is a word w'_n that is not in L_{bi}. So, by our assumption, Theorem 3.3 can be applied on this word: w'_n has two consecutive subwords which can be interchanged to get another word of L_d. Let us find the subwords x and y. It is obvious that xy must be either in the first half or the second half of w'_n. (If it intersected the middle of w'_n, then after the interchange the letters e, f and g, h and/or their signed versions e', f', g', h' would be in a wrong order.) The subword xy cannot intersect the first quarter of w'_n, because then the word obtained by mapping M_2 will not be of the desired form. Similar argument works for the third quarter. Then, the length of x and y must be both divisible by 3. (If it is not fulfilled, then there will be a problem by the order of a, b, c, d and e, f, g, h and e', f', g', h' letters, i.e., they will not be only on 2 modulo 3, 0 modulo 3 and 1 modulo 3 positions, respectively.) If xy is inside a quarter of w'_n and both x and y has length divisible by 3, then by interchanging the subwords x and y the same word w'_n is obtained. Therefore there is no way to partition w'_n to four parts as $uxyv$ such that $uyxv \in L_d$ and $uxyv \neq uyxv$.

By Theorem 3.2 and the previous example we have proved that the inclusion $\mathbf{L}_{perm_2} \subsetneq \mathbf{L}_{perm_3}$ is strict. The previous results can be summarized in the following way:

Theorem 3.4. *The place of these new language classes in the Chomsky-hierarchy:*

$$\mathbf{L}_{CF} \subsetneq \mathbf{L}_{perm_2} \subsetneq \mathbf{L}_{perm_3} \subsetneq \mathbf{L}_{CS}.$$

Now an interchange lemma is presented for other families of $\mathbf{L}_{perm}$ by generalizing Theorem 3.3.

Theorem 3.5. *Let $L \in \mathbf{L}_{perm_n}$ and let L_b be any of its basis languages. For any word $w \in L$, $w \notin L_b$ there exists a word $w' \in L$ such that there exist $u, v, x_1, ..., x_n \in T^*$: $w = ux_1...x_nv$, $w' = ux_{p(1)}...x_{p(n)}v$ and $w \neq w'$, where $p(1), ..., p(n)$ is a permutation of $1, ..., n$.*

Proof. Let G be a grammar having permutation rules of length n such that $L(G) = L$ and L_b is the basis language of L with respect to the grammar G. Since $w \notin L_b$ there exists a/last rule $A_{p(1)}...A_{p(n)} \to A_1...A_n$ that is applied in a derivation of w for some non-terminals A_i. Then there is a derivation, such that $S \Rightarrow^* uA_{p(1)}...A_{p(n)}v \Rightarrow uA_1...A_nv \Rightarrow^* ux_1...x_nv = w$ in which w is generated in context-free way from $uA_1...A_nv$. Obviously

$w' = ux_{p(1)}...x_{p(n)}v \in L$ generated without the last application of the interchange rule $A_{p(1)}...A_{p(n)} \to A_1...A_n$. □

Example 3.4. The language $L_{abc} = \{a^nb^nc^n \mid n \in \mathbb{N}\}$ is not a context-free language. Moreover, this language does not satisfy the condition in Theorem 3.5. Hence this language is not even in $\mathbf{L}_{perm_m}$ for any value of m, and so it is not in $\mathbf{L}_{perm}$.

The language $\{a^nb^nc^n|n \in \mathbb{N}\}$ is an example for semi-linear context-sensitive languages that is not in $\mathbf{L}_{perm_m}$ for any value of $m \in \mathbb{N}$, $m \geq 2$.

The language L_{abc} is important from linguistical point of view. It is a well known language belonging to mildly context-sensitive language families. It can be obtained by the intersection of the language of Example 3.1 and the regular set $a^*b^*c^*$, or by intersection of the language generated in Example 3.2 and $a^*b^*c^*$. By Lemma 3.1 it is clear that a context-free grammar with interchange rule without any other help cannot generate L_{abc}. With a similar method one can have another important elements of the mildly context-sensitive language families, such as the language $L_{abcd} = \{a^nb^mc^nd^m|n, m \in \mathbb{N}\}$.[10]

Over the 1-letter terminal alphabet the interchange rules do not add anything to the generating power of context-free grammars. Moreover in this case only regular languages can be generated since it is known that in this case all semi-linear languages are regular.

In the derivations of these grammars we drop the basic property of the (context-free) derivation trees. We allow to change the order of letters in the sentential form and therefore in the derivation (tree) as Fig. 1 shows.

Now the closure properties of $\mathbf{L}_{perm}$ and its subfamilies will be analysed.

Proposition 3.2. *The language families $\mathbf{L}_{perm}$ and $\mathbf{L}_{perm_n}$ $(n \geq 2)$ are closed under the regular operations (union, concatenation, Kleene-closure).*

The closure under regular operations can be proved by the usual technique. As a consequence, they are also closed under n-th power for any $n \in \mathbb{N}$. Now closure under some other operations are shown.

Theorem 3.6. *The language families $\mathbf{L}_{perm}$ and $\mathbf{L}_{perm_n}$ $(n \geq 2)$ are closed under shuffle, commutative-closure, homomorphism and mirror image.*

Proof. Let $G = (N, T, S, H)$ be a permutation grammar such that terminals occur only in rules of type $A \to a$ with $A \in N, a \in T$. (This can be done by substituting every occurrence of each terminal a by newly introduced

non-terminal A_a in every rule that is not of the form $A \to a$ and adding the new rule $A_a \to a$ to the grammar for every terminal a.) We present a grammar $G_{com} = (N, T, S, H \cup H_{com})$ which provides the commutative closure L_{com} of the language $L(G)$. So, let $H_{com} = \{AB \to BA | A, B \in N\}$. There is a derivation of $w \in L(G)$ such that the sentential form $u \in N^*$ can be obtained from which w is derived by only rules of type $A \to a$. From u all of its permutations can be obtained in G_{com} by the help of the rules in H_{com}. In this way one can get any kind of permutations of the originally derived word w.
Now for closure under shuffle we use the grammars $G_1 = (N_1, T, S_1, H_1)$ and $G_2 = (N_2; T, S_2, H_2)$ in the same kind of normal form generating $L(G_1)$ and $L(G_2)$, respectively (with conditions $N_1 \cap N_2 = \emptyset$ – that can be obtained by renaming non-terminals; and terminals can be introduced in both the grammars only by rules of type $A \to a$). Now we give a grammar that generates the shuffle L_{shuf} of the languages $L(G_1)$ and $L(G_2)$. Let $G_{shuf} = (N_1 \cup N_2 \cup \{S\}, T, S, H_1 \cup H_2 \cup \{S \to S_1 S_2\} \cup H_{shuf})$ be a grammar with $H_{shuf} = \{AB \to BA | A \in N_1, B \in N_2\}$, where $S \notin N_1 \cup N_2$. It generates the shuffle of $L(G_1)$ and $L(G_2)$.
Observe that only context-free rules and permutation rules of length 2 are needed to add to the rule set. Therefore if $n \geq 2$ is the length of the longest permutation rule of H, then the generated language L_{com} is in $\mathbf{L}_{perm_n}$. Moreover, it is true that L_{com} is in $\mathbf{L}_{perm_2}$, since the original permutation rules can be omitted, it is enough to use only the permutation rules of H_{com} to obtain L_{com}. If G_1 is a permutation grammar of order m_1 and G_2 is a permutation grammar of order m_2, then L_{shuf} belongs to $\mathbf{L}_{perm_n}$ where $n = \max(m_1, m_2, 2)$.
Now we prove the closure under homomorphism. Let G be a permutation grammar such that terminals appear only in rules type $A \to a$. Let $h : T^* \to (T')^*$ be a homomorphism. Let $G_{hom} = (N, T', S, H')$ be the grammar obtained from G by changing every rule $A \to a$ (where $a \in T$) to the rule $A \to v$ where $v = h(a)$. It is clear that the generated language is exactly $h(L(G))$, moreover G_{hom} has the same order as G, because only context-free rules are modified.
Finally, let $G = (N, T, S, H)$ be a permutation grammar. For the closure under mirror image the elements of the set of rules H must be changed into their reverse, i.e., $A \to u$ ($u \in (N \cup T)^*$) is in H_m if and only if $(A \to u^T) \in H$; and the permutation rule $(v \to u) \in H_m$ if and only if $(v^T \to u^T) \in H$. Obviously for each word $w \in L(G)$ the mirror of its 'derivation tree' gives a real derivation in $G_m = (N, T, S, H_m)$ and vice-

versa. Therefore G_m generates exactly the mirror of $L(G)$. Obviously the order of G_m is the same as the order of G. □

Theorem 3.7. *The language families $\mathbf{L}_{perm}$ and $\mathbf{L}_{perm_n}$ are not closed under intersection by regular languages and hence under intersection. They are not closed under complement.*

Proof. As we have already shown (see Example 3.4) the language L_{abc} is not in $\mathbf{L}_{perm}$ and not in $\mathbf{L}_{perm_n}$, but can be obtained by the intersection of a regular and a permutation language. Moreover the permutation language of Example 3.1 can be used, and that is especially belonging to $\mathbf{L}_{perm_2}$, and so to every class $\mathbf{L}_{perm_n}$. Therefore none of the families of permutation languages is closed under intersection (and under intersection by regular languages).
The class $\mathbf{L}_{CF}$ is not closed under complement: the complement of L_{abc} is context-free, while L_{abc} is not. Hence $\mathbf{L}_{CF} \subsetneq \mathbf{L}_{perm}$ and $\mathbf{L}_{CF} \subsetneq \mathbf{L}_{perm_n}$ (for any $n > 2$), the complement of L_{abc} belongs also to every class of permutation languages. Opposite to this, as we already proved, L_{abc} is not belonging to any of them. In this way the theorem is proved. □

Concluding the previous two theorems:

Corollary 3.2. *$\mathbf{L}_{perm}$ and $\mathbf{L}_{perm_n}$ $(n \geq 2)$ are not trios, not AFL's and not anti-AFL's.*

So, our language families are similar to the context-free family in the fact that they are not closed under complement and intersection. But opposite the context-free class the classes of permutation languages ($\mathbf{L}_{perm}$ and $\mathbf{L}_{perm_n}$) are closed under commutative closure and shuffle product. Note, that the family $\mathbf{L}_{CS}$ is also closed under shuffle and commutative closure. It is interesting that the smaller families of the Chomsky-hierarchy such as regular sets and context-free languages are not closed under commutative closure, but their commutative closure are in $\mathbf{L}_{perm}$, moreover they are in $\mathbf{L}_{perm_2}$. Since every commutative semi-linear language is a commutative closure of a regular language, all commutative semi-linear languages are in $\mathbf{L}_{perm_2}$ (and so in every other family of permutation languages). Opposite to $\mathbf{L}_{CF}$ and the families of permutation languages, $\mathbf{L}_{CS}$ is closed under intersection and complement.

4. Conclusions, Further Remarks

In context free grammars the branches of the derivations are independent of each other. In this paper context-free derivations with branch-interchange were presented. The language families $\mathbf{L}_{perm}$ and $\mathbf{L}_{perm_n}$ $(n \geq 2)$ strictly contain the context-free class, and they contain only semi-linear languages. They are strictly inside the context-sensitive class. Moreover $\mathbf{L}_{perm_2}$ is strictly smaller than $\mathbf{L}_{perm_3}$. We left open the problem whether the hierarchies between $\mathbf{L}_{perm_n}$ and $\mathbf{L}_{perm_{n+1}}$ and between $\mathbf{L}_{perm_n}$ and $\mathbf{L}_{perm}$ (with $n > 2$) are strict. Closure properties under several operations (such as, for instance, shuffle, commutative closure and complement) are analysed. It was shown that all commutative semi-linear languages are in $\mathbf{L}_{perm_2}$.

Now we present further open problems related to the permutation languages. The solution of the parsing problem of theses new language families is an important open problem. It is also an interesting task to analyse the effect of other controlling mechanisms in the derivations. For instance, what is the case when priority can be used among various (types of) rules? It is also a subject of future research to establish the connection of permutation grammars and the field of shuffles of trajectories.[16] They seem to be highly related to each other.

Acknowledgements

Useful comments of the participants of the International Workshop on Automata, Formal Languages and Algebraic Systems (AFLAS 2008) are gratefully acknowledged.

The research is party supported by the programme Öveges of the Hungarian National Office for Research and Technology (NKTH) and by a Japanese-Hungarian bilateral project of the Hungarian Science and Technology Foundation (TÉT).

References

1. J. E. Hopcroft and J. D. Ullmann, *Introduction to Automata Theory, Languages, and Computation,* (Addison-Wesley, Reading, 1979).
2. C. Martin-Vide, V. Mitrana and Gh. Paun (eds.), *Formal Languages and Applications*, (Springer-Verlag, Berlin, Heidelberg, 2004).
3. Gy. E. Révész, *Introduction to Formal Languages*, (McGraw-Hill, New York, 1983).
4. A. Salomaa, *Formal Languages,* (Academic Press, New York, 1973).
5. J. Dassow and Gh. Paun, *Regulated Rewriting in Formal Language Theory,* (Springer-Verlag, Berlin, 1989).

6. M. Penttonen, One-sided and two-sided context in formal grammars, *Information and Control* **25**, (1974), pp. 371–392.
7. E. Mäkinen, On permutative grammars generating context-free languages. *BIT* **25**, (1985), pp. 604–610.
8. B. Nagy, Languages generated by context-free grammars extended by type $AB \to BA$ rules, *Journal of Automata, Languages and Combinatorics* **14**/2 (2009), an earlier version is presented as *Languages generated by context-free and type $AB \to BA$ rules,* in *Proc. 8th Int. Symposium of Hungarian Researchers on Computational Intelligence and Informatics (CINTI 2007)*, (Budapest, Hungary, 2007) pp. 563–572.
9. R. V. Book, On the structure of context-sensitive grammars, *Int. Journal of Computer and Information Sciences* **2** (1973), pp. 129–139.
10. B. Nagy, Permutation languages in formal linguistics, in *Proc. IWANN 2009, Lecture Notes in Computer Science* **5517** (2009), pp. 504–511.
11. R. J. Parikh, On context-free languages, *J. ACM*, **18** (1966), pp. 570–581.
12. Y. Bar-Hillel, M. Perles and E. Shamir, On formal properties of simple phrase strucuture grammars, *Z. Phonetik. Sprachwiss. Kommunikationsforsch.* **14** (1961), pp. 143–172.
13. W. Ogden, A helpful result for proving inherent ambiguity, *Math. Systems Theory* **2** (1968), pp. 191–194.
14. C. Bader and A. Moura, A generalization of Ogden's lemma, *J. ACM* **29** (1982), pp. 404–407.
15. P. Dömösi, M. Ito, M. Katsura and C. Nehaniv, New pumping lemma for context-free languages, in *Proc. DMTCS'96*, eds. D. S. Bridges, C. S. Calude, J. Gibbons, S. Reeves and I.H. Witten, (Springer-Verlag, 1997), pp. 187–193.
16. A. Mateescu, G. Rozenberg and A. Salomaa, Shuffle on trajectories: syntactic constraints, *Theoretical Computer Science* **197** (1998), pp. 1–56.

Received: May 7, 2009

Revised: March 22, 2010

DERIVATION TREES FOR CONTEXT-SENSITIVE GRAMMARS

BENEDEK NAGY

Department of Computer Science, Faculty of Informatics, University of Debrecen, 4010 Debrecen, PO Box. 12, Hungary,
E-mail: nbenedek@inf.unideb.hu
www.inf.unideb.hu/~nbenedek

One of the main reasons that context-free grammars are widely used is the fact that the concept of derivation trees fits to them very well. The left-most derivations play important role both in theory and practice. Unfortunately with left-most derivations only context-free languages can be derived even if the rules of the grammar are not context-free. In this paper we investigate derivation trees for context-sensitive grammars based on Penttonen's one-sided normal form. The concept of the presented derivation graphs are a kind of extension of the well-known context-free derivation-trees. Moreover it allows to define left-most derivations in context-sensitive grammars without loosing the efficiency of the derivations. These left-most derivations are not derivations in the usual sentential form sense. They are the generalizations of the classical (context-free) left-most derivations, in the way of constructing a derivation tree. Some examples and a new type of ambiguity are also shown.

Keywords: Chomsky hierarchy; Context-sensitive languages; Derivation tree; Canonical derivation; Left-most derivation; Ambiguity.

1. Introduction

Context-free grammars and languages are well known, their theory is well developed, they are widely used; however the world is not-context-free.[1] There are well-known phenomena proved to be context-sensitive, non context-free, such as language of logical tautologies, programming and natural languages etc. Two of the most important reasons that context-free grammars and languages are widely spread and used in practice are that the derivations can be represented by tree-graphs and left-most derivations are sufficient. 'No image general about the way in which a sentence is generated following the grammar's productions has been obtained' in context-sensitive case, therefore these languages 'are not so much studied'

— claimed Atanasiu.[2] For this reason tree-like derivation graphs are recommended for grammars in Kuroda normal form, moreover by dividing the set of non-terminals to two disjoint sets (cd-Kuroda normal-form) a left-most derivation is also defined in Ref. 2. Unfortunately even if every words of the language can be derived by this left-most derivation, it is generally not sufficient, it works by blocking some branches of the derivation graph. We note here that graphical representations of derivations are not only nice visualizations, but also important for scientific reasons. They help in analysis and give a tool for — both theoretical and practical — further research including relations to complexity classes, relations to other formalisms, parsing techniques etc. We also note that Brandenburg can represent the derivations of phrase-structure grammars by trees,[3] but the nodes of his tree are complex derivations starting from a part of the sentential form till a part in which a derived terminal (or in case of non-context-sensitive grammars, the empty string) can be found. The complexity (diameter) of the nodes is measured, and these trees are used to describe relations to complexity classes.

In the next section we recall some basic definitions and facts that we need later on. In Sec. 3 based on Penttonen's old result[4,5] we build the derivation graphs for context-sensitive grammars in a tree-like form. The branches of these derivation trees need some synchronization points. These synchronization points cause a new type of ambiguity that can also be important in linguistics. Based on the newly introduced derivation-trees the concept of left-most derivation can also be extended. Our left-most derivation gives back the classical concept in context-free case, however it does not coincide with the concept of derivation generally. One of our most important results that it works in context-sensitive case as well, i.e., all context-sensitive derivation-trees can be constructed by our left-most derivation.

2. Preliminaries

In this section we recall some basic concepts and facts about formal languages. First the definitions of the Chomsky-type grammars and the Chomsky hierarchy are shown.

2.1. *Chomsky-hierarchy: basic notions and definitions*

We assume that the reader is familiar with the basic concepts of formal language theory. We only fix our notation and briefly recall some facts that are needed later. (See Refs. 6,7 for more details.)

The length of the empty word is zero, i.e., $|\lambda| = 0$. A *grammar* is a construct $G = (N, T, S, H)$, where N, T are the non-terminal and terminal alphabets, with $N \cap T = \emptyset$; they are finite sets. $S \in N$ is a special symbol, called the start symbol (initial letter). H is a finite set of production rules, where a rule uses to be written in the form $v \to w$ with $v \in (N \cup T)^* N (N \cup T)^*$ and $w \in (N \cup T)^*$. Throughout the paper capital letters $(A, B, C, D, ...)$ stand for nonterminals, while lower case letters $(a, b, c, ...)$ stand for terminals. Let G be a grammar and $v, w \in (N \cup T)^*$. Then $v \Rightarrow w$ is a *direct derivation* if there exist $v_1, v_2, v', w' \in (N \cup T)^*$ such that $v = v_1 v' v_2$, $w = v_1 w' v_2$ and $v' \to w' \in H$. We will use the term *underlined (or signed) derivation* for derivations in which the left-side of the used rule is underlined at the place of the application: $v_1 \underline{v'} v_2 \Rightarrow w = v_1 w' v_2$. The *derivation* is the reflexive and transitive closure of the direct derivation and also of the signed derivation. There is a special derivation that plays a very important role in this paper: A derivation is called *left-most* if in every derivation step the subword $v_1 \in T^*$. The string $v \in (N \cup T)^*$ (it is a word over $N \cup T$) is called a sentential form if it has a derivation from S. The language generated by a grammar G is the set of (terminal) words can be derived from the initial letter: $L(G) = \{w | S \Rightarrow^* w \wedge w \in T^*\}$.

If $\lambda \notin L$, then we say that L is λ-free. Two grammars are *equivalent* if they generate the same language up to the empty word. In this paper, from now on, we do not care about whether λ is in the language or not. Depending on the possible structures of the derivation rules we have the following classes. They are the grammars of the Chomsky-hierarchy.

- every grammar is phrase-structured
- monotonous grammars: for every derivation rule $v \to w$, $|v| \leq |w|$
- type 1 (context-sensitive) grammars: each rule is in the form $v_1 A v_2 \to v_1 w v_2$, with $v_1, v_2 \in (N \cup T)^*$, $A \in N$ and $w \in (N \cup T)^* \setminus \{\lambda\}$
- type 2 (context-free) grammars: each rule is in the form $A \to v$ with $A \in N$ and $v \in (N \cup T)^*$.
- type 3, or regular grammars: each derivation rule is one of the following forms: $A \to a$, $A \to aB$, $A \to \lambda$; where $A, B \in N$ and $a \in T$.

The generated language is regular $(\mathbf{L}_{reg})$/ context-free $(\mathbf{L}_{CF})$/ context-sensitive $(\mathbf{L}_{CS})$/ monotonous $(\mathbf{L}_{mon})$/ recursive enumerable $(\mathbf{L}_{RE})$ if it is generated by a regular/ context-free /context-sensitive/ monotonous/ phrase-structure grammar, respectively. These classes form the following (so-called Chomsky) hierarchy: $\mathbf{L}_{reg} \subsetneq \mathbf{L}_{CF} \subsetneq \mathbf{L}_{CS} = \mathbf{L}_{mon} \subsetneq \mathbf{L}_{RE}$.

It is known, that any type of grammar has an equivalent one that is the same type and there is no terminal rewriting (i.e., every rule has only non-terminal(s) in the left-hand-side). Moreover every non-regular grammar has a same type equivalent grammar such that terminals occur only rules of type $A \to a$.

A derivation is a left-most derivation if $v_1 \in T^*$ in each signed derivation step. In this way the first non-terminal of the sentential form is rewritten by an applicable rule in each derivation step.

It is a very important property of left-most derivations (see Ref. 7) that they provide a context-free language independently of the type of the grammar. In context-free case the (originally) generated language coincides with the language generated by left-most derivations. The left-most derivations play important role, for instance, at pushdown machines.

For the Chomsky-type grammars there are so-called normal forms, in which the form of the used derivation rules are more restricted than in the original definition. Moreover it is well-known, that using only such restricted-form rules the generating power remains the same. Now, we recall possible normal-forms for context-free and context-sensitive grammars.

For each context-free grammar there is an equivalent grammar in which all derivation rules are in one of the forms $A \to BC, A \to a$ $(A, B, C \in N, a \in T)$. A grammar has only these kinds of rules is in *Chomsky normal form*.

In *Kuroda normal form* the rules can be in the following forms: $AB \to CD, A \to BC, A \to B, A \to a$ (where $A, B, C, D \in N, a \in T$). For every monotonous grammar there is an equivalent one in Kuroda normal form. There is a strong restriction for the length of the rules. Any of the left and right-hand-side of a rule has maximum length two. Moreover these rules are monotonous.
By a trick from Révész every rule of type $AB \to CD$ can be replaced by four context-sensitive rules:[8] $AB \to AZ, AZ \to WZ, WZ \to WD, WD \to CD$, where W, Z are newly introduced non-terminals of the grammar. (This trick also gives a proof of the equivalence between monotonous and context-sensitive grammars.) Every (λ-free) context-sensitive language can be generated by rules of the forms $AB \to AC, AB \to CB, A \to BC, A \to B, A \to a$ (where $A, B, C \in N, a \in T$). This is the so-called *Révész normal form* for context-sensitive grammars.

There was an open question (stated, for instance in Ref. 9) whether one-sided context-sensitivity is enough to generate all context-sensitive languages or one-sided context-sensitive languages are strictly included in the context-sensitive language class. The next normal form gives the answer for this question. The *Penttonen normal form* was introduced in Ref. 4, where it is called one-sided normal form: Every context-sensitive language can be generated by a grammar whose production rules are of the forms $A \to BC$, $AB \to AC$, $A \to a$, where A, B and C are nonterminals and a is a terminal.

2.2. *Derivation graphs for phrase-structure grammars*

The derivations have graphical representations (see, for instance, Refs. 8, 10,11, where they are called syntactical graphs and they were described informally). In this paper we deal with derivation graphs and trees; we start with these graphs, therefore we give a formal definition:
The *derivation graphs* can have two kinds of nodes: *symbol-labeled nodes* have label from $T \cup N \cup \{\lambda\}$, while *rule-labeled (or cross) nodes* have labels from H. Let these directed graphs be defined in the following inductive way. Let the *start-graph* be the one-node graph without any edges with node labeled by S. The start-graph is a derivation graph corresponding to the sentential form S.
Inductive step: In each (signed) derivation step a new rule-labeled node α (having label $u \to v \in H$) is constructed with in-edges from the nodes corresponding to the underlined symbols of the sentential form (the underlined string must be u). The out-edges from α are going to new symbol-labeled nodes, in a correct order, according to the right-hand-side v of the used production. In this new derivation graph, the symbol-labeled nodes that are not having out-edges (yet) represent the symbols of the actual sentential form (in left to right order).
A derivation graph is finished if it represents a derivation of a terminal word.

Note that every derivation graph, but the start-graph, has the following properties. It is a bipartite graph: its nodes are labeled from H and by the set of symbols $(T \cup N \cup \{\lambda\})$, respectively. We allow to use the symbol λ to label a node generated by a rule of type $u \to \lambda$ for some appropriate u. The 'root' is a node labeled by S having only an out-edge and no in-edge, as the derivation starts from S. Every symbol-labeled node, but the root, has exactly 1 in-edge (representing the derivation step in which this symbol is generated), and can have at most 1 out-edge. If it was already used in the derivation as a symbol of the left-hand-side of an applied rule, then there is an out-edge. The other (rule-labeled) nodes can have several in and

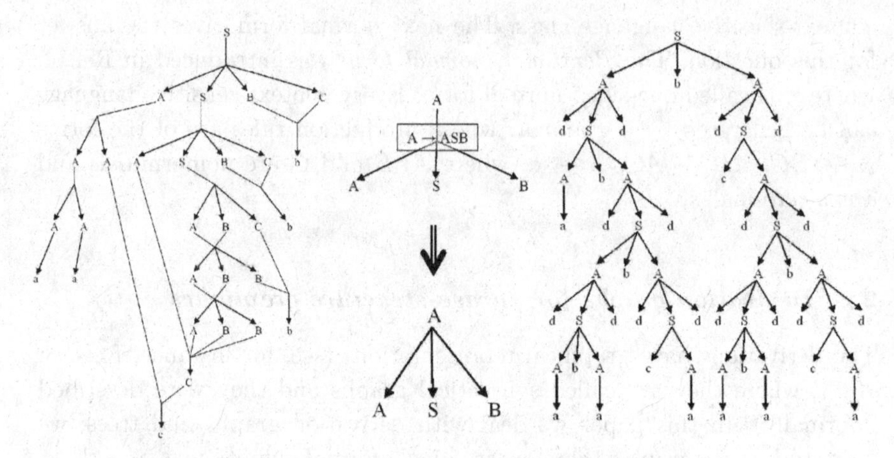

Fig. 1. Derivation graph in a phrase-structure grammar (left) and Modification of the graph (middle) to usual Derivation tree in context-free case (right)

several out-edges. These nodes represent the derivation steps. Let us see an example.

Example 2.1. Let $G_0 = (\{S, A, B, C\}, \{a, b, c\}, S, \{S \to ABC, A \to AA, S \to ASBC, ABC \to \lambda, AABB \to C, AC \to c, BC \to b, AB \to ABB, A \to a\})$ be a phrase structure grammar. Figure 1 (left) shows a finished derivation-graph in this system. In the figure only the symbol-labeled nodes are labeled, the labels of the rule-labeled nodes are not marked, but they are uniquely determined by their connections to labeled nodes.

Note that the widely used graphical representations of derivations are trees for context-free grammars, these trees are usually called *derivation-trees*, they are leaf-ordered trees, as we detail in the next subsection.

2.3. *Derivation trees for context-free case*

Graphical representation is more frequently used for derivations in context-free grammars. If only context-free rules are used, then the graphs can be simplified. In these graphs each rule-labeled node has exactly 1 in-edge. The graph holds the same information about the derivation excluding the rule-labeled nodes (see Fig. 1 middle).

labels of the corresponding	edges	of
the original graph		the tree
$A \to \alpha$		
$\alpha \to v_1$		$A \to v_1$
...	$\Rightarrow$	...
$\alpha \to v_i$		$A \to v_i$

In Fig. 1 (right) a derivation-tree is shown. In these trees the symbols of the leaves give the sentential form. Terminated derivation trees have not any nonterminal labeled leaves. In these graphs every leaf is labeled by a terminal symbol (or sometimes by the empty word λ). All other nodes (they are labeled by non-terminals) must have some (at least one) successor(s).

The context-free grammars are very popular because the concept of derivation trees fits very well to their derivations. For a signed derivation the derivation tree is uniquely determined, which is not true without signing the derivation. Applying the rule $S \to SS$ to S one obtains SS, and SSS by the second application without knowing which S of SS is rewritten. Thus, two different derivation trees can represent this derivation. It is a kind of ambiguity of context-free grammars.

Using the Chomsky normal form a finished tree is a binary tree. Each non-terminal labeled node has two successor nodes labeled by non-terminals or only one successor node labeled by a terminal. In these derivation graphs, the subtrees rooted at the children of a node v having branching factor two are called *left and right subtrees* of v. The set of nodes that are obtained with a path from the root in which the first (left-most) branch is used at every node is the *left-branch* of the tree. Similarly the *right-branch* is also defined. In trees there is a partial ordering relation, called *dominance.* A node v dominates the node v' if v' is in the subtree rooted by v. In this case it is said that v is an ancestor of v'. Every two nodes have a uniquely determined *latest common ancestor* in the tree. If v', v'' are not in dominating relation in any order, then their latest common ancestor node v is the root of a subtree with the following property. One of the nodes (of v', v'') are in the left subtree, the other is in the right subtree of v. We say, that two nodes v', v'' are *in neighbor branches* of the tree if v' is in the right-branch of the left-subtree of v and v'' is in the left-branch of the right-subtree of the same node v. We also can say that v' is on the left-neighbor branch of the node v'' if v' and v'' are in neighbor branches and v' is in the left subtree of v. In context-free grammars there is a derivation corresponding to the given derivation tree, in which the symbols of the nodes v' and v'' are neighbors in the sentential form.

In context-free case the left-most derivations are those derivations in which the first non-terminal of the sentential form is rewritten by an applicable rule in each derivation step. Regarding the derivation tree, the left-most derivation gives the same way as the depth-first graph-search algorithm discover a tree.[12] Moreover the following facts are known.

Fact 2.1. For each signed derivation there is a unique derivation tree. The possible signed derivations in context-free grammars form equivalent classes. Each class can be represented by a uniquely determined derivation tree. Each derivation tree can be represented by a unique derivation: by the left-most derivation.

In the next sections we show that the concept of derivation trees and left-most derivations can be extended to context-sensitive grammars as well.

2.4. *Derivation graphs for context-sensitive languages*

In this subsection we analyse the form of derivation graphs in monotonous and context-sensitive grammars. Although there were several attempts to describe the derivations of context-sensitive grammars by tree-like structures, in general, the result was not satisfactory. We will recall Atanasiu's approach in next subsection; and solve the problem in Sec. 3.

The monotonous and context-sensitive cases are between the general phrase-structure and context-free cases. Can we use special graphs to represent the derivations? Let us start with the derivations of monotonous grammars. In every monotonous rule the right-hand side has at least the same length as the left-hand side has. Since these rules are more related to the rules of a phrase-structure grammar than the rules of a context-free grammar, using them the derivation graphs are very different from trees as it can be seen, for instance, in the next example.

Example 2.2. Let $G_m = (\{S, A, B, C, D, E, F, G\}, \{a, b\}, S, H_m)$ be a monotonous grammar with rule set $H_m = \{S \to DABE$, $S \to DABEF$, $F \to GG, F \to GGF, G \to b, aG \to Ga, BE \to aa, Aa \to aa, Da \to aa$, $AB \to BAA, DB \to DC, CA \to AAC, CE \to BE\}$. Figure 2 shows a possible derivation-graph in this system.

As we can see, the graph in Fig. 2 is not a tree. In these derivation graphs there are two kinds of nodes as we defined earlier. Each cross-node has

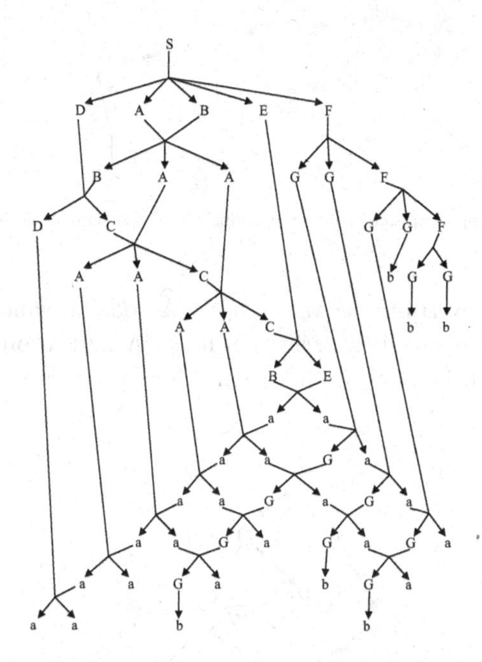

Fig. 2. Derivation in a monotonous grammar

in-edges and at least the same number of out-edges due to the monotone property. Moreover λ is not used in these graphs. These are the only differences comparing these graphs to the derivation graphs of phrase-structure grammars.

Observing our example derivation, one can see that the derivation is continued form the node labeled by D in the left-hand-side of the figure (in 'level' 4) after a long waiting for several steps in the middle of the graph.

2.4.1. *Atanasiu's approach for monotonous grammars*

A variation of the previously mentioned derivation graphs are presented in Ref. 2 based on Kuroda normal form.

Note, that a main difference between the derivation graphs and derivation trees is the existence of rule-labeled nodes. A new type of edge was introduced, see Fig. 3, and this difference was removed. The rules type $AB \to CD$ are represented by these special types of edges from the node labeled by A to the node labeled by B and the latter node has two children labeled by C and D. The node labeled by A is called context-node, while

Fig. 3. Transforming graphical representations of rules of type $AB \to CD$ by Atanasiu

the node labeled by B is a derivation node. A node for which a context-free rule is applied is also called derivation node. A derivation graph (tree) is shown in Fig. 4 using this modification. Moreover it is proved that every

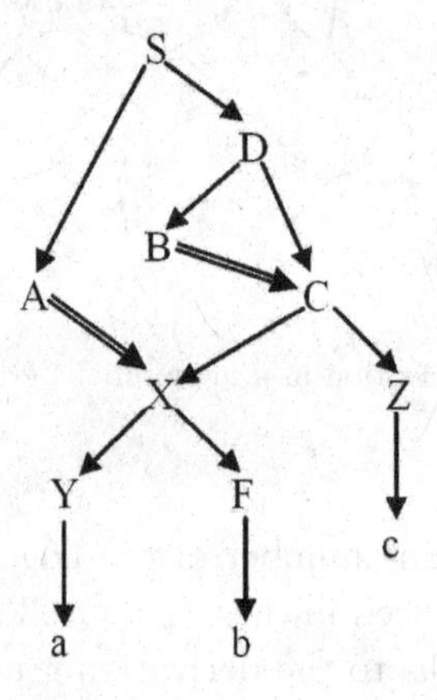

Fig. 4. Atanasiu's derivation tree for a grammar in Kuroda normal form

monotonous grammar has an equivalent one in which the set of possible context and derivation non-terminals are disjoint (these grammars are called cd Kuroda normal form). Further, the concept of left-most derivation was modified such that in each step the left-most derivation non-terminal of the actual sentential form must be rewritten. It was shown that in this way every monotonous language can be generated in left-most way. However one may feel, that these derivations are not really left-most derivations, since some branches of the derivation graphs are blocked by context-non-terminals and waiting for their turn, in which they are rewritten.

Now we are going to define derivation graphs that are more similar to trees. We will exclude the cross-nodes of the derivation graphs by a new method.

3. A New Concept of Derivation Tree in Context-Sensitive Case

We will show, that in context-sensitive grammars, and specially in Penttonen normal form the derivations can be represented by graphs that are very close to trees.

Now we are using the original definition of the context-sensitive grammars to give the new approach. In the rules the context is the same in both sides of a rule. Now, considering the derivation graphs in these grammars, one can observe, that the context symbols are repeated. We do not need to repeat them in the graph, but it should be marked that they are needed for these derivation steps. Figure 5 shows a part of the original form of the derivation graph (left), and the new way of the representation of this part (right). We note here that if the replaced non-terminal and the context are not uniquely defined (such as, for instance, at $ABC \to ASBC$) we can use the first non-terminal that can be the replaced one, without any problem.

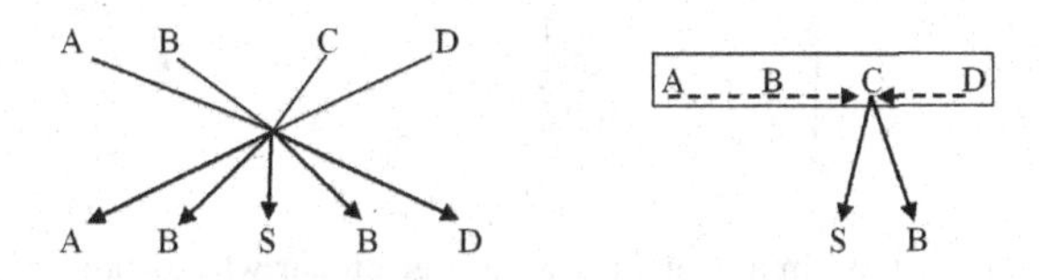

Fig. 5. Graphical representations of the context-sensitive rule $ABCD \to ABSBD$.

The concept of derivation-trees does not work in pure form for context-sensitive grammars and derivations. The neighborhood of a non-terminal can also be important at an application of a replacing rule. We use two kinds of edges in these derivation graphs to represent the context-sensitive derivation steps. The original, derivation edges are coming from the replaced non-terminal and going to the new (terminal or nonterminal) symbol(s) given in the right hand side of the used derivation rule. Each new type of edge (represented by box and broken arrows) shows the neighborhood of the replaced non-terminal as it is required in the used rule. We will use the names context-box and context-edge.

Example 3.1. Let $G_{cs} = (\{S, A, B, C, D, E, F, G, I, J, K, L, M, O, P\}, \{a, b, c\}, S, H_{cs})$ be a context-sensitive grammar, with rule set $H_{cs} = \{S \to aSA, S \to bSB, abS \to abCE, baSA \to baDFA, EA \to EG,$

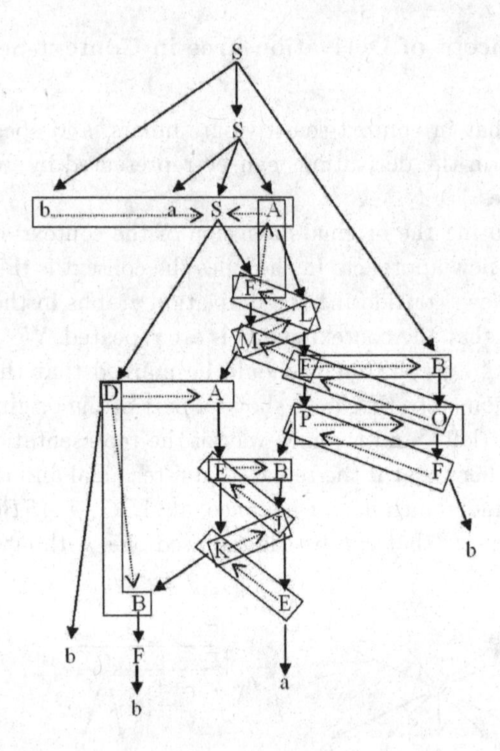

Fig. 6. 'Derivation-tree' in a context sensitive grammar with context-boxes

$EG \to IG, IG \to IE, IE \to AE, EB \to EJ, EJ \to KJ, KJ \to KE,$ $KE \to BE, FA \to FL, FL \to ML$, $ML \to MF, MF \to AF, FB \to FO,$ $FO \to PO, PO \to PF, PF \to BF, CA \to CE, CB \to CF, DA \to DE,$ $DB \to DF, C \to a, D \to b, E \to a, F \to b\}$. Figure 6 shows a possible derivation-graph in this system.

Opposite to Atanasiu's approach the context symbols are not rewritten in the derivation step (in which they are used as context).

The derivation graphs are more similar to trees if grammars in Penttonen normal form are applied. The derivation 'tree' will be simpler; each context-box contains only a left-neighbor non-terminal. (Therefore we will use only context-edges without context-boxes in these cases.) For these situations we define the derivation trees for context-sensitive case in a formal way.

Definition 3.1. Let $G = (N, T, S, H)$ be a grammar in Penttonen normal form. The derivation tree for a sentential form v is a directed graph build up

from labeled nodes (labels from $N \cup T$ and two kinds of edges (derivation and context) as follows. The nodes and the derivation edges form a tree graph (as in context-free case): The root is labeled by S. Every interior node is labeled by a non-terminal. The sequence of labels of leaves reading from left to right yields v.
Let A be the label of an interior node, and let u be its child(ren)'s label(s reading from left to right). Then one of the following conditions is fulfilled.

(i) If there is no context-edge that ends at that interior node, then $A \to u \in H$.
(ii) If exactly one context edge ends at that interior node (coming from a node labeled by $C \in N$), then $CA \to Cu \in H$.

Every context-edge connects two neighbor branches of the tree (directed from left to right), i.e., a context edge can go from v' to v'' if their latest common ancestor is v and v' is in the right-branch from the first child of v and v'' is in the left-branch from the second child of v. There are no context-edges crossing each other, i.e., a node v' which is an ancestor of a node v'' (with $v' \neq v''$) must not have an out-context-edge to a node v''' if there is a context-edge from v'' to any of the ancestors of v'''. A node can have several out-context-edges and at most 1 in-context-edge. A derivation tree is said to be finished if it has only terminal-labeled leaves.

Regarding the structure of our trees a non-terminal labeled node can have successor(s) according to a rule in H. Let the label of an interior node be $A \in N$, then it can have one or two successor nodes. In the first case this child can be labeled by a terminal (based on a rule $A \to a$ where $a \in T$) or can be a non-terminal C corresponding to a rule $BA \to BC$ with a non-terminal B; while in the latter case these nodes (and their order) must coincide with a rule in the form $A \to BC$ with $B, C \in N$.
Besides the representation of context-free steps, the graph also contains context-edges representing the applied context-sensitive rules. When a node v labeled by a non-terminal A has only one successor node which is also non-terminal labeled (let its label be C), then v has exactly one in-context-edge and it is from a node v' labeled by B (satisfying $BA \to BC \in H$) such that v and v' are in neighbor branches. Moreover there are some restrictions for context-edges, namely none of them can cross any other edges (nor derivation, nor context).

In Ref. 13 the language $a^n b^n c^n$ is generated by a left context-sensitive grammar. Now a similar example is shown in Penttonen normal form.

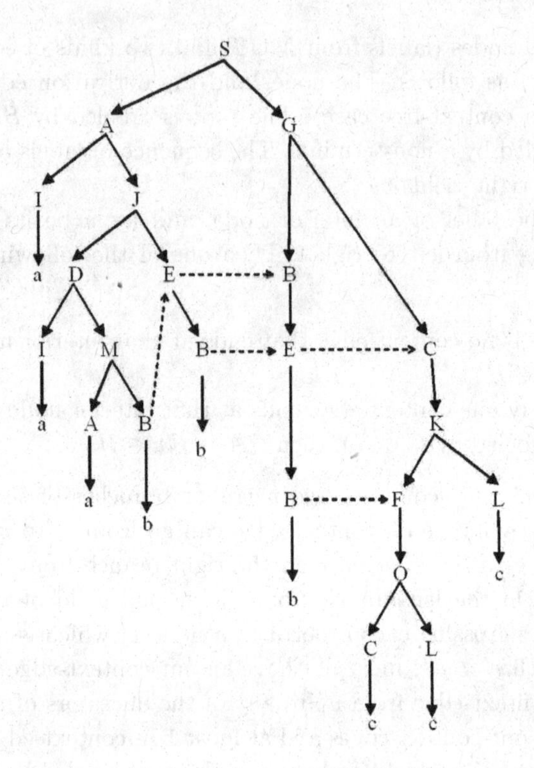

Fig. 7. Derivation-tree in a grammar in Penttonen normal form

Example 3.2. Let $G_p = (\{S, A, B, C, D, E, F, G, I, J, K, L, M, O\}, \{a, b, c\}, S, H_p)$ be a context-sensitive grammar in Penttonen normal form, with rule set $H_p = \{S \to AG, G \to BC, A \to IJ, J \to DE, EB \to EE, EC \to EK, K \to FL, D \to IM, M \to AB, BE \to BB, BF \to BO, O \to CL, A \to a, B \to b, C \to c, D \to a, E \to b, F \to c, I \to a, L \to c\}$. Figure 7 shows a possible derivation tree in this system. Observe that the structure of the graph is simple and easily readable.

The new concept (derivation-tree for context-sensitive grammars) has a strict relation to derivations in the same way as derivation trees relate to derivations in context-free case.

Theorem 3.1. *For a grammar G there exists a derivation tree for a string w if and only if* $S \Rightarrow^* w$ *(i.e., w is a sentential form).*

Proof. We give a relation between signed derivations and derivation trees. The proof goes by induction on the number of derivation steps. Trivially

the tree having only a node labeled by S is the initial sentential form. Let us assume that there is a derivation tree and its leaves give the sentential form w. There are three cases based on the possible derivation steps. Applying a context-free rule it is trivial that the same method works as in context-free case, the new tree corresponds to the new sentential form. Applying a rule in the form $AB \to AC$ in the derivation both nodes labeled by A and B must be leaves of the tree, moreover they are neighbors. The context edge between them can be added and a new derivation edge can connect the node labeled by B to the new node labeled by C.
In the opposite direction a derivation tree represents real derivations as the next construction shows. Let a context-sensitive derivation tree be given, it is the 'target-tree'. Our aim is to prove that it represents a derivation of a sentential form, i.e., it can be generated in a way presented in the previous part of the proof. For this purpose we construct a sequence of derivation-trees. This sequence starts with the graph with the only node labeled by S having no edges; and it finishes with the target-tree. Moreover every two consecutive elements of the sequence represent trees in which the derivation is continued by exactly 1 step. In the construction of the sequence the term 'actual derivation-tree' will be used to represent the element from which we are continuing the construction. It is clear that the root of the target-tree: a node labeled by S corresponds to the symbol S from which the derivation starts and there is no context-edge from the root.
We call a leaf node v of the actual derivation-tree 'continuable' if it is not leaf of the target-tree and one of the following conditions is fulfilled:

(i) there are no context-edges starting/ending at this node in the target-tree; or
(ii) if there is an in-context-edge to the node v in the target-tree from a node u, then u is a leaf of the actual derivation-tree; and if there is an out-context-edge to a node u from v in the target-tree, then u is in the actual derivation-tree and u is not a leaf of this tree.

Let us start to construct the sequence of trees. In every step the derivation will be continued by derivation edge(s) from one of the continuable nodes of the actual tree. The place of the applied derivation step is given by the place of the node of the tree, while the used derivation rule is given by the structure of the target-tree. (If there is an in-context-edge to the actual node, then a context-sensitive rule is applied, and a context-free rule is applied in other cases.) By the previous notion of continuable nodes, it is easy to see that the derivation process can be continued from the continuable nodes. Moreover, until the actual derivation-tree equals to the target-tree, there is at least 1 continuable node in the actual derivation-tree. Therefore derivation trees correspond exactly to sentential forms that can be derived in the grammar. □

A leaf node can be continuable if it is already used as a context in the derivation steps that needs it as a context. The previous theorem immediately implies the following statement.

Corollary 3.1. *For a grammar G there exists a finished derivation tree for a word w if and only if $w \in L(G)$.*

Remark 3.1. The branches of the derivation tree in a context-sensitive grammar are not independent, communication (synchronization) among them is needed by the context-edges.

A derivation from a non-terminal can be continued when all branches are after the points where this non-terminal was needed as a (part of a) context, i.e., this non-terminal has been used at all context-edges which contain it.

As we can see, using a kind of synchronization (communication, or appearance check) among the branches of the derivation the generating power of the grammar is increasing.

4. Left-Most Derivations in Context-Sensitive Case

Since left-most derivations play very important role in context-free grammars, there were several attempts to define left-most derivations in context-sensitive case keeping the generative power. As we already mentioned the proposed solutions are not satisfactory in general. In this section we solve this problem. Using the classical definition of left-most derivation only context-free languages can be obtained even if the grammar is phrase-structured.[7] Therefore we need another kind of left-most derivation, which does not coincide with the left-most derivation in the original (sentential form) sense. We are considering left-most derivations in derivation-tree sense, which differs from the classical concept of left-most derivations for non-context-free derivation trees. The new type of left-most derivation is universal, in the sense, that all words of the language can be obtained by its help. For context-sensitive grammars the left-most derivation is defined in (derivation) graph sense opposite to the context-free case in which it is left-most in sentential-form sense as well:

Definition 4.1. Let $G = (N, T, S, H)$ a grammar in Penttonen normal form. The left-most derivation for a word w is the method of the construction of a derivation tree for w in the following way. In each step the left-most non-terminal labeled leaf of the tree is used to provide new node(s) by derivation edge(s). It is possible that this step goes by using an in-context-edge from one of its neighbor nodes (the derivation has already continued from these neighbor non-terminal labeled nodes).

The left-most derivation is exactly the same order of construction as a left-most traversal goes in the tree (using the order given by the tree with the derivation edges). In context-free case, since the branches of the tree are independent, the left-most traversal of the tree gives the left-most (sentential) derivation. There is a significant difference between context-free and context-sensitive cases. In both cases it can be considered as a construction of the derivation tree, but while in context-free case it is a real (sentential) derivation in the same time, in context-sensitive case it is usually not a real derivation: using the sentential form, the rules cannot be applied in this order. Now we give a recursive algorithm which can provide a left-most derivation.

Algorithm 4.1.
input: *a context-sensitive grammar G in Penttonen normal form*
output: *a derivation tree in this grammar*
Step 1. *Let the start-graph: a node labeled by S be given. Let the node v be this node.*
Step 2. *Call the method* `continue` *with parameter v.*

Method: `continue`
input: *a node v of the derivation tree*
Step 1. *Let A be the label of v.*
Step 2. *Apply an appropriate derivation rule of G to extend the tree:*
– ***a.*** *any context-free rule of the form $A \to a$ can be applied with $a \in T$.*
If such a rule is applied then
Extend the tree by a new node labeled by a and a derivation edge from v to the new node.
Return.
– ***b.*** *any context-free rule of the form $A \to BC$ can be applied with $B, C \in N$.*
If such a rule is applied then
Extend the tree by two new nodes labeled by B and C and two derivation

edges from v to the new nodes, respectively.

Call the method **continue** *with the (left-)child node labeled by B.*

Call the method **continue** *with the (right-)child node labeled by C.*

Return.

– **c.** *context-sensitive rule $BA \to BC$ can be applied if there is a node v' with label B in the left-neighbor branch of v such that there is no context-edge starting from any node v'' dominated by v' (with $v' \neq v''$)*

If such a rule is applied then

Extend the tree by a context edge from v' to v, a new node with label C and a derivation edge from v to the new node.

Call the method **continue** *with the child node labeled by C.*

Return.

The algorithm gives a left-most traversal of the tree (using derivation edges). The left-most derivation (i.e., the construction of the tree) can go in a left to right order. The leftmost branch does not depend on other branches. The next branch may need context somewhere, and one can find it on the finished left neighbor branch. The left-hand-side part of the graph is never changed, but the right-most non-terminals of the graph may needed as contexts to build up the remaining part. It is important that the context edges cannot cross each-other.

For the example (Fig. 7) the left-most derivation is the following. The graph is built up in the next order: $S \to AG$, $A \to IJ$, $I \to a$, it is the left-most branch of the derivation-graph. It is independent of all the remaining part of the graph. The construction is continuing from the left-most non-terminal: $J \to DE$, $D \to IM$, $I \to a$, another branch is finished. The left-most non-terminal leaf is M, so $M \to AB$, $A \to a$. Now $B \to b$. Then E is the left-most non-terminal leaf, and a context B is used: from the already given right-hand-side of the actual branch of the graph: $BE \to BB$. Then $B \to b$, $G \to BC$, and contexts are needed: $EB \to EE$, $BE \to BB$ and $B \to b$ finishes this branch. Now, $EC \to EK$, $K \to FL$, and a context-edge again: $BF \to BO$. Then $O \to CL$, $C \to c$, and $L \to c$ finish these branches, while and $L \to c$ finish the construction of the derivation-graph.

The left-most derivation in context-sensitive case usually is not a real derivation in traditional (sentential form) sense, but it is unique for every derivation tree. It is a way to construct the derivation tree with context-edges, but it relates to sentential derivations.

Proposition 4.1. *Having the derivation tree any real (sentential) derivations can be constructed with the following condition. In a derivation step* *every non-terminal must remain from which a context edge starts to a non-terminal that is not already rewritten in the derivation.*

The previous proposition is closely related to the proof of Theorem 3.1, where the term continuable was used for those non-terminals from which the (sentential) derivation can be continued. Observe that usually a derivation graph represents more than one (signed) derivations. From the graph one can obtain the possible left-most real sentential derivation of a word by replacing the left-most continuable non-terminal at each derivation step. In our derivation of Example 3.2 (Fig. 7), a real (and left-most as possible) derivation is: $S \Rightarrow AG \Rightarrow IJG \Rightarrow aJG \Rightarrow aDEG \Rightarrow aIMEG \Rightarrow aaMEG \Rightarrow aaABEG \Rightarrow aaaBEG \Rightarrow aaaBEBC \Rightarrow aaaBEEC \Rightarrow aaaBBEC \Rightarrow aaabBEC \Rightarrow aaabBEK \Rightarrow aaabBBK \Rightarrow aaabbBK \Rightarrow aaabbBFL \Rightarrow aaabbBOL \Rightarrow aaabbbOL \Rightarrow aaabbbCLL \Rightarrow aaabbbcLL \Rightarrow aaabbbccL \Rightarrow aaabbbccc$. This derivation is a real derivation and usually does not coincide with our newly defined and analysed left-most derivation.

Now a new type of ambiguity is shown.

Example 4.1. Let $G_{amb} = (\{S, F, U, T, I, L, M\}, \{if, then, else, a < b, b < c, a = a + 1, c = c + 1\}, S, H_{if})$ be a context-sensitive grammar in Penttonen normal form, with rule set $H_{if} = \{S \to SS, S \to FU, U \to TS, F \to IL, E \to MS, US \to UE, I \to if, T \to then, M \to else, L \to a < b, L \to b < c, S \to a = a + 1, S \to c = c + 1\}$. Figure 8 shows two examples of context-sensitive derivation trees in this system.

In the left-most derivation the same rules are applied in the same order, the only difference is the place of the used context in rule $US \to UE$. The two derivation trees differ by the origin of the context-edge. A version

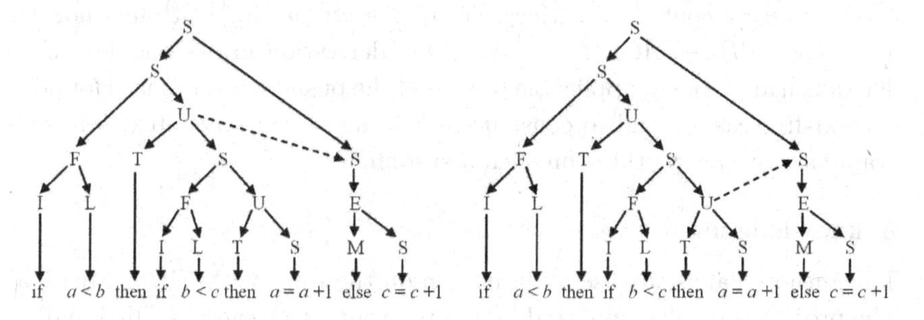

Fig. 8. Ambiguity caused by various places of the applied context

of dangling-else ambiguity is presented with two possible structure of the conditions.

5. Conclusions

Derivation graphs are analysed. The concept of derivation trees was extended to the context-sensitive case. Our solution differs from Atanasiu's approach. He defined his derivation trees for monotonous grammars in Kuroda normal form, and used his left-most and right-most derivations in cryptographical applications. Contrary, in our approach the derivation tree is extended by the help of context-edges based on Penttonen normal form. In our derivation trees the context nodes are not rewritten in steps in which they are used as context. Every non-terminal is rewritten exactly once in a finished derivation tree. Moreover, the left-most derivation is extended in the sense of constructing the derivation tree in the left-most way. It is uniquely defined for every derivation tree. Moreover, since in every step the left-most non-terminal labeled node of the tree is used (i.e. rewritten in this step), one can easily built any derivation tree in this way. It is not necessary to wait some steps or block some nodes since the new parts of the tree have no influence to the already generated left part. A new form of ambiguity is obtained from this approach, as within a derivation various context relations may become applicable. To have right-most derivations the symmetric form of Penttonen normal form is needed: it can be defined only for grammars having rules of the forms $A \to a, A \to BC, AB \to CB$. In context free grammars the branches of the derivation tree are independent of each other. In context-sensitive case it is enough to use only a node of the left-neighbor branch in the synchronization. The derivation trees and the left-most derivations can easily be extended to phrase-structure grammars. We note here that in Ref. 14 a new normal form was presented for context-sensitive grammars excluding the possible iterations in the derivation caused by context-sensitive rules of a grammar in Penttonen normal form, e.g., $AB \to AC, AC \to AB$. Our derivation graph and left-most derivation approach is applicable to extend the pushdown automata for non-context-free case,[15] and to construct a parsing-system to context-sensitive grammars, we are working on such a system.

Acknowledgements

The author thanks the useful discussion on the topic with Victor Mitrana. The project is partly supported by the programme Öveges (NKTH) and by a bilateral Japanese-Hungarian project (TéT).

References

1. J. Dassow and Gh. Paun, *Regulated Rewriting in Formal Language Theory,* (Springer-Verlag, Berlin, 1989).
2. A. Atanasiu and V. Mihalache, About derivations in context-sensitive grammars, report (1993) 21 pages, an earlier shorter version is appeared as: Derivation graphs for context-sensitive grammars, in *Salodays in Theoretical Computer Science,* (Hyperion Press, 1993) 5 pages.
3. F.-J. Brandenburg, On the tranformation of derivation graphs to derivation trees (Preliminary report), in *Proc. of MFCS'81, LNCS* **118** (Springer, 1981), pp. 224–233.
4. M. Penttonen, One-sided and two-sided context in formal grammars, *Information and Control* **25**, (1974), pp. 371–392.
5. A. Mateescu, 2003, On context-sensitive grammars, Lecture Notes in 2nd Int. PhD School on Formal Languages and Applications, (Tarragona, Spain, 2003), in *Formal Languages and Applications,* eds. C. Martin-Vide, V. Mitrana and Gh. Paun (Springer-Verlag, Berlin, Heidelberg, 2004), pp. 139–161.
6. J. E. Hopcroft and J. D. Ullmann, *Introduction to Automata Theory, Languages, and Computation,* (Addison-Wesley, Reading, 1979).
7. G. Rozenberg and A. Salomaa (eds.), *Handbook of Formal Languages,* 3 volumes (Springer, Berlin, 1997).
8. Gy. E. Révész, *Introduction to Formal Languages,* (McGraw-Hill, New York, 1983).
9. A. Salomaa, *Formal Languages,* (Academic Press, New York, 1973).
10. G. Buntrock and F. Otto, Growing context-sensitive languages and Church-Rosser languages, *Information and Computation* **141**, (1998), pp. 1–36.
11. J. Loeckx, 1970, The parsing for general phrase-structure grammars, *Information and Control* **16**, (1970), pp. 443–464.
12. S. J. Russel and P. Norvig, *Artificial Intelligence: A Modern Approach* (Prentice Hall, New Jersey, 1995).
13. C. Martin-Vide, Formal language theory: classical and non-classical machineries, Lecture Notes in 2nd Int. PhD School on Formal Languages and Applications, (Tarragona, Spain, 2003), a shorter version can be found in: *The Oxford Handbook of Computational Linguistics,* ed. R. Mitkov (Oxford University Press, Oxford, 2003), pp. 157–177.
14. B. Nagy and P. Varga, A new normal form for context-sensitive grammars, in *Proc. of (35th Conf. on Current Trends in) Theory and Practice of Computer Science (SOFSEM 2009), Volume II,* (Spindleruv Mlyn, Czech Repulic, 2009), pp. 60–71.
15. B. Nagy, An automata-theoretic characterization of the Chomsky-hierarchy, in *Proc. of 7th Annual Conf. on Theory and Applications of Models of Computation (TAMC 2010), LNCS* (Springer, 2010), accepted for publication

Received: June 21, 2009

Revised: March 23, 2010

ON PROPER LANGUAGES AND TRANSFORMATIONS OF LEXICALIZED TYPES OF AUTOMATA

FRIEDRICH OTTO

Fachbereich Elektrotechnik/Informatik, Universität Kassel,
D-34109 Kassel, Germany
E-mail: otto@theory.informatik.uni-kassel.de

Motivated by the way in which sentences of natural languages are analyzed in linguistics, types of automata are studied that work on extended alphabets which, in addition to the input symbols, also contain certain *auxiliary symbols*. The latter model the use of (morphological, syntactical, and semantical) categories in the process of analyzing sentences. The automata we consider work on so-called *characteristic languages*, that is, on languages that include auxiliary symbols. The *proper language* is obtained from a characteristic language by removing all occurrences of auxiliary symbols. By requiring that the automata are *lexicalized*, we restrict the lengths of blocks of auxiliary symbols that are admitted. We study the classes of proper languages for *deterministic* finite-state acceptors, pushdown automata, two-pushdown automata, and freely rewriting restarting automata that are lexicalized. In addition, we use a generalization of the notion of proper language to associate a (binary) transformation with a characteristic language. This leads to the study of transformations of a certain form for the above types of automata.

Keywords: Proper Language; Lexicalized Type of Automaton; Transformation.

1. Introduction

Automata with a restart operation were introduced originally to describe a method of grammar-checking for the Czech language.[1] These automata started the investigation of restarting automata as a suitable tool for modeling the so-called *analysis by reduction*, which is a technique that is often used (implicitly) for developing formal descriptions of natural languages based on the notion of *dependency*.[2,3] In particular, the Functional Generative Description[4] (FGD) for the Czech language is based on this method.

FGD is a dependency based system, which translates given sentences into their underlying tectogrammatical representations, which are (at least in principle) disambiguated. Let w be a sentence of a natural language that is to be analyzed. First w is split into a sequence of tokens, which can then be taken as the input symbols of the proper analyzer. So we assume that w is already given as a sequence of tokens $w_1, w_2, \ldots, w_n$. Now the process we are interested in consists of three main phases:

(1) In the first phase, called *lexical analysis*, each token is annotated with all tags (that is, categories) that could possibly apply to this particular token using a dictionary. Thus, in this phase a sequence of nonterminals of bounded length is inserted after each token w_i. This annotation contains morphological, syntactical, and possibly also some semantical information. It describes all possibilities for classifying the token w_i.
(2) If the annotation for a token w_i gives several possible classifications, then one tries to delete all classifications that contradict the actual situation given by the context. This process, which is called *disambiguation*, ends with either a unique classification for all tokens, with the detection of an error (in case all classifications are removed from a token based on context information), or with a small number of remaining classifications.[5] In the first case the disambiguation process is successful, while in the latter case the remaining ambiguities are resolved by generating all possible completely disambiguated sequences, which are then considered separately.
(3) Finally, analysis by reduction is applied to the disambiguated sequence of tokens and their remaining annotations. It consists in stepwise simplifications (that is, reductions), which continue until the so-called core predicative structure of the sentence is reached. Each simplification replaces a small part of the sentence by an even shorter phrase.

Here we formalize the latter step of this process by studying various deterministic types of automata for *proper languages*. These automata work on so-called *characteristic languages*, that is, on languages that include auxiliary symbols (categories) in addition to the input symbols. The *proper language* is obtained from the characteristic language by removing all auxiliary symbols from its words (sentences). By requiring that the automata considered are *lexicalized* we restrict the lengths of blocks of auxiliary symbols that are allowed by a constant. This restriction is quite natural from a linguistic point of view, as these blocks of auxiliary symbols model the meta-language categories from individual linguistic layers with which an input string is being enriched when its disambiguated form is being produced (see above). We use *deterministic* types of automata only in order to ensure the *correctness preserving property* for the analysis.

This paper is structured as follows. In Section 2 we shortly restate the hierarchy results that have been obtained recently for lexicalized restarting automata.[6–8] In fact, for freely rewriting restarting automata various two-dimensional hierarchies of proper languages have been derived, where the one dimension is governed by the number of rewrites that may be executed in any one cycle, and the other dimension is parametrized by the number of auxiliary symbols that may occur concurrently on the tape in any valid computation. In Section 3 we then consider the proper languages of deterministic finite-state acceptors and deterministic pushdown automata. We will see that the class of proper languages of finite-state acceptors is just the class REG of regular languages, while the class of proper languages of deterministic pushdown automata is the class CFL of all context-free languages. In Section 4 we study the proper languages of deterministic two-pushdown automata that are *shrinking* or even *length-reducing*. As we will see they are universal, that is, each recursively enumerable language is the proper language of such an automaton. Therefore, we turn to *lexicalized* two-pushdown automata, showing that the class of proper languages of deterministic two-pushdown automata that are shrinking (or length-reducing) and lexicalized coincides with the class GCSL of growing context-sensitive languages. Finally, in Section 5 we turn to transformations defined by characteristic languages. We define the concept in general terms, and we present a few preliminary results for deterministic finite-state acceptors, pushdown automata, and shrinking/length-reducing two-pushdown automata. The paper closes with a short summary and some open problems.

Notation. Throughout the paper we will use λ to denote the empty word. Further, $|w|$ will denote the *length* of the word w, and if a is an element of the underlying alphabet, then $|w|_a$ denotes the *a-length* of w, that is, the number of occurrences of the letter a in w. Further, $\mathbb{N}_+$ will denote the set of all positive integers.

If Σ is a subalphabet of Γ, then by Pr^Σ we denote the projection from Γ^* onto Σ^*, that is, Pr^Σ is the morphism defined by $a \mapsto a$ $(a \in \Sigma)$ and $A \mapsto \lambda$ $(A \in \Gamma \smallsetminus \Sigma)$. If $v = \mathsf{Pr}^\Sigma(w)$, then v is the *Σ-projection* of w, and w is an *expanded version* of v. For a language $L \subseteq \Gamma^*$, $\mathsf{Pr}^\Sigma(L) = \{\, \mathsf{Pr}^\Sigma(w) \mid w \in L \,\}$.

If M is an automaton (finite-state, pushdown, etc.) with input alphabet Σ and tape alphabet Γ containing Σ, then $L_C(M)$ will denote the *characteristic language* of M, which consists of all words w over Γ that are accepted by M. Now $L(M) = L_C(M) \cap \Sigma^*$ is the *input language* of M, and

$L_P(M) = \mathsf{Pr}^{\Sigma}(L_C(M))$ is the *proper language* of M. Thus, a word $u \in \Sigma^*$ belongs to the proper language of M if and only if an extended version $v \in \Gamma^*$ of u is in the characteristic language of M.

For any class A of automata, $\mathcal{L}_C(\mathsf{A})$ will denote the class of characteristic languages recognizable by automata from A, $\mathcal{L}(\mathsf{A})$ will denote the class of input languages recognizable by automata from A, and $\mathcal{L}_P(\mathsf{A})$ will denote the class of proper languages of automata from A. By DCFL we denote the class of deterministic context-free languages. Occasionally we will use regular expressions instead of the corresponding regular languages.

2. Restarting Automata

The restarting automaton has been designed specifically as a formal model for the analysis by reduction. In fact, a large variety of types of restarting automata has been developed over the years. Here we are particularly interested in the *freely rewriting restarting automaton*, FRR-automaton for short, as described in Ref. 8.

An FRR-automaton is a (nondeterministic) machine that consists of a finite-state control, a single flexible tape with end markers, and a read/write window of fixed size. Formally, it is described by an 8-tuple $M = (Q, \Sigma, \Gamma, ¢, \$, q_0, k, \delta)$, where Q is a finite set of states, Σ is a finite input alphabet, Γ is a finite tape alphabet containing Σ, the symbols $¢, \$ \notin \Gamma$ are used as markers for the left and right border of the work space, respectively, $q_0 \in Q$ is the initial state, $k \geq 1$ is the size of the *read/write window*, and δ is the *transition relation* that associates to each pair (q, w) consisting of a state q and a possible content w of the read/write window a finite set of possible transition steps. There are four types of transition steps:

1. A *move-right step* (MVR) causes M to shift the read/write window one position to the right and to change the state. However, the read/write window cannot move across the right sentinel \$.
2. A *rewrite step* causes M to replace a non-empty prefix u of the content w of the read/write window by a shorter string v, thereby reducing the length of the tape, and to change the state. Further, the read/write window is placed immediately to the right of the string v. However, occurrences of the delimiters ¢ and \$ can neither be deleted nor newly created by a rewrite step.
3. A *restart step* causes M to place its read/write window over the left end of the tape, so that the first symbol it sees is the left sentinel ¢, and to reenter the initial state q_0.
4. An *accept step* causes M to halt and accept.

If $\delta(q, w) = \emptyset$ for some pair (q, w), then M necessarily halts, and we say that M *rejects* in this situation. If $\delta(q, w)$ contains at most a single transition for each pair (q, w), then M is a *deterministic* FRR-automaton. We use the prefix det- to denote deterministic types of restarting automata.

Observe that the rewrite steps of an FRR-automaton differ slightly from those for a *classical* restarting automaton like the RRWW-automaton.[9] A rewrite step of an RRWW-automaton replaces the complete content w of the read/write window by a shorter word v, and then the read/write window is moved to the right of the newly written word. For an FRR-automaton, however, a rewrite step replaces a non-empty prefix u of the content $w = uz$ of the read/write window by a shorter word v, producing the factor vz, and then the read/write window is moved just to the right of the factor v. Hence, after executing this rewrite step, the suffix z is still inside the read/write window. This change in the definition of the rewrite step is caused by the following observation: When an FRR-automaton is to rewrite a factor u by a word v, a certain finite look-ahead z may be needed to determine the correct occurrence of the factor u to be rewritten. However, this very factor z (or a suffix thereof) might be used in the next rewrite step, and so the read/write window must not skip across it.

Observe further that the model of the FRR-automaton presented here differs from the model studied in Ref. 10. Our model has length-reducing rewrite steps only, while the rewrite steps of the model considered in Ref. 10 are just required to be weight-reducing with respect to some weight function, that is, that model is a generalization of the *shrinking* restarting automaton.[11]

A *configuration* of an FRR-automaton M is a string $\alpha q \beta$, where $q \in Q$, and either $\alpha = \lambda$ and $\beta \in \{¢\} \cdot \Gamma^* \cdot \{\$\}$ or $\alpha \in \{¢\} \cdot \Gamma^*$ and $\beta \in \Gamma^* \cdot \{\$\}$; here q represents the current state, $\alpha\beta$ is the current content of the tape, and it is understood that the window contains the first k symbols of β or all of β when $|\beta| \leq k$. A *restarting configuration* is of the form $q_0 ¢ w \$$. If $w \in \Sigma^*$, then $q_0 ¢ w \$$ is an *initial configuration.*

We observe that any computation of M consists of certain phases. A phase, called a *cycle*, starts in a restarting configuration, the head moves along the tape performing move-right and rewrite operations until a restart operation is performed and thus a new restarting configuration is reached. If no further restart operation is performed, the computation necessarily finishes in a halting configuration, which is either *rejecting* or *accepting.* Such a phase is called a (rejecting or an accepting) *tail.* It is required that in each cycle M performs at least one rewrite step – thus each cycle strictly reduces the length of the tape. We use the notation $u \vdash^c_M v$ to denote a cycle of M that begins with the restarting configuration $q_0 ¢ u \$$ and ends with the restarting configuration $q_0 ¢ v \$$; the relation $\vdash^{c^*}_M$ is the reflexive and transitive closure of $\vdash^c_M$.

A *sentential form* $w \in \Gamma^*$ *is accepted by* M, if there is an accepting computation which starts from the restarting configuration $q_0 ¢ w \$$. By $L_C(M)$ we denote the language consisting of all sentential forms accepted by M; we say that $L_C(M)$ is the *characteristic language* of M, while the set $L(M) = L_C(M) \cap \Sigma^*$ of all input sentences accepted by M is called the *input language recognized* by M.

We emphasize the following basic properties[9,12] of restarting automata, which are often used implicitly in proofs.

Proposition 2.1 (Error Preserving Property). *Let M be an FRR-automaton, and let $x, y \in \Gamma^*$. If $x \vdash^{c^*}_M y$ and $x \notin L_C(M)$, then $y \notin L_C(M)$, either.*

Proposition 2.2 (Correctness Preserving Property). *Let M be an FRR-automaton, and let $x, y \in \Gamma^*$. If $x \in L_C(M)$, and if $x \vdash^{c^*}_M y$ is part of an accepting computation of M, then $y \in L_C(M)$, too.*

Observe that the latter property does in general not hold for input languages, as apart from the initial configuration, each restarting configuration in an accepting computation may contain some auxiliary (that is, non-input) symbols.

Finally we come to the notion of *monotonicity.* Let $C = \alpha q \beta$ be a *rewrite configuration* of an FRR-automaton M, that is, a configuration in which a rewrite step is to be applied. Then $|\beta|$ is called the *right distance* of C, which is denoted by $D_r(C)$. A *sequence of rewrite configurations* $S = (C_1, C_2, \ldots, C_n)$ is called *monotone* if

$$D_r(C_1) \geq D_r(C_2) \geq \cdots \geq D_r(C_n).$$

Let j be a positive integer. We say that a sequence of rewrite configurations $S = (C_1, C_2, \ldots, C_n)$ is *j-monotone* if there is a partition of S into j subsequences

$$S_1 = (C_{1,1}, C_{1,2}, \ldots, C_{1,n_1}), \ldots, S_j = (C_{j,1}, C_{j,2}, \ldots, C_{j,n_j})$$

such that each S_i, $1 \leq i \leq j$, is monotone. Observe that it is not required that the subsequences $S_1, \ldots, S_j$ follow sequentially one after another in the original sequence. Instead they are in general all scattered throughout the original sequence. Hence, a sequence of rewrite configurations $(C_1, C_2, \ldots, C_n)$ is *not* j-monotone if and only if there exist indices $1 \le i_1 < i_2 < \cdots < i_{j+1} \le n$ such that $D_r(C_{i_1}) < D_r(C_{i_2}) < \cdots < D_r(C_{i_{j+1}})$.

A *computation* of an FRR-automaton M is called *j-monotone* if the sequence of rewrite configurations that is obtained from the cycles of that computation is j-monotone. Observe that here those rewrite configurations are not taken into account that correspond to the rewrite steps that are executed in the tail of that computation. A *computation* is *j-rewriting* if none of its cycles contains more than j rewrite steps. Finally, a *computation* is *j-constrained* if it is both j-rewriting and j-monotone, and the FRR-automaton M is called *j-constrained* if each of its computations is j-constrained. We use the prefix j-constr- to denote j-constrained types of FRR-automata.

Above we introduced FRR-automata as acceptors for characteristic languages and input languages. Now we turn to proper languages of FRR-automata. Here we will only consider FRR-automata $M = (Q, \Sigma, \Gamma, ¢, \$, q_0, k, \delta)$ that are deterministic. Recall that the *proper language of* M is defined as the set of words $L_P(M) = \mathsf{Pr}^{\Sigma}(L_C(M))$, that is, a word $v \in \Sigma^*$ belongs to $L_P(M)$ if and only if there exists an expanded version u of v such that $u \in L_C(M)$. As a det-FRR-automaton M can easily be simulated by a deterministic Turing machine in quadratic time, we see that the membership problems for the languages $L_C(M)$ and $L(M)$ are solvable in quadratic time.

The class CRL of *Church-Rosser languages* is a *basis* for the class RE of recursively enumerable languages,[13] that is, for each recursively enumerable language $L \subseteq \Sigma^*$, there exists a Church-Rosser language B on some alphabet Δ strictly containing Σ such that $\mathsf{Pr}^{\Sigma}(B) = L$. As CRL coincides with the class of input languages of deterministic RRWW-automata,[14] there exists a deterministic RRWW-automaton M' with input alphabet Δ and tape alphabet Γ such that $L(M') = B$. Hence, $L = \mathsf{Pr}^{\Sigma}(B) \subseteq \mathsf{Pr}^{\Sigma}(L_C(M'))$. However, the language $L_C(M')$ will in general also contain words for which the projection onto Σ does not belong to the language L, that is, the above inclusion is in general a strict one. Nevertheless, using a technically more involved construction the following result has been derived,[6] where $\Sigma_0 = \{a, b\}$, $\Sigma_1 = \Sigma_0 \cup \{c\}$, $\varphi_0 : \Sigma_0^* \to \Sigma_0^*$ is the injective morphism that is defined by $a \mapsto aa$ and $b \mapsto bb$, and $\varphi : \Sigma_0^* \to \Sigma_1^*$ denotes the mapping that is defined by $\varphi(w) = \varphi_0(w) \cdot c$.

Proposition 2.3. *For each recursively enumerable language* $L \subseteq \Sigma_0^+$, *there exists a 1-rewriting det-FRR-automaton* M *such that* $L_P(M) \cap \Sigma_0^* \cdot c = \varphi(L)$.

Thus, a word $w \in \Sigma_0^*$ belongs to the recursively enumerable language L if and only if its image $\varphi(w)$ belongs to the proper language $L_P(M)$. This yields the following result.

Corollary 2.1. *There exists a deterministic FRR-automaton* M *such that the language* $L_P(M)$ *is non-recursive.*

Thus, the proper languages of deterministic FRR-automata are in general far more complex than the corresponding input and characteristic languages. Therefore we now turn our attention to deterministic FRR-automata for which the use of auxiliary symbols is restricted as in Refs. 6 and 15.

Definition 2.1. Let $M = (Q, \Sigma, \Gamma, ¢, \$, q_0, k, \delta)$ be a det-FRR-automaton.

(a) A word $w \in \Gamma^*$ is *not immediately rejected by* M if, starting from the restarting configuration $q_0¢w\$$, M either performs a cycle of the form $w \vdash_M^c z$ for some word $z \in \Gamma^*$, or M accepts w in a tail computation. By $\mathsf{NIR}(M)$ we denote the set of all words that are not immediately rejected by M.
(b) The det-FRR-automaton M is called *lexicalized* if there exists a constant $j \in \mathbb{N}_+$ such that, whenever $v \in (\Gamma \smallsetminus \Sigma)^*$ is a factor of a word $w \in \mathsf{NIR}(M)$, then $|v| \leq j$.
(c) M is called *strongly lexicalized* if it is lexicalized, and if each of its rewrite operations just deletes some symbols.

Strong lexicalization is a technique that is used in dependency based formal descriptions of natural languages.[4] Below we are interested in (strongly) lexicalized FRR-automata and their proper languages. By LRR (SLRR) we denote the class of (strongly) lexicalized FRR-automata, and by t-LRR (t-SLRR) we denote the class of (strongly) lexicalized FRR-automata which execute at most t rewrite steps in any cycle. Further, by j-constr-LRR (j-constr-SLRR) we denote the class of (strongly) lexicalized FRR-automata that are j-constrained. Recall that lexicalized FRR-automata are deterministic.

If M is a lexicalized FRR-automaton, and if $w \in \Gamma^*$ is an extended version of an input word $v = \mathsf{Pr}^\Sigma(w)$ such that w is not immediately rejected by M, then $|w| \leq (j+1) \cdot |v| + j$ for some constant $j > 0$. Accordingly we have the following result.

Corollary 2.2. *If M is a lexicalized FRR-automaton, then the proper language $L_{\mathrm{P}}(M)$ is context-sensitive.*

On the other hand, we have the following negative result.

Proposition 2.4. *The Church-Rosser language $L_{\mathrm{e}} = \{ a^{2^n} \mid n \in \mathbb{N} \}$ is not contained in $\mathcal{L}_{\mathrm{P}}(\text{1-LRR})$.*

Proof. Assume that $L_{\mathrm{e}} = L_{\mathrm{P}}(M)$ for a 1-rewriting LRR-automaton $M = (Q, \{a\}, \Gamma, ¢, \$, q_0, k, \delta)$, and let $z = a^{2^n} \in L_{\mathrm{e}}$, where n is a large integer. Then there exists an extended version $w \in \Gamma^*$ of z such that $w \in L_{\mathrm{C}}(M)$. Thus, the computation of M with input w is accepting. From the Pumping Lemma for restarting automata,[14] it is easily seen that this computation cannot just consist of an accepting tail computation, that is, it begins with a cycle of the form $w \vdash^c_M w'$. From the Correctness Preserving Property it follows that $w' \in L_{\mathrm{C}}(M)$, which in turn implies that $\mathsf{Pr}^{\{a\}}(w') \in L_{\mathrm{e}}$. Thus, $\mathsf{Pr}^{\{a\}}(w') = a^m$ for some integer m satisfying $2^n - k \le m < 2^n + k$. As m must be a power of two, it follows from the choice of z that $m = 2^n$, that is, w' is obtained from w by rewriting some auxiliary symbols only. We can repeat this argument until eventually M either rewrites some occurrences of the symbol a, which will then yield a word $\hat{w} \in L_{\mathrm{C}}(M)$ for which the projection $\mathsf{Pr}^{\{a\}}(\hat{w})$ does not belong to the language L_{e} anymore, or until M accepts a word $\tilde{w}$ in a tail computation for which $\mathsf{Pr}^{\{a\}}(\tilde{w}) = a^{2^n}$ holds. In the latter case the Pumping Lemma can be applied to show that $L_{\mathrm{P}}(M)$ contains words that do not belong to the language L_{e}. In either case it follows that $L_{\mathrm{P}}(M) \neq L_{\mathrm{e}}$, contradicting our assumption above. Thus, L_{e} is not the proper language of any 1-rewriting LRR-automaton. □

Finally we consider the following static complexity measure for LRR-automata.

Definition 2.2. Let $M = (Q, \Sigma, \Gamma, ¢, \$, q_0, k, \delta)$ be an LRR-automaton, and let $m \in \mathbb{N}$. The automaton M has *word-expansion* m, denoted by $\mathsf{W}(M) = m$, if each word from $\mathsf{NIR}(M)$ contains at most m occurrences of auxiliary symbols, that is, if $w \in \Gamma^*$ is not immediately rejected by M, then $|\mathsf{Pr}^{\Gamma \smallsetminus \Sigma}(w)| \le m$.

We use the prefix $\mathrm{W}(m)$- to denote classes of deterministic FRR-automata that have word-expansion m. The following result taken from Ref. 8 is a generalization of a result for lexicalized RRWW-automata.[6,15]

Theorem 2.1. *If M is a $W(m)$-LRR-automaton for some $m \in \mathbb{N}$, then the membership problem for $L_P(M)$ is solvable deterministically in time $O(n^{m+2})$.*

As the 1-(S)LRR-automaton is almost identical to the (strongly) lexicalized RRWW-automaton,[6,15] we have the following result.

Theorem 2.2. *The class* CFL *of context-free languages coincides with the class of proper languages of 1-constrained (strongly) lexicalized FRR-automata, that is,*

$$\mathcal{L}_P(\text{1-constr-LRR}) = \mathcal{L}_P(\text{1-constr-SLRR}) = \text{CFL}.$$

Actually, based on the parameter j of constrainability and the word expansion factor $W(m)$ several two-dimensional hierarchies of classes of proper languages have been derived.[6,8]

Theorem 2.3. *For all $m \geq 0$, all $i \geq 1$, and all* $\text{X} \in \{\text{LRR}, \text{SLRR}\}$,

(a) $\mathcal{L}_P(\text{W}(m)\text{-}i\text{-constr-X}) \subsetneq \mathcal{L}_P(\text{W}(m)\text{-}(i+1)\text{-constr-X})$.

(b) $\mathcal{L}_P(\text{W}(m)\text{-}i\text{-constr-X}) \subsetneq \mathcal{L}_P(\text{W}(m+1)\text{-}i\text{-constr-X})$.

(c) $\mathcal{L}_P(\text{W}(m)\text{-}i\text{-X}) \subsetneq \mathcal{L}_P(\text{W}(m)\text{-}(i+1)\text{-X})$.

(d) $\mathcal{L}_P(\text{W}(m)\text{-}i\text{-X}) \subsetneq \mathcal{L}_P(\text{W}(m+1)\text{-}i\text{-X})$.

Some further results relating these language classes to some other classes of proper languages of certain restricted types of restarting automata can also be found in Refs. 6 and 8.

3. Finite-State Acceptors and Pushdown Automata

Here we study proper languages of finite-state acceptors and pushdown automata. Let $A = (Q, \Sigma, \Gamma, q_0, F, \delta)$ be a *deterministic finite-state acceptor*, DFA for short, where Q is a finite set of states, Σ is a finite input alphabet, Γ is a finite tape alphabet containing Σ, $q_0 \in Q$ is the initial state, $F \subseteq Q$ is the set of final states, and $\delta : Q \times \Gamma \to Q$ is the (partial) transition function. Then $L_C(A) = \{ w \in \Gamma^* \mid A \text{ accepts on input } w \}$ is the *characteristic language* of A, while $L_P(A) = \text{Pr}^{\Sigma}(L_C(A))$ is the *proper language* of A. The DFA A is called *lexicalized* if there exists a constant $c \in \mathbb{N}_+$ such that, for each $w \in L_C(A)$ and each factor $v \in (\Gamma \smallsetminus \Sigma)^*$ of w, $|v| \leq c$ holds. We use the prefix lex- to denote lexicalized types of automata.

It is well-known that $\mathcal{L}_C(\text{DFA}) = \text{REG}$, the class of regular languages. As REG is closed under arbitrary morphisms, it follows that with $L_C(A)$, also $L_P(A) = \mathsf{Pr}^{\Sigma}(L_C(A))$ is regular. On the other hand, if $\Sigma = \Gamma$, then $L_C(A) = L_P(A)$. Thus, we have the following trivial observation.

Observation 3.1. $\mathcal{L}_P(\mathsf{DFA}) = \mathcal{L}_P(\text{lex-}\mathsf{DFA}) = ?\mathsf{REG}$.

Thus, in the case of finite-state acceptors proper languages give just another characterization for the class of regular languages.

A *deterministic pushdown automaton*, DPDA for short, is given through a tuple $M = (Q, \Sigma, \Gamma, \Delta, q_0, \bot, F, \delta)$, where Q is a finite set of states, Σ is a finite input alphabet, Γ is a finite tape alphabet containing Σ, Δ is a finite pushdown alphabet containing the bottom marker $\bot$, $q_0 \in Q$ is the initial state, $F \subseteq Q$ is the set of final states, and $\delta : Q \times (\Gamma \cup \{\lambda\}) \times \Delta \to Q \times \Delta^*$ is the (partial) transition function. A *configuration* of M is written as a triple (q, u, α), where $q \in Q$ is the current state, $u \in \Gamma^*$ is the remaining input with the input head on the first symbol of u (or on λ, if $u = \lambda$), and $\alpha \in \Delta^*$ is the current content of the pushdown store with the first letter of α at the bottom and the last letter of α at the top. A word $w \in \Gamma^*$ is accepted by M, if the computation of M that starts with the initial configuration $(q_0, w, \bot)$ reaches a final configuration of the form (q, λ, α), where $q \in F$ and $\alpha \in \Delta^*$. Then $L_C(M) = \{ w \in \Gamma^* \mid M \text{ accepts on input } w \}$ is the *characteristic language* of M, while $L_P(M) = \mathsf{Pr}^{\Sigma}(L_C(M))$ is the *proper language* of M.

It is well-known that $\mathcal{L}_C(\mathsf{DPDA}) = \mathsf{DCFL}$. However, as the language class DCFL is not closed under arbitrary morphisms, we obtain a strictly larger class of languages by considering proper languages of DPDAs.

Theorem 3.1. $\mathcal{L}_P(\mathsf{DPDA}) = \mathcal{L}_P(\text{lex-}\mathsf{DPDA}) = \mathsf{CFL}$.

Proof. As the class of context-free languages is closed under arbitrary morphisms, we see that $\mathcal{L}_P(\mathsf{DPDA}) \subseteq \mathcal{L}_P(\mathsf{PDA}) \subseteq \mathsf{CFL}$, where PDA denotes the class of (nondeterministic) pushdown automata.

Conversely, assume that $L \subseteq \Sigma^*$ is a context-free language. Without loss of generality we may assume that L does not contain the empty word. Thus, there exists a context-free grammar $G = (N, \Sigma, S, P)$ for L that is in *Greibach normal form*,[16] that is, each rule of P has the form $A \to \alpha$ for some string $\alpha \in \Sigma \cdot N^*$. For the following construction we assume that the rules of G are numbered from 1 to m.

From G we construct a new grammar $G' = (N, \Sigma \cup B, S, P')$, where $B = \{ \nabla_i \mid 1 \leq i \leq m \}$ is a set of new terminal symbols that are in one-to-one correspondence to the rules of G, and

$$P' = \{ A \to \nabla_i \alpha \mid (A \to \alpha) \text{ is the } i\text{-th rule of } G,\ 1 \le i \le m \}.$$

Obviously, a word $\omega \in (\Sigma \cup B)^*$ belongs to $L(G')$ if and only if ω has the form $\omega = \nabla_{i_1} a_1 \nabla_{i_2} a_2 \cdots \nabla_{i_n} a_n$ for some integer $n > 0$, where $a_1, \ldots, a_n \in \Sigma$, $i_1, \ldots, i_n \in \{1, \ldots, m\}$, and these indices describe a (left-most) derivation of $w = a_1 a_2 \cdots a_n$ from S in G. Thus, $\mathsf{Pr}^{\Sigma}(L(G')) = L(G) = L$. From ω this derivation can be reconstructed deterministically. In fact, the language $L(G')$ is deterministic context-free. Hence, there exists a DPDA M for this language. By interpreting the symbols of B as auxiliary symbols, we obtain a DPDA M' such that $\mathsf{Pr}^{\Sigma}(L_{\mathrm{C}}(M')) = \mathsf{Pr}^{\Sigma}(L(M)) = \mathsf{Pr}^{\Sigma}(L(G')) = L$. It follows that $\mathcal{L}_{\mathrm{P}}(\mathsf{DPDA}) = \mathsf{CFL}$.

Observe from the proof above that within each word from $L_{\mathrm{C}}(M')$, auxiliary letters and input letters alternate. Thus, the DPDA M' is even lexicalized. □

4. Two-Pushdown Automata

Finally we turn to the proper languages of two-pushdown automata. A *two-pushdown automaton*, TPDA for short, is a nondeterministic automaton with two pushdown stores. Formally, it is defined as an 8-tuple $M = (Q, \Sigma, \Gamma, k, \delta, q_0, \bot, F)$, where Q is a finite set of states, Σ is a finite input alphabet, Γ is a finite pushdown alphabet with $\Gamma \supsetneq \Sigma$ and $\Gamma \cap Q = \emptyset$, $k \ge 1$ is the size of the pushdown windows, $q_0 \in Q$ is the initial state, $\bot \in \Gamma \setminus \Sigma$ is the bottom marker of the pushdown stores, and $\delta\colon Q \times \Gamma^{\le k} \times \Gamma^{\le k} \to \mathcal{P}_{\mathsf{fin}}(Q \times \Gamma^* \times \Gamma^*)$ is a transition relation. Here $\Gamma^{\le k} = \{ u \in \Gamma^+ \mid |u| \le k \}$, and $\mathcal{P}_{\mathsf{fin}}(Q \times \Gamma^* \times \Gamma^*)$ denotes the set of finite subsets of $Q \times \Gamma^* \times \Gamma^*$. In addition, we require that the special symbol $\bot$ can only occur at the bottom of a pushdown store, and that no other letter can occur at that place. The automaton M is a *deterministic two-pushdown automaton* (DTPDA), if δ is a (partial) function from $Q \times \Gamma^{\le k} \times \Gamma^{\le k}$ into $Q \times \Gamma^* \times \Gamma^*$.

Following Ref. 17, a configuration of a (D)TPDA is described by a word uqv, where $q \in Q$ is the current state, $u \in \Gamma^*$ is the content of the first pushdown store with the first letter of u at the bottom and the last letter of u at the top, and $v \in \Gamma^*$ is the content of the second pushdown store with the last letter of v at the bottom and the first letter of v at the top. For an input string $w \in \Sigma^*$, the corresponding initial configuration is $\bot q_0 w \bot$. The (D)TPDA M induces a computation relation $\vdash_M^*$ on the set of configurations, which is the reflexive transitive closure of the single-step computation relation $\vdash_M$. The (D)TPDA M accepts with empty pushdown stores, that is, $L_C(M) = \{ w \in \Gamma^* \mid \bot q_0 w \bot \vdash_M^* q \text{ for some } q \in F \}$ is the *characteristic language accepted by* M, and $L(M) = L_C(M) \cap \Sigma^*$ is its *input language.*

Definition 4.1.

(a) A (D)TPDA is called *shrinking*, if there exists a weight function $\varphi : Q \cup \Gamma \rightarrow \mathbb{N}_+$ such that, for all $q \in Q$ and all $u, v \in \Gamma^{\leq k}$, $(p, u', v') \in \delta(q, u, v)$ implies that $\varphi(u'pv') < \varphi(uqv)$. Here φ is extended to a morphism from $(Q \cup \Gamma)^*$ into $\mathbb{N}$ by taking $\varphi(\lambda) = 0$ and $\varphi(wa) = \varphi(w) + \varphi(a)$ for all words $w \in (Q \cup \Gamma)^*$ and all letters $a \in Q \cup \Gamma$.
(b) A (D)TPDA is called *length-reducing*, if, for all $q \in Q$ and all $u, v \in \Gamma^{\leq k}$, $(p, u', v') \in \delta(q, u, v)$ implies that $|u'v'| < |uv|$ holds.

Obviously, the length-reducing TPDA is a special case of the shrinking TPDA. Observe that the input is provided to a TPDA as the initial contents of its second pushdown store, and that in order to accept a TPDA is required to empty its pushdown stores. Thus, it is forced to consume its input completely.

From the definition of the transition relation δ we see that M halts immediately whenever one of its pushdown stores is emptied. Because of the above property this happens if and only if a transition of the form $(q, u, v\bot) \mapsto (q', u', \lambda)$ or $(q, \bot u, v) \mapsto (q', \lambda, v')$ is performed. Thus, we can assume without loss of generality that M has a single halting state q_f, and that all the halting and accepting configurations of M are of the form q_f.

The following results have been obtained on the descriptive power of shrinking and length-reducing (D)TPDAs.[17,18]

Proposition 4.1.

(a) *A language is growing context-sensitive, if and only if if it is the input language of a length-reducing TPDA, if and only if it is the input language of a shrinking TPDA.*
(b) *A language is Church-Rosser, if and only if if it is the input language of a length-reducing deterministic TPDA, if and only if it is the input language of a shrinking deterministic TPDA.*

What can we say about the proper languages of shrinking/length-reducing DTPDAs? As mentioned before the class CRL of Church-Rosser languages is a basis for the class RE of recursively enumerable languages. Hence, if $L \subseteq \Sigma^*$ is a recursively enumerable language, then there exists a

Church-Rosser language B on some alphabet Δ strictly containing Σ such that $\mathsf{Pr}^{\Sigma}(B) = L$. Thus, there exists a shrinking/length-reducing DTPDA M' with input alphabet Δ and tape alphabet Γ such that $L(M') = B$. As CRL is closed under intersection with regular languages,[18] we can even assume that $L_{\mathrm{C}}(M') \subseteq \Delta^*$ holds, implying that $L_{\mathrm{C}}(M') = B$. If we now interpret M' as a shrinking/length-reducing DTPDA with input alphabet Σ, then we obtain that $L_{\mathrm{P}}(M') = \mathsf{Pr}^{\Sigma}(B) = L$, that is, we have the following result. Here we use the prefixes sh- and lr- to denote shrinking and length-reducing TPDAs, respectively.

Proposition 4.2. $\mathcal{L}_{\mathrm{P}}(\textsf{sh-DTPDA}) = \mathcal{L}_{\mathrm{P}}(\textsf{lr-DTPDA}) = \textsf{RE}$.

Thus, we see that the class of proper languages of shrinking/length-reducing DTPDAs is already universal. Accordingly, we need to restrict the use of auxiliary symbols for these DTPDAs. Therefore, we consider proper languages of shrinking/length-reducing DTPDAs that are lexicalized.

Proposition 4.3. *If M is a shrinking/length-reducing DTPDA that is lexicalized, then the proper language $L_{\mathrm{P}}(M)$ is growing context-sensitive.*

Proof. Let M be a shrinking/length-reducing DTPDA with input alphabet Σ and tape alphabet Γ, and assume that M is lexicalized with constant $j \in \mathbb{N}$. Then no word $w \in L_{\mathrm{C}}(M)$ contains any factor from $(\Gamma \setminus \Sigma)^*$ of length exceeding j. Thus, the morphism $\mathsf{Pr}^{\Sigma} : \Gamma^* \to \Sigma^*$ has *j-limited erasing*[16] on $L_{\mathrm{C}}(M)$. As $L_{\mathrm{C}}(M)$ is a Church-Rosser language, it belongs to the class GCSL of growing context-sensitive languages.[17] This in turn implies that $L_{\mathrm{P}}(M) = \mathsf{Pr}^{\Sigma}(L_{\mathrm{C}}(M))$ is also growing context-sensitive, as this class is closed under limited erasing.[19] □

Actually each growing context-sensitive language is the proper language of a shrinking/length-reducing DTPDA that is lexicalized. As remarked by Buntrock,[19] each growing context-sensitive language $L \subseteq \Sigma^*$ is the image of a Church-Rosser language $L_1 \subseteq \Sigma_1^*$ under some λ-free morphism $\psi : \Sigma_1^* \to \Sigma^*$. Without loss of generality we can assume that Σ_1 and Σ are disjoint. Define $\Gamma = \Sigma_1 \cup \Sigma$, and define a new morphism $\Psi : \Sigma_1^* \to \Gamma^*$ by mapping $s \mapsto \psi(s)s$ for all $s \in \Sigma_1$. It is easily seen that with L_1, also $\Psi(L_1)$ is a Church-Rosser language. We can now construct a shrinking/length-reducing DTPDA M_1 with input alphabet Σ and tape alphabet Γ' containing Γ such that $L_{\mathrm{C}}(M_1) = \Psi(L_1)$ and $L_{\mathrm{P}}(M_1) = \mathsf{Pr}^{\Sigma}(\Psi(L_1)) = \psi(L_1) = L$. Thus, we have the following characterization.

Corollary 4.1. $\mathcal{L}_{\mathrm{P}}(\textsf{lex-sh-DTPDA}) = \mathcal{L}_{\mathrm{P}}(\textsf{lex-lr-DTPDA}) = \textsf{GCSL}$.

5. Transformations Computed by Deterministic Types of Automata

The real goal of performing analysis by reduction on (the enriched form of) an input sentence is not simply to accept or reject this sentence, but to extract information from that sentence and to translate it into another form (be it in another natural language or a formal representation). Therefore, we want to interpret various types of automata as "transducers," that is, we study (binary) relations that are computed by various types of automata. The general setting will be as follows.

Let Γ be a finite alphabet, and let Σ_1 and Σ_2 be two disjoint subalphabets of Γ, where Σ_1 is interpreted as an *input alphabet*, and Σ_2 is seen as an *output alphabet*. With an automaton M on Γ one can now associate the relation $\mathrm{Rel}(M) \subseteq \Sigma_1^* \times \Sigma_2^*$ that is defined as follows:

$$\mathrm{Rel}(M) = \{\, (u,v) \mid \exists w \in L_{\mathrm{C}}(M) : \mathsf{Pr}^{\Sigma_1}(w) = u \text{ and } \mathsf{Pr}^{\Sigma_2}(w) = v \,\},$$

where Pr^{Σ_i} denotes the projection from Γ^* onto Σ_i^*, $i = 1, 2$. Thus, a pair $(u, v) \in \Sigma_1^* \times \Sigma_2^*$ belongs to the relation $\mathrm{Rel}(M)$ if and only if there exists a word x in the shuffle of u and v such that an extended version w of x belongs to the characteristic language $L_{\mathrm{C}}(M)$. We say that M *recognizes* (or *computes*) the relation $\mathrm{Rel}(M)$. For $u \in \Sigma_1^*$, the *image* of u under $\mathrm{Rel}(M)$ is defined by

$$\mathrm{Rel}(M)(u) = \{\, v \in \Sigma_2^* \mid (u,v) \in \mathrm{Rel}(M) \,\},$$

and for $v \in \Sigma_2^*$, the *preimage* of v with respect to $\mathrm{Rel}(M)$ is defined by

$$\mathrm{Rel}^{-1}(M)(v) = \{\, u \in \Sigma_1^* \mid (u,v) \in \mathrm{Rel}(M) \,\}.$$

A relation $R \subseteq \Sigma_1^* \times \Sigma_2^*$ is a *transduction of type* X if there exists an automaton M of type X such that $\mathrm{Rel}(M) = R$ holds. By $\mathcal{R}el(\mathsf{X})$ we denote the class of all transductions of type X.

Now, for any type X of automaton, the following questions are of interest:

- What are typical examples of transductions of type X?
- What closure and non-closure properties does the class of transductions of type X have?
- What are the algorithmical and complexity-theoretical properties of the class of transductions of type X?
- Is there a characterization of the class of transductions of type X in terms of more classical language families or automata?

In analogy to the situation for proper languages we are in particular interested in relations that are computed by automata which are lexicalized.

We first study binary relations of deterministic finite-state acceptors. To this end we need the following notions and definitions.

A *rational transducer* is defined as $T = (Q, \Sigma, \Delta, q_0, F, E)$, where Q is a finite set of internal states, Σ is a finite input alphabet, Δ is a finite output alphabet, $q_0 \in Q$ is the initial state, $F \subseteq Q$ is the set of final states, and $E \subset Q \times \Sigma^* \times \Delta^* \times Q$ is a finite set of transitions.

If $e = (p_1, u_1, v_1, q_1)(p_2, u_2, v_2, q_2) \cdots (p_n, u_n, v_n, q_n) \in E^*$ is a sequence of transitions, then its *label* is the pair $\ell(e) = (u_1 u_2 \cdots u_n, v_1 v_2 \cdots v_n) \in \Sigma^* \times \Delta^*$. By $\ell_{\text{in}}(e)$ we denote the first component $u_1 u_2 \cdots u_n \in \Sigma^*$, and by $\ell_{\text{out}}(e)$ we denote the second component $v_1 v_2 \cdots v_n \in \Delta^*$. The sequence e above is called a *path* from p_1 to q_n, if $p_{i+1} = q_i$ for all $i = 1, \ldots, n-1$. It is called *successful* if p_1 is the initial state q_0, and if q_n is a final state. By $\Lambda(p, q)$ we denote the set of all paths from $p \in Q$ to $q \in Q$, and we define $\Lambda(p, Q') = \bigcup_{q \in Q'} \Lambda(p, q)$ for all subsets $Q' \subseteq Q$. Finally, $T(p, q) = \{\, \ell(e) \mid e \in \Lambda(p, q) \,\}$ and $T(p, Q') = \{\, \ell(e) \mid e \in \Lambda(p, Q') \,\}$. Thus, $\Lambda(q_0, F)$ is the set of all successful paths, and $T(q_0, F)$ is the set of labels of all successful paths. Then $\text{Rel}(T) = T(q_0, F)$ is called the *relation* defined by T. For $u \in \Sigma^*$ and $v \in \Delta^*$, $T(u) = \{\, v \in \Delta^* \mid (u, v) \in T(q_0, F) \,\}$, and $T^{-1}(v) = \{\, u \in \Sigma^* \mid (u, v) \in T(q_0, F) \,\}$. Obviously, the *domain* of $\text{Rel}(T)$ is the language $L(T) = \{\, u \in \Sigma^* \mid T(u) \neq \emptyset \,\}$, which is the set of all input words for which T has an accepting computation.

The relations defined by rational transducers are just the so-called *rational relations*,[20] that is, the rational subsets of the monoid $\Sigma^* \times \Delta^*$. We denote the class of rational relations over $\Sigma^* \times \Delta^*$ by $\mathsf{Rat}(\Sigma, \Delta)$. Actually we will make use of the following characterization of rational relations.[20]

Theorem 5.1. *A relation $R \subseteq \Sigma^* \times \Delta^*$ is rational if and only if there exist a finite alphabet Ω, a regular language K over Ω, and two morphisms $f : \Omega^* \to \Sigma^*$ and $g : \Omega^* \to \Delta^*$ such that $R = \{\, (f(w), g(w)) \mid w \in K \,\}$.*

Based on this characterization we can easily derive the following result.

Proposition 5.1.

$$\mathcal{R}el(\mathsf{DFA}) = \bigcup_{\substack{\Sigma, \Delta \\ \Sigma \cap \Delta = \emptyset}} \mathsf{Rat}(\Sigma, \Delta).$$

Proof. Let $R \subseteq \Sigma^* \times \Delta^*$ be a rational relation. Then by Theorem 5.1 there exist a finite alphabet Ω, a regular language K over Ω, and two morphisms $f : \Omega^* \to \Sigma^*$ and $g : \Omega^* \to \Delta^*$ such that $R = \{ (f(w), g(w)) \mid w \in K \}$. Assume that the three alphabets Σ, Δ, and Ω are pairwise disjoint. As K is a regular language over Ω, there exists a complete DFA $A = (Q, \Omega, q_0, F, \delta)$ for this language. From A we now design a DFA B with tape alphabet $\Theta = \Sigma \cup \Delta \cup \Omega$ as follows: For all $p, q \in Q$ and all $a \in \Omega$, if $\delta(p, a) = q$, then B has a sequence of transition steps that takes B from p to q, while reading the word $\psi(a) := af(a)g(a)$. It is easily seen that B can be constructed in such a way that $L_C(B) = \{ \psi(w) \mid w \in K \}$. Hence,

$$\begin{aligned} \mathrm{Rel}(B) &= \{ (\mathsf{Pr}^{\Sigma}(\psi(w)), \mathsf{Pr}^{\Delta}(\psi(w))) \mid w \in K \} \\ &= \{ (f(w), g(w)) \mid w \in K \} \qquad = R. \end{aligned}$$

Conversely, let $A = (Q, \Sigma_1, \Sigma_2, \Gamma, q_0, F, \delta)$ be a DFA with tape alphabet Γ, input alphabet $\Sigma_1 \subset \Gamma$, and output alphabet $\Sigma_2 \subset \Gamma$, where Σ_1 and Σ_2 are disjoint. Then $\mathrm{Rel}(A) = \{ (\mathsf{Pr}^{\Sigma_1}(w), \mathsf{Pr}^{\Sigma_2}(w)) \mid w \in L_C(A) \}$. As $L_C(A)$ is a regular language over Γ, and as $\mathsf{Pr}^{\Sigma_1} : \Gamma^* \to \Sigma_1^*$ and $\mathsf{Pr}^{\Sigma_2} : \Gamma^* \to \Sigma_2^*$ are morphisms, we see from Theorem 5.1 that the relation $\mathrm{Rel}(A)$ is rational. This completes the proof of Proposition 5.1. □

Next we turn to the so-called *pushdown relations*. A *pushdown transducer* (PDT for short) is defined as $T = (Q, \Sigma, \Delta, X, q_0, Z_0, F, E)$, where Q is a finite set of internal states, Σ is a finite input alphabet, Δ is a finite output alphabet, X is a finite pushdown alphabet, $q_0 \in Q$ is the initial state, $Z_0 \in X$ is the initial symbol on the pushdown, $F \subseteq Q$ is the set of final states, and $E \subset Q \times (\Sigma \cup \{\lambda\}) \times X \times Q \times X^* \times \Delta^*$ is a finite set of transitions.[21] A *configuration* of T is written as (q, u, α, v), where $q \in Q$ is a state, $u \in \Sigma^*$ is the still unread part of the input, $\alpha \in X^*$ is the contents of the pushdown store with the first letter of α at the bottom and the last letter at the top, and $v \in \Delta^*$ is the output produced so far. If $(q, au, \alpha x, v)$ is a configuration, where $a \in \Sigma \cup \{\lambda\}$ and $x \in X$, and $(q, a, x, p, y, z) \in E$, then T can perform the transition step $(q, au, \alpha x, v) \vdash (p, u, \alpha y, vz)$, that is, a is read from the input, the topmost symbol x on the pushdown is replaced by the string y, the word z is appended to the output, and the internal state changes from q to p. The relation $\mathrm{Rel}(T)$ computed by T is defined as

$$\begin{aligned} \mathrm{Rel}(T) = \{ (u, v) \in \Sigma^* \times \Delta^* \mid\ & (q_0, u, Z_0, \lambda) \vdash^* (q, \lambda, \alpha, v) \\ & \text{for some } q \in F \text{ and } \alpha \in X^* \}. \end{aligned}$$

A relation $R \subseteq \Sigma^* \times \Delta^*$ is called a *pushdown relation* if $R = \mathrm{Rel}(T)$ holds for some PDT T. By $\mathsf{PDR}(\Sigma, \Delta)$ we denote the class of all pushdown relations over $\Sigma^* \times \Delta^*$.

Proposition 5.2.

$$\mathcal{R}el(\mathsf{DPDA}) = \bigcup_{\substack{\Sigma,\Delta \\ \Sigma\cap\Delta=\emptyset}} \mathsf{PDR}(\Sigma, \Delta).$$

Proof. Let $T = (Q, \Sigma, \Delta, X, q_0, Z_0, F, E)$ be a PDT such that $\Sigma \cap \Delta = \emptyset$. We present a DPDA $M = (Q', \Sigma, \Delta, \Gamma, X, q_0', \bot, F', \delta')$ such that

$$\mathrm{Rel}(M) = \{ (\mathsf{Pr}^{\Sigma}(w), \mathsf{Pr}^{\Delta}(w)) \mid w \in L_{\mathrm{C}}(M) \} = \mathrm{Rel}(T)$$

holds. With each possible transition $e \in E$ of T, we associate a new auxiliary letter γ_e, that is, we take $\Gamma = \Sigma \cup \Delta \cup \{ \gamma_e \mid e \in E \}$. As usual we assume without loss of generality that the various subalphabets of Γ are pairwise disjoint. For $e = (p, a, x, q, y, \beta)$, where $p, q \in Q$, $a \in \Sigma \cup \{\lambda\}$, $x \in X$, $y \in X^*$, and $\beta \in \Delta^*$, let $\psi(e)$ denote the word $\psi(e) := \gamma_e a \beta$. Now M is designed in such a way that it expects an input of the form $\psi(w)$ for some $w \in E^+$ such that the sequence of transition steps described by w forms an accepting computation of T. Obviously, during this computation T consumes the input $\iota(w)$ and produces the output $\omega(w)$, where $\iota : E^* \to \Sigma^*$ and $\omega : E^* \to \Delta^*$ are the morphisms induced by mapping $e = (p, a, x, q, y, \beta)$ onto a and onto β, respectively. Then it is easily verified that $L_{\mathrm{C}}(M) = \{ \psi(w) \mid w$ describes an accepting computation of $T \}$, and accordingly, $\mathrm{Rel}(M) = \mathrm{Rel}(T)$ follows.

Conversely, let $M = (Q, \Sigma, \Delta, \Gamma, X, q_0, \bot, F, \delta)$ be a DPDA that computes a relation $\mathrm{Rel}(M) = \{ (\mathsf{Pr}^{\Sigma}(w), \mathsf{Pr}^{\Delta}(w)) \mid w \in L_{\mathrm{C}}(M) \}$. We need to describe a PDT $T = (Q, \Sigma, \Delta, X, q_0, Z_0, F, E)$ that realizes this very binary relation. We define the transition relation E of T as follows, where $p, q \in Q$, $x \in X$, and $y \in X^*$:

$$\begin{aligned}
(p, a, x, q, y, \lambda) \in E \text{ if } \delta(p, a, x) = (q, y) \quad & \text{for all } a \in \Sigma \cup \{\lambda\}, \\
(p, \lambda, x, q, y, c) \in E \text{ if } \delta(p, c, x) = (q, y) \quad & \text{for all } c \in \Delta, \\
(p, \lambda, x, q, y, \lambda) \in E \text{ if } \delta(p, b, x) = (q, y) \quad & \text{for all } b \in \Gamma \smallsetminus (\Sigma \cup \Delta).
\end{aligned}$$

Thus, given a word $u \in \Sigma^*$ as input, T nondeterministically guesses an expanded version $w \in \Gamma^*$ of u and tries to simulate an accepting computation of M on input w. If T is successful, then $w \in L_{\mathrm{C}}(M)$ holds, and we see that while processing w, T produces the output $\mathsf{Pr}^{\Delta}(w)$. Hence, $\mathrm{Rel}(T) \subseteq \mathrm{Rel}(M)$ holds. Conversely, if $(u, v) \in \mathrm{Rel}(M)$, then there exists an expanded version $w \in \Gamma^*$ of the shuffle of u and v such that $w \in L_{\mathrm{C}}(M)$

holds. Obviously, T has an accepting computation that simulates the accepting computation of M on input w. It follows that $\text{Rel}(M) = \text{Rel}(T)$ holds. □

Finally we turn to relations that are computed by deterministic two-pushdown automata.

Proposition 5.3. *Let Σ and Δ be two finite disjoint alphabets. Then a binary relation $R \subseteq \Sigma^* \times \Delta^*$ can be computed by a shrinking/length-reducing DTPDA if and only if it is recursively enumerable.*

Proof. If $R \subseteq \Sigma^* \times \Delta^*$ is recursively enumerable, then also the language $L_R = \{ uv \mid (u, v) \in R \}$ is recursively enumerable. Hence, by Proposition 4.2 there exists a shrinking/length-reducing DTPDA M such that $L_P(M) = L_R$. It follows that $\text{Rel}(M) = \{ (\mathsf{Pr}^\Sigma(w), \mathsf{Pr}^\Delta(w)) \mid w \in L_C(M) \} = R$ holds.

Conversely, if $R = \text{Rel}(M)$ for a shrinking/length-reducing DTPDA M, then obviously the relation R is recursively enumerable. □

As the proof of Proposition 5.3 uses Proposition 4.2 in an essential way, it follows that words from $L_C(M)$ may contain a very large number of occurrences of auxiliary symbols. A DTPDA $M = (Q, \Sigma, \Delta, \Gamma, k, \delta, q_0, \perp, F)$ is called *lexicalized* if there exists a constant $j \in \mathbb{N}_+$ such that, for all $w \in L_C(M)$, $|w| \leq j \cdot (|w|_\Sigma + |w|_\Delta)$ holds, that is, the number of occurrences of auxiliary symbols in $w \in L_C(M)$ is bounded from above by a fixed linear multiple of the combined number of input and output symbols in w.

Proposition 5.4. *Let Σ and Δ be two disjoint finite alphabets. Then a binary relation $R \subseteq \Sigma^* \times \Delta^*$ can be computed by a shrinking/length-reducing DTPDA that is lexicalized, if the language $L_R = \{ uv \mid (u, v) \in R \}$ is growing context-sensitive.*

Proof. Assume that $R \subseteq \Sigma^* \times \Delta^*$ is a binary relation for which the language L_R is growing context-sensitive. Then it follows from Corollary 4.1 that $L_R = L_P(M)$ for a lexicalized shrinking/length-reducing DTPDA. But then $R = \text{Rel}(M)$ follows immediately. □

However, the converse of Proposition 5.4 does not hold in general. Let $\Sigma = \{a, b\}$, let $\overline{\Sigma} = \{\bar{a}, \bar{b}\}$ such that $\Sigma \cap \overline{\Sigma} = \emptyset$, and let $^- : \Sigma^* \to \overline{\Sigma}^*$ be the morphism induced by $a \mapsto \bar{a}$ and $b \mapsto \bar{b}$. The *marked copy language* $L_{\text{mc}} = \{ w\overline{w} \mid w \in \Sigma^* \}$ is not growing context-sensitive,[19,22] and so by

Corollary 4.1, it is not the proper language of any lexicalized shrinking DTPDA. On the other hand, if $h : \Sigma^* \to (\Sigma \cup \overline{\Sigma})^*$ denotes the morphism induced by $a \mapsto a\bar{a}$ and $b \mapsto b\bar{b}$, then the language $L_{\text{md}} = \{ h(w) \mid w \in \Sigma^* \}$ is regular, that is, it is in particular the proper language of a shrinking DTPDA M. If we now interpret Σ as input alphabet and $\overline{\Sigma}$ as output alphabet of M, then the relation $\text{Rel}(M)$ coincides with the relation $R = \{ (w, \overline{w}) \mid w \in \Sigma^* \}$, which implies that the language L_R coincides with the language L_{mc}. Thus, although the relation R is computed by a shrinking DTPDA that is lexicalized, the language L_R is not growing context-sensitive.

6. Conclusion

We have investigated the classes of proper languages of various types of (lexicalized) automata. First we have considered classes of proper languages of (strongly) lexicalized restarting automata with multiple rewrites. We have seen the influence of two parameters on the expressive power of these automata: the number of rewrites per cycle, and the number of auxiliary symbols that may appear on the tape at the same time. Then for finite-state acceptors we have seen that we just obtain another characterization for the class of regular languages. More interestingly, the proper languages of (lexicalized) deterministic pushdown automata yield another description of the class of context-free languages, and for shrinking/length-reducing deterministic two-pushdown automata, the class of proper languages is universal, while in the case of lexicalized automata of this type, we obtain another characterization of the class of growing context-sensitive languages.

Finally we have considered transductions, that is, binary relations, that are computed by deterministic finite-state acceptors, deterministic pushdown automata, and deterministic two-pushdown automata that are shrinking/length-reducing. As it turned out the former are just the class of rational relations and the class of pushdown relations, respectively. Which binary relations are computed by shrinking/length-reducing deterministic two-pushdown automata that are lexicalized? What can we say about the classes of transductions that are computed by the various types of deterministic restarting automata?

References

1. V. Kuboň and M. Plátek. A grammar based approach to a grammar checking of free word order languages. In: *COLING'94, Proc., Vol. II*, Kyoto, Japan, 1994, 906–910.

2. M. Lopatková, M. Plátek, and V. Kuboň. Modeling syntax of free word-order languages: Dependency analysis by reduction. In: V. Matoušek, P. Mautner, and T. Pavelka (eds.), *TSD 2005, Proc.*, *Lecture Notes in Computer Science* 3658, Springer, Berlin, 2005, 140–147.
3. P. Sgall, E. Hajičová, and J. Panevová. *The Meaning of the Sentence in Its Semantic and Pragmatic Aspects.* Reidel Publishing Company, Dordrecht, 1986.
4. M. Lopatková, M. Plátek, and P. Sgall. Towards a formal model for functional generative description: Analysis by reduction and restarting automata. *The Prague Bulletin of Mathematical Linguistics* 87 (2007) 7–26.
5. P. Květoň. Rule-based morphological desambiguation. *The Prague Bulletin of Mathematical Linguistics* 85 (2006) 57–71.
6. F. Mráz, F. Otto, and M. Plátek. The degree of word-expansion of lexicalized RRWW-automata – A new measure for the degree of nondeterminism of (context-free) languages. *Theoretical Computer Science* 410 (2009) 3530–3538.
7. F. Otto and M. Plátek. A two-dimensional taxonomy of proper languages of lexicalized FRR-automata. In: C. Martin-Vide, F. Otto, and H. Fernau (eds.), *LATA 2008, Proc.*, *Lecture Notes in Computer Science* 5196, Springer, Berlin, 2008, 409–420.
8. M. Plátek, F. Otto, and F. Mráz. Two-dimensional hierarchies of proper languages of lexicalized FRR-automata. *Information and Computation* 207 (2009) 1300–1314.
9. P. Jančar, F. Mráz, M. Plátek, and J. Vogel. On monotonic automata with a restart operation. *Journal of Automata, Languages and Combinatorics* 4 (1999) 283-292.
10. F. Mráz, F. Otto, and M. Plátek. Free word order and restarting automata. In: R. Loos, S.Z. Fazekas, and C. Martín-Vide (eds.), *LATA 2007, Preproc.*, Report 35/07, Research Group on Math. Linguistics, Universitat Rovira i Virguli, Tarragona, 2007, 425–436.
11. T. Jurdziński and F. Otto. Shrinking restarting automata. *International Journal of Foundations of Computer Science* 18 (2007) 361–385. An extended abstract appeared in: J. Jędrzejowicz and A. Szepietowski (eds.), *MFCS 2005, Proc.*, *Lecture Notes in Computer Science* 3618, Springer, Berlin, 2005, 532–543.
12. F. Otto. Restarting automata and their relations to the Chomsky hierarchy. In: Z. Ésik and Z. Fülöp (eds.), *Developments in Language Theory, DLT'2003, Proc.*, *Lecture Notes in Computer Science* 2710, Springer, Berlin, 2003, 55–74.
13. F. Otto, M. Katsura, and Y. Kobayashi. Infinite convergent string-rewriting systems and cross-sections for finitely presented monoids. *Journal of Symbolic Computation* 26 (1998) 621–648.
14. F. Otto. Restarting automata. In: Z. Ésik, C. Martin-Vide, and V. Mitrana (eds.), *Recent Advances in Formal Languages and Applications*, Studies in Computational Intelligence, Vol. 25, Springer, Berlin, 2006, 269–303.
15. F. Mráz, M. Plátek, and F. Otto. A measure for the degree of nondeterminism

of context-free languages. In: J. Holub and J. Žd'árek (eds.), *CIAA 2007, Proc., Lecture Notes in Computer Science* 4783, Springer, Berlin, 2007, 192–202.

16. J. Hopcroft and J. Ullman. *Introduction to Automata Theory, Languages, and Computation.* Addison-Wesley, Reading, MA, 1979.
17. G. Buntrock and F. Otto. Growing context-sensitive languages and Church-Rosser languages. *Information and Computation* 141 (1998) 1–36.
18. G. Niemann, F. Otto. The Church-Rosser languages are the deterministic variants of the growing context-sensitive languages. *Information and Computation* 197 (2005) 1–21.
19. G. Buntrock. *Wachsende kontextsensitive Sprachen.* Habilitationsschrift, Universität Würzburg, 1996.
20. J. Berstel. *Transductions and Context-free Languages.* Teubner Studienbücher, Teubner, Stuttgart, 1979.
21. K. Culik and C. Choffrut. Classes of Transducers and Their Properties. *Research Report CS-81-04*, Department of Computer Science, University of Waterloo, 1981.
22. C. Lautemann. One pushdown and a small tape. In: K.W. Wagner (ed.), *Dirk Siefkes zum 50. Geburtstag*, Technische Universität Berlin and Universität Augsburg, 1988, 42–47.

Received: June 21, 2009

Revised: April 28, 2010

INITIAL LITERAL SHUFFLES OF UNIFORM CODES

GENJIRO TANAKA and YOSHIYUKI KUNIMOCHI

Dept. of Computer Science, Shizuoka Institute of Science and Technology, Fukuroi-shi, 437-8555 Japan

E-mail: {tanaka,kunimoti}@cs.sist.ac.jp

It is known that the family of prefix codes is closed under the initial literal shuffle operation. However, it has not been known whether or not other families of codes are closed under this operation. In this paper we investigate the initial literal shuffles of subsets of full uniform codes. We shall show that the initial literal shuffle operation preserves various properties of uniform codes under certain conditions. From this fact, we can construct more complicated uniform codes from simpler uniform codes by using the initial literal shuffle operation.

Keywords: Initial literal shuffle; Code; Uniform code; Circular code; Limited code; Extractable submonoid.

Mathematics Subject Classification: 68Q70

1. Introduction

Let A be an alphabet, A^* the free monoid over A, and 1 the empty word. Let $A^+ = A^* - \{1\}$. A word $v \in A^*$ is a *right factor* of a word $u \in A^*$ if there is a word $w \in A^*$ such that $u = wv$. For a word $w \in A^*$ and a letter $x \in A$ we let $|w|_x$ denote the number of x in w. The length $|w|$ of w is the number of letters in w. Therefore, $A^n = \{w \in A^* |\ |w| = n\}$, $n \geq 1$. Two words x, y are said to be *conjugate* if there exist words u, v such that $x = uv$, $y = vu$. The conjugacy relation is an equivalence relation. By $Cl(x)$ we denote the class of x of this equivalence relation. We define the permutation on A^* by

$$\Gamma(1) = 1 \quad \text{and} \quad \Gamma(av) = va \quad \text{for all } a \in A, v \in A^*.$$

Then words x and y are conjugate if and only if $\Gamma^n(x) = y$ for some $n \geq 0$. A word w is called a *primitive* word if w is not a power of another word.

A nonempty subset C of A^+ is said to be a *code* if for $x_1, ..., x_p, y_1, ..., y_q \in C$, $p, q \geq 1$,

$$x_1 \cdots x_p = y_1 \cdots y_q \implies p = q,\ x_1 = y_1,\ \ldots,\ x_p = y_p.$$

A subset M of A^* is a *submonoid* of A^* if $M^2 \subseteq M$ and $1 \in M$. Every submonoid M of a free monoid has a unique minimal set of generators $C = (M - \{1\}) - (M - \{1\})^2$. C is called the *base* of M.

A submonoid M is *right unitary* in A^* if for all $u, v \in A^*$,

$$u,\ uv \in M \Longrightarrow v \in M.$$

M is *left unitary* in A^* if it satisfies the dual condition. A submonoid M is *biunitary* if it is both left and right unitary. Let M be a submonoid of a free monoid A^*, and C its base. If $CA^+ \cap C = \emptyset$ (resp. $A^+C \cap C = \emptyset$), then C is called a *prefix* (resp. *suffix*) code over A. C is called a *bifix* code if it is a prefix and suffix code. A submonoid M of A^* is right unitary (resp. biunitary) if and only if its minimal set of generators is a prefix code (resp. bifix code) ([1, p.46],[5, p.108]).
Let C be a nonempty subset of A^*. If $|x| = |y|$ for all $x, y \in C$, then C is a bifix code. We call such a code a *uniform code.* The uniform code A^n, $n \geq 1$, is called a *full uniform code.*

Definition 1. Let $x, y{\in}A^*$. Then the *initial literal shuffle* $x \circ y$ of x and y is defined as follows:
(1) If either $x = 1$ or $y = 1$, then $x \circ y = xy$.
(2) Let $x = a_1a_2 \cdots a_m$ and let $y = b_1b_2 \cdots b_n$, a_i, $b_j \in A$. Then

$$x \circ y = \begin{cases} a_1b_1a_2b_2 \cdots a_nb_na_{n+1}a_{n+2} \cdots a_m & \text{if } m \geq n, \\ a_1b_1a_2b_2 \cdots a_mb_mb_{m+1}b_{m+2} \cdots b_n & \text{if } m < n. \end{cases}$$

Let C_1, $C_2 \subset A^*$. Then the *initial literal shuffle* $C_1 \circ C_2$ of C_1 and C_2 is defined as $C_1 \circ C_2 = \{x \circ y \,|\, x \in C_1,\ y \in C_2\}$.

Now we list the fundamental properties of the initial literal shuffle.

Let w, u, v, u' $v' \in A^+$.
(1) If $|u| = |v|$ and $|u'| = |v'|$, then $(u \circ v)(u' \circ v') = uu' \circ vv'$.
(2) If $w = u \circ v$ and $|u| = |v|$, then $w^k = u^k \circ v^k$, $k \geq 1$.
(3) If $|u| = |v| + 1$ and $|u'| = |v'|$, then $(u \circ v)(u' \circ v') = uv' \circ vu'$.
(4) If $|u| = |v| + 1$ and $|u'| + 1 = |v'|$, then $(u \circ v)(u' \circ v') = uv' \circ vu'$.
(5) If $w = u{\circ}v$ and $|u| = |v|{+}1$, then $w^2 = uv{\circ}vu$ and $w^{2m} = (uv)^m{\circ}(vu)^m$.
Let C_1 and C_2 be subsets of A^n.
(6) If $c_i \in C_1$, $d_i \in C_2$, $1 \leq i \leq p$, then

$$(c_1 \circ d_1)(c_2 \circ d_2) \cdots (c_p \circ d_p) = (c_1c_2 \cdots c_p) \circ (d_1d_2 \cdots d_p).$$

(7) If $u \in C_1^*$, $v \in C_2^*$ and $|u| = |v|$, then $u \circ v \in (C_1 \circ C_2)^*$.
(8) If $|u| = |v|$ and $u \circ v \in (C_1 \circ C_2)^*$, then $u \in C_1^*$ and $v \in C_2^*$.

(9) Let x and y be elements in A^* such that $|x| = |y|$. Then, $x \in C_1^*$ and $y \in C_2^*$ if and only if $x \circ y \in (C_1 \circ C_2)^*$.
(10) $(C_1 \circ C_2)^* = \bigcup_{k=0}^{\infty}(C_1^k \circ C_2^k)$.

2. Initial Literal Shuffles of Uniform Codes

A submonoid M of A^* is said to be *pure* ([6]) if for all $x \in A^*$ and $n \geq 1$,

$$x^n \in M \Longrightarrow x \in M.$$

A submonoid M of A^* is *very pure* if for all $u, v \in A^*$,

$$uv,\ vu \in M \Rightarrow u,\ v \in M.$$

It is obvious that a very pure submonoid is a pure submonoid.

Definition 2. A subset C of A^* is called a *circular* code if for all $n, m \geq 1$ and $x_1, \cdots, x_n,\ y_1, \cdots, y_m \in C$, and $p \in A^*$ and $s \in A^+$, the following implication holds:

$$sx_2 \cdots x_n p = y_1 \cdots y_m, \text{ and } x_1 = ps \quad \Longrightarrow \quad n = m,\ p = 1, \text{ and } x_i = y_i\ (1 \leq i \leq n).$$

Any nonempty subset of a circular code is also a circular code. A submonoid of A^* is very pure if and only if its minimal set of generators is a circular code([1], p.323).

Definition 3. Let $p, q \geq 0$ be two integers. A code C is said to be $(p,\ q)$-*limited* if for any sequence $u_0, u_1, \cdots, u_{p+q}$ of words in A^*, the assumptions $u_{i-1}u_i \in C^*$ $(1 \leq i \leq p+q)$ imply $u_p \in C^*$.

C is called a limited code if it is a (p,q)-*limited* for some integers $p, q \geq 0$.

Let A be a finite set, and let $C \subset A^n$. If C is limited, then C is circular([1, p330]). Any subset D of a circular code C is also circular. Therefore the finite circular code D is limited([1,p333]). If D is (p,q)-limited, then for arbitrary sequence $u_0, u_1, \cdots, u_{p+q} \in A^*$ such that $u_iu_{i+1} \in D^*$ we have $u_p \in D$. Since D^* is biunitary and $u_p, u_{p-1}u_p, u_pu_{p+1} \in D^*$, we have $u_{p-1}, u_{p+1} \in D^*$. It follows that $u_i \in D^*$ for all $0 \leq i \leq p+q$. Therefore, D is (s,t)-limited for all s, t with $s + t = p + q$.

Proposition 1. Let n be an even number, and let $C_1,\ C_2 \subset A^n$. If both C_1^* and C_2^* are pure, then $(C_1 \circ C_2)^*$ is pure.

Proof. Suppose that $w^m \in (C_1 \circ C_2)^*$, $w \in A^+$.
Case 1. $|w|$ is odd, and $w = x_1y_1 \cdots x_py_px_{p+1}$, $x_i, y_i \in A$, $p \geq 0$.
We show that Case 1 cannot occur. That is, we prove that if $|w|$ is odd, then $w^m \notin (C_1 \circ C_2)^*$ for any $m \geq 1$. Put $x = x_1x_2 \cdots x_{p+1}$, $y = y_1y_2 \cdots y_p$, then $w = x \circ y$. Since $|w^m| = m|w|$ is even, $m = 2m_0$ for some integer $m_0 \geq 1$. From $w^2 = (x \circ y)(x \circ y) = (xy) \circ (yx)$ we have $w^m = (xy)^{m_0} \circ (yx)^{m_0} \in (C_1 \circ C_2)^*$. Thus $(xy)^{m_0} \in C_1^*$, and $(yx)^{m_0} \in C_2^*$. Since C_1^* is pure, we have $xy \in C_1^*$. Thus C_1 contains a word of odd length. This contradicts our assumption. Thus Case 1 does not occur.
Case 2. $|w|$ is even, and $w = x_1y_1 \cdots x_py_p$, $x_i, y_i \in A$.
Put $x = x_1x_2 \cdots x_p$, $y = y_1y_2 \cdots y_p$, then we have $w^m = x^m \circ y^m \in (C_1 \circ C_2)^*$. Thus $x \in C_1^*$, $y \in C_2^*$ and $|x| = |y|$. Consequently $w = x \circ y \in (C_1 \circ C_2)^*$. Therefore, $(C_1 \circ C_2)^*$ is pure. Q.E.D.

For the case $C_1 = C_2$ we have the following:

Corollary 1. Let n be an even number, and let $C \subset A^n$. If C^* is pure, then $(C \circ C)^*$ is pure.

Proposition 2. Let $C \subset A^n$. If $(C \circ C)^*$ is pure, then C^* is pure.

Proof. Suppose that $w^m \in C^*$. Then $w^m \circ w^m = (w \circ w)^m \in (C \circ C)^*$. Consequently $w \circ w \in (C \circ C)^*$. Therefore, $w \in C^*$. Q.E.D.

Example 1. Let $C_1 = \{ab, ba\}$, $C_2 = \{b^2\}$. Then C_1^* is pure, and C_2^* is not pure.
(1) $C_1 \circ C_1 = \{a^2b^2, ab^2a, ba^2b, b^2a^2\}$. $(C_1 \circ C_1)^*$ is a pure submonoid.
(2) $C_1 \circ C_2 = \{ab^3, b^2ab\}$. $(C_1 \circ C_2)^*$ is pure. Therefore, the converse of Proposition 1 does not hold.
(3) $\{b\}^*$ is pure, but $(\{b\} \circ \{b\})^* = \{b^2\}^*$ is not pure.

Proposition 3. Let C_1, C_2, and C be subsets of A^n $(n \geq 2)$ such that $(C_1 \cup C_2) \subset C$. If C is circular, then $C_1 \circ C_2$ is circular.

Proof. Note that both C_1 and C_2 are circular since they are subsets of the circular code C.
Let uv, $vu \in (C_1 \circ C_2)^*$, $u, v \in A^*$. If either $u = 1$ or $v = 1$, then u, $v \in (C_1 \circ C_2)^*$. Consequently we consider other cases.

Case 1. $uv \in (C_1 \circ C_2)^+ - (C_1 \circ C_2)$. **Case 2**. $uv \in C_1 \circ C_2$.

Case 1. Let $uv = \alpha_1\alpha_2 \cdots \alpha_k \cdots \alpha_m$, $\alpha_i = (c_i \circ d_i) \in C_1 \circ C_2$, $1 \le i \le m$, and $\alpha_k = (x_1x_2 \cdots x_n) \circ (y_1y_2 \cdots y_n)$, $x_j, y_j \in A$, $1 \le j \le n$.

Case 1-i. If $u = \alpha_1 \cdots \alpha_k$ and $v = \alpha_{k+1} \cdots \alpha_m$, then $u, v \in (C_1 \circ C_2)^*$.

Case 1-ii. $u = \alpha_1 \cdots \alpha_{k-1}x_1y_1 \cdots x_py_p$, $v = x_{p+1}y_{p+1} \cdots x_ny_n\alpha_{k+1} \cdots \alpha_m$, $1 \le p < n$. In this case we have

$$\begin{aligned} vu &= x_{p+1}y_{p+1} \cdots x_ny_n\alpha_{k+1} \cdots \alpha_m\alpha_1 \cdots \alpha_{k-1}x_1y_1 \cdots x_py_p \\ &= (e_1 \circ f_1)(e_2 \circ f_2) \cdots (e_m \circ f_m) \end{aligned}$$

for some $e_k \in C_1$, $f_k \in C_2$, $1 \le k \le m$. It follows that $(x_{p+1} \cdots x_n c_{k+1} \cdots c_mc_1 \cdots c_{k-1}x_1 \cdots x_p) \circ (y_{p+1} \cdots y_nd_{k+1} \cdots d_md_1 \cdots d_{k-1}y_1 \cdots y_p) = (e_1e_2 \cdots e_m) \circ (f_1f_2 \cdots f_m)$. Thus

$$x_{p+1} \cdots x_nc_{k+1} \cdots c_mc_1 \cdots c_{k-1}x_1 \cdots x_p = e_1e_2 \cdots e_m.$$

This contradicts the fact that C_1 is circular. Thus Case 1-ii cannot occur.

Case 1-iii. $u = \alpha_1 \cdots \alpha_{k-1}x_1y_1 \cdots x_py_px_{p+1}, v = y_{p+1} \cdots x_ny_n\alpha_{k+1} \cdots \alpha_m$, $0 \le p < n$. In this case we have

$$\begin{aligned} vu &= y_{p+1} \cdots x_ny_n\alpha_{k+1} \cdots \alpha_m\alpha_1 \cdots \alpha_{k-1}x_1y_1 \cdots y_px_{p+1} \\ &= (e_1 \circ f_1)(e_2 \circ f_2) \cdots (e_m \circ f_m) \end{aligned}$$

for some $(e_j \circ f_j) \in C_1 \circ C_2$, $1 \le j \le m$. It follows that $(y_{p+1} \cdots y_n d_{k+1} \cdots d_md_1 \cdots d_{k-1}y_1 \cdots y_p) \circ (x_{p+2} \cdots x_nc_{k+1} \cdots c_mc_1 \cdots c_{k-1}x_1 \cdots x_{p+1}) = (e_1 \cdots e_m) \circ (f_1 \cdots f_m)$. Thus

$$y_{p+1} \cdots y_nd_{k+1} \cdots d_md_1 \cdots d_{k-1}y_1 \cdots y_p = e_1 \cdots e_m, \tag{1}$$

$$x_{p+2} \cdots x_nc_{k+1} \cdots c_mc_1 \cdots c_{k-1}x_1 \cdots x_{p+1} = f_1 \cdots f_m. \tag{2}$$

Note that $c_i, d_i, e_i, f_i \in C$ for all i $(1 \le i \le m)$ and $n \ge 2$. If $p > 1$, then the equality (1) contradicts our assumption that C is circular. If $p = 0$, then the equality (2) also contradicts our assumption. Therefore, Case 1-iii cannot occur.

Case 2-i. $uv = x_1y_1 \cdots x_ny_n$, $x_1x_2 \cdots x_n \in C_1, y_1y_2 \cdots y_n \in C_2$, and $u = x_1y_1 \cdots x_py_p$, $p \ge 1$. In this case we have

$$\begin{aligned} vu &= x_{p+1}y_{p+1} \cdots x_ny_nx_1y_1 \cdots x_py_p \\ &= (x_{p+1} \cdots x_nx_1 \cdots x_p) \circ (y_{p+1} \cdots y_ny_1 \cdots y_p) \in C_1 \circ C_2. \end{aligned}$$

Thus $x_1x_2 \cdots x_n$, $x_{p+1} \cdots x_nx_1 \cdots x_p \in C_1$. That is, C_1 contains nontrivial conjugate elements. This contradicts the fact that C_1 is circular.

Case 2-ii. $uv = x_1y_1 \cdots x_ny_n$, $x_1x_2 \cdots x_n \in C_1$, $y_1y_2 \cdots y_n \in C_2$, and $u = x_1y_1 \cdots x_py_px_{p+1}$, $0 \le p < n$. In this case we have

$$\begin{aligned} vu &= y_{p+1} \cdots x_ny_nx_1y_1 \cdots x_py_px_{p+1} \\ &= (y_{p+1} \cdots y_ny_1 \cdots y_p) \circ (x_{p+2} \cdots x_nx_1 \cdots x_{p+1}) \in C_1 \circ C_2. \end{aligned}$$

Thus $x_1x_2\cdots x_n \in C_1$, $x_{p+1}\cdots x_nx_1\cdots x_p \in C_2$. That is, C contains two distinct conjugate words. This contradicts our assumption. Therefore, Case 2 can not occur. Consequently only Case 1-i can occur. Q.E.D.

Since $C = C \cup C$, we have the following:

Corollary 2. Let $C \subset A^n$, $n \geq 2$. If C is circular, then $C \circ C$ is circular.

Proposition 4. Let $C \subset A^n$. If $C \circ C$ is circular, then C is circular.

Proof. Suppose that there exist u, $v \in A^*$ such that uv, $vu \in C^*$ and $u \notin C^*$. For the two elements $u \circ u$, $v \circ v \in A^*$ we have

$$(u \circ u)(v \circ v) = uv \circ uv \in (C \circ C)^*, \quad (v \circ v)(u \circ u) = vu \circ vu \in (C \circ C)^*.$$

However $u \circ u \notin (C \circ C)^*$. Thus $(C \circ C)^*$ is not very pure, that is, $C \circ C$ is not circular. Q.E.D.

Example 2. (1) Let $C = \{ab, ba\}$. Then C^* is pure, but it is not very pure. $(C \circ C)^*$ is pure. However, by Proposition 4, $(C \circ C)^*$ is not very pure. Thus $C \circ C$ is not circular.
(2) Let $C_1 = \{aba\}$ and $C_2 = \{a^2b\}$. Then both C_1 and C_2 are circular. However, $C_1 \cup C_2$ is not circular. $C_1 \circ C_2 = \{(a^2b)^2\}$ and $a^2b \notin (C_1 \circ C_2)^*$. Thus $(C_1 \circ C_2)^*$ is not pure. Therefore, $C_1 \circ C_2$ is not circular.
(3) Let $A = \{a_i \,|\, i \geq 0\}$ (an infinite set). Let $C = \{\, a_ia_{i+1} \,|\, i \geq 0\} \subset A^2$. Then C is a circular code([1, p.330]). Thus $C \circ C = \{a_ia_ja_{i+1}a_{j+1} \,|\, i, j \geq 0\}$ is a circular code. Since $D = \{a_i^2a_{i+1}^2 | i \geq 0\}$ is a subset of a circular code $C \circ C$, D is a circular code.

Proposition 5. Let C_1, C_2, and C be nonempty subsets of A^n $(n \geq 2)$ such that $(C_1 \cup C_2) \subset C$. If C is (p, q)-limited, then $C_1 \circ C_2$ is (p, q)-limited.

Proof. Let $D = C_1 \circ C_2$. Assume that $u_0, u_1, \cdots, u_{p+q} \in A^*$ and $u_{k-1}u_k \in D^*$, $1 \leq k \leq p+q$.
Case 1. If $u_i \in D^*$ for some $i\,(0 \leq i \leq p+q)$, then $u_k \in D^*$ for all $k\,(0 \leq k \leq p+q)$ since D^* is biunitary.
Case 2. So we consider the other case that $u_i \notin D^*$ for all $i\,(0 \leq i \leq p+q)$. From $u_{i-1}u_i \in D^+$ we have

$$u_{i-1}u_i = (c_1 \circ d_1)(c_2 \circ d_2)\cdots(c_{t_i} \circ d_{t_i})$$

for some integer t_i and $(c_k \circ d_k) \in C_1 \circ C_2$, $c_k \in C_1$, $d_k \in C_2$ for $1 \le k \le t_i$. Then there exist some $m\,(1 \le m \le t_i)$ and $w_{i-1}, w'_{i-1} \in A^+$ such that $c_m \circ d_m = w_{i-1}w'_{i-1}$ and

$$u_{i-1} = (c_1 \circ d_1) \cdots (c_{m-1} \circ d_{m-1}) w_{i-1},\ u_i = w'_{i-1}(c_{m+1} \circ d_{m+1}) \cdots (c_{t_i} \circ d_{t_i}).$$

Let $c_m = x_{i-1,1}x_{i-1,2} \cdots x_{i-1,n}$, $d_m = y_{i-1,1}y_{i-1,2} \cdots y_{i-1,n}$, $x_{i-1,j}, y_{i-1,j} \in A, 1 \le j \le n$, then w_{i-1}(resp. w'_{i-1}) is a left factor (resp. a right factor) of $x_{i-1,1}y_{i-1,1}x_{i-1,2}y_{i-1,2} \cdots x_{i-1,n}y_{i-1,n}$. It follows that

$$\begin{aligned} u_{i-1}u_i &= (c_1 \cdots c_{m-1} \circ d_1 \cdots d_{m-1})(c_m \circ d_m)(c_{m+1} \cdots c_{t_i} \circ d_{m+1} \cdots d_{t_i}) \\ &= (\alpha_{i-1} \circ \beta_{i-1}) x_{i-1,1}y_{i-1,1}x_{i-1,2}y_{i-1,2} \cdots x_{i-1,n}y_{i-1,n}(\gamma_{i-1} \circ \delta_{i-1}), \end{aligned}$$

where $\alpha_{i-1} = c_1 \cdots c_{m-1}$, $\beta_{i-1} = d_1 \cdots d_{m-1}$, $c_m \circ d_m = x_{i-1,1}y_{i-1,1}x_{i-1,2}y_{i-1,2} \cdots x_{i-1,n}y_{i-1,n}$,
$\gamma_{i-1} = c_{m+1} \cdots c_{t_i}$, $\delta_{i-1} = d_{m+1} \cdots d_{t_i}$. We note that if $m = 1$ (resp. $m = t_i$), then $\alpha_{i-1} \circ \beta_{i-1} = 1$ (resp. $\gamma_{i-1} \circ \delta_{i-1} = 1$).

Case 2-i. $|u_0|$ is even.
$u_0 = (\alpha_0 \circ \beta_0)x_{01}y_{01} \cdots x_{0r}y_{0r}$, $u_1 = x_{0,r+1}y_{0,r+1} \cdots x_{0n}y_{0n}(\gamma_0 \circ \delta_0)$, $r \ge 1$. In this case, all u_i $(0 \le i \le p+q)$ have even lengths since $u_{i-1}u_i \in (A^{2n})^*$. From

$$u_1u_2 = (\alpha_1 \circ \beta_1)x_{11}y_{11}x_{12}y_{12} \cdots x_{1n}y_{1n}(\gamma_1 \circ \delta_1),$$

we have

$$\begin{aligned} u_1 &= (\alpha_1 \circ \beta_1)x_{11}y_{11} \cdots x_{1s}y_{1s},\ s \ge 1, \\ u_2 &= x_{1,s+1}y_{1,s+1} \cdots x_{1n}y_{1n}(\gamma_1 \circ \delta_1), \end{aligned}$$

for some s. Consequently

$$u_1 = x_{0,r+1}y_{0,r+1} \cdots x_{0n}y_{0n}(\gamma_0 \circ \delta_0) = (\alpha_1 \circ \beta_1)x_{11}y_{11} \cdots x_{1s}y_{1s}.$$

Note that $|\gamma_0 \circ \delta_0| \equiv 0 \bmod 2n$, $|\alpha_1 \circ \beta_1| \equiv 0 \bmod 2n$, and

$$|u_1| = |\gamma_0 \circ \delta_0| + 2(n-r) = |\alpha_1 \circ \beta_1| + 2s.$$

If $|\gamma_0 \circ \delta_0| \ne |\alpha_1 \circ \beta_1|$, then we have either $n - r > n$ or $s > n$. This is a contradiction. Thus we have $|\gamma_0 \circ \delta_0| = |\alpha_1 \circ \beta_1|$ and $n - r = s$. It follows that

$$\begin{aligned} u_1 &= x_{0,r+1}y_{0,r+1} \cdots x_{0n}y_{0n}(\gamma_0 \circ \delta_0) = (\alpha_1 \circ \beta_1)x_{11}y_{11} \cdots x_{1,n-r}y_{1,n-r}, \\ u_2 &= x_{1,n-r+1}y_{1,n-r+1} \cdots x_{1n}y_{1n}(\gamma_1 \circ \delta_1). \end{aligned}$$

From

$$u_2u_3 = (\alpha_2 \circ \beta_2)x_{21}y_{21}x_{22}y_{22}\cdots x_{2n}y_{2n}(\gamma_2 \circ \delta_2),$$

we have

$$u_2 = (\alpha_2 \circ \beta_2)x_{21}y_{21}\cdots x_{2t}y_{2t},$$
$$u_3 = x_{2,t+1}y_{2,t+1}\cdots x_{2n}y_{2n}(\gamma_2 \circ \delta_2),$$

for some t. It follows that

$$u_2 = x_{1,n-r+1}y_{1,n-r+1}\cdots x_{1n}y_{1n}(\gamma_1 \circ \delta_1) = (\alpha_2 \circ \beta_2)x_{21}y_{21}\cdots x_{2t}y_{2t}.$$

Note that $|\gamma_1 \circ \delta_1| \equiv 0 \bmod 2n$, $|\alpha_2 \circ \beta_2| \equiv 0 \bmod 2n$, and

$$|u_2| = |\gamma_1 \circ \delta_1| + 2r = |\alpha_2 \circ \beta_2| + 2t.$$

Then we have $t = r$. That is,

$$u_2 = x_{1,n-r+1}y_{1,n-r+1}\cdots x_{1n}y_{1n}(\gamma_1 \circ \delta_1) = (\alpha_2 \circ \beta_2)x_{21}y_{21}\cdots x_{2r}y_{2r},$$
$$u_3 = x_{2,r+1}y_{2,r+1}\cdots x_{2n}y_{2n}(\gamma_2 \circ \delta_2).$$

By repeating this argument we have

$$\begin{aligned}
u_0 &= (\alpha_0 \circ \beta_0)x_{01}y_{01}\cdots x_{0r}y_{0r} = (\alpha_0 x_{01}\cdots x_{0r}) \circ (\beta_0 y_{01}\cdots y_{0r}),\\
u_{2f-1} &= x_{2f-2,r+1}y_{2f-2,r+1}\cdots x_{2f-2,n}y_{2f-2,n}(\gamma_{2f-2} \circ \delta_{2f-2})\\
&= (\alpha_{2f-1} \circ \beta_{2f-1})x_{2f-1,1}y_{2f-1,1}\cdots x_{2f-1,n-r}y_{2f-1,n-r},\ f \geq 1,\\
u_{2f} &= x_{2f-1,n-r+1}y_{2f-1,n-r+1}\cdots x_{2f-1,n}y_{2f-1,n}(\gamma_{2f-1} \circ \delta_{2f-1})\\
&= (\alpha_{2f} \circ \beta_{2f})x_{2f,1}y_{2f,1}\cdots x_{2f,r}y_{2f,r},\ f \geq 1,\\
u_{p+q} &= x_{p+q-1,r+1}y_{p+q-1,r+1}\cdots x_{p+q-1,n}y_{p+q-1,n}(\gamma_{p+q-1} \circ \delta_{p+q-1}),\\
&\qquad\qquad \text{if } p+q \text{ is odd},\\
u_{p+q} &= x_{p+q-1,n-r+1}y_{p+q-1,n-r+1}\cdots x_{p+q-1,n}y_{p+q-1,n}(\gamma_{p+q-1} \circ \delta_{p+q-1}),\\
&\qquad\qquad \text{if } p+q \text{ is even}.
\end{aligned}$$

Note that

$$\begin{aligned}
u_{2f-1} &= (x_{2f-2,r+1}\cdots x_{2f-2,n}\gamma_{2f-2}) \circ (y_{2f-2,r+1}\cdots y_{2f-2,n}\delta_{2f-2})\\
&= (\alpha_{2f-1}x_{2f-1,1}\cdots x_{2f-1,n-r}) \circ (\beta_{2f-1}y_{2f-1,1}\cdots y_{2f-1,n-r}).\\
u_{2f} &= (x_{2f-1,n-r+1}\cdots x_{2f-1,n}\gamma_{2f-1}) \circ (y_{2f-1,n-r+1}\cdots y_{2f-1,n}\delta_{2f-1})\\
&= (\alpha_{2f}x_{2f,1}\cdots x_{2f,r}) \circ (\beta_{2f}y_{2f,1}\cdots y_{2f,r}),
\end{aligned}$$

Let

$$v_{2i} = \alpha_{2i}x_{2i,1}\cdots x_{2i,r}, \quad i \geq 0,$$

$$v_{2i+1} = \alpha_{2i+1}x_{2i+1,1}\cdots x_{2i+1,n-r}, \quad i \geq 0,$$

$$v_{p+q} = x_{p+q-1,n-r+1}\cdots x_{p+q-1,n}\gamma_{p+q-1}, \quad \text{if } p+q \text{ is even},$$

$$v_{p+q} = x_{p+q-1,r+1}\cdots x_{p+q-1,n}\gamma_{p+q-1}, \quad \text{if } p+q \text{ is odd}.$$

For $v_0, v_1, \cdots, v_{p+q} \in A^*$ we have $v_j v_{j+1} \in C_1^* \subset C^*$. Since C is a bifix (p,q)-limited code, we have $v_j \in C^*$ for all $0 \leq j \leq p+q$. However $|v_0|$ is not multiple of n. This is a contradiction. Thus Case 2-i cannot occur.
Case 2-ii. $|u_0|$ is odd. In this case all u_i have odd length since $u_{i-1}u_i \in (A^{2n})^*$.
If $u_0{=}(\alpha_0{\circ}\beta_0)x_{01}y_{01}\cdots x_{0r}y_{0r}x_{0,r+1}$ and $u_1{=}y_{0,r+1}x_{0,r+2}\cdots x_{0n}y_{0n}(\gamma_0{\circ}\delta_0)$, then we have that for $f \geq 1$,

$$\begin{aligned} u_{2f-1} &= y_{2f-2,r+1}x_{2f-2,r+2}\cdots x_{2f-2,n}y_{2f-2,n}(\gamma_{2f-2}\circ\delta_{2f-2}) \\ &= (\alpha_{2f-1}\circ\beta_{2f-1})x_{2f-1,1}y_{2f-1,1}\cdots y_{2f-1,n-r-1}x_{2f-1,n-r}, \\ u_{2f} &= y_{2f-1,n-r}x_{2f-1,n-r+1}\cdots x_{2f-1,n}y_{2f-1,n}(\gamma_{2f-1}\circ\delta_{2f-1}) \\ &= (\alpha_{2f}\circ\beta_{2f})x_{2f,1}y_{2f,1}\cdots x_{2f,r}y_{2f,r}x_{2f,r+1}, \\ u_{p+q} &= y_{p+q-1,n-r}\cdots x_{p+q-1,n}y_{p+q-1,n}(\gamma_{p+q-1}\circ\delta_{p+q-1}), \quad \text{if } p+q \text{ is even}, \\ u_{p+q} &= y_{p+q-1,r+1}\cdots x_{p+q-1,n}y_{p+q-1,n}(\gamma_{p+q-1}\circ\delta_{p+q-1}), \quad \text{if } p+q \text{ is odd}. \end{aligned}$$

Therefore,

$$\beta_{2f-1}y_{2f-1,1}\cdots y_{2f-1,n-r-1} = x_{2f-2,r+2}\cdots x_{2f-2,n}\gamma_{2f-2},$$

$$\alpha_{2f}x_{2f,1}\cdots x_{2f,r+1} = y_{2f-1,n-r}\cdots y_{2f-1,n}\delta_{2f-1}.$$

If $r+1 < n$, then we set

$$v_0 = \alpha_0 x_{01}\cdots x_{0,r+1},$$

$$v_{2f-1} = \beta_{2f-1}y_{2f-1,1}\cdots y_{2f-1,n-r-1},$$

$$v_{2f} = \alpha_{2f}x_{2f,1}\cdots x_{2f,r+1},$$

$$v_{p+q} = x_{p+q-1,r+2}\cdots x_{p+q-1,n}\gamma_{p+q-1}, \quad \text{if } p+q \text{ is odd},$$

$$v_{p+q} = y_{p+q-1,n-r}\cdots y_{p+q-1,n}\delta_{p+q-1}, \quad \text{if } p+q \text{ is even}.$$

If $r+1=n$, then we put

$$\begin{aligned} v_0 &= \beta_0 y_{01} \cdots y_{0,n-1}, \\ v_{2f-1} &= \alpha_{2f-1} x_{2f-1,1}, \\ v_{2f} &= \beta_{2f} y_{2f,1} \cdots y_{2f,n-1}, \\ v_{p+q} &= y_{p+q-1,n} \delta_{p+q-1}, \quad \text{if } p+q \text{ is odd}, \\ v_{p+q} &= x_{p+q-1,2} \cdots x_{p+q-1,n} \gamma_{p+q-1}, \quad \text{if } p+q \text{ is even}. \end{aligned}$$

Then $v_{i-1}v_i \in C^*$ for all $i = 1, \cdots p+q$. Since C is a bifix (p,q)-limited code, we have $v_j \in C_1^*$ for any $0 \leq j \leq p+q$. However the length of v_0 is not a multiple of n. This is a contradiction. Therefore, Case 2-ii cannot occur.
Hence only Case 1 is possible to occur. Thus $C_1 \circ C_2$ is (p,q)-limited. Q.E.D.

For the case that $C_1 = C_2 = C$, we have the following:

Corollary 3. Let $C \subset A^n$ $(n \geq 2)$. If C is (p, q)-limited, then $C \circ C$ is (p, q)-limited.

Proposition 6. Let $C \subset A^n$. If $C \circ C$ is (p,q)-limited, then C is (p,q)-limited.

Proof. Let $u_0, u_1, \cdots, u_{p+q}$ be a sequence of words in A^* where $u_{i-1}u_i \in C^*$ for $1 \leq i \leq p+q$. Consider the following sequence of words in A^*; $u_0 \circ u_0, u_1 \circ u_1, \cdots, u_{p+q} \circ u_{p+q}$. Then

$$(u_{i-1} \circ u_{i-1})(u_i \circ u_i) = (u_{i-1} \circ u_i) \circ (u_{i-1} \circ u_i) \in (C \circ C)^*.$$

Since $C \circ C$ is (p,q)-limited, we have $u_p \circ u_p \in (C \circ C)^*$. Therefore, $u_p \in C^*$. Q.E.D.

A subset $C \subset A^+$ is said to be *infix* if for all $x, y, z \in A^*$ the assumptions $z \in C$, $xzy \in C$ imply $x = y = 1$. For example a subset C of A^n is an infix code. A nonempty subset $C \subset A^+$ is said to be an *intercode* of index m, $m \geq 1$, if $C^{m+1} \cap A^+ C^m A^+ = \emptyset$.

An intercode of index m is (p, q)-limited for all p and q with $p+q = 2m+1$. In general, if an infix code L is (p, q)-limited for some $p, q \geq 0$ with $p+q = 2m+1$, then L is an intercode of index m([9]).

Let $C \subset A^n$, $n \geq 2$. If C is an intercode of index m, then C is (p, q)-limited for some p, $q \geq 0$ with $p + q = 2m + 1$. Therefore, by Proposition 5, $C \circ C$ is an infix (p, q)-limited code. Thus $C \circ C$ is an intercode of index m. That is, Proposition 5 implies the next proposition. However we give a direct proof for the sake of clarity and completeness.

Proposition 7. Let $C \subset A^n$, $n \geq 2$, and let C_1 and C_2 be nonempty subsets of C. If C is an intercode of index m, then $C_1 \circ C_2$ is an intercode of index m.

Proof. Assume that $w \in (C_1 \circ C_2)^{m+1} \cap A^+(C_1 \circ C_2)^m A^+ \neq \emptyset$. Then

$$w = (\alpha_1\alpha_2 \cdots \alpha_{m+1}) \circ (\beta_1\beta_2 \cdots \beta_{m+1}) = u((\gamma_1\gamma_2 \cdots \gamma_m) \circ (\delta_1\delta_2 \cdots \delta_m))v$$

for some u, $v \in A^+$ and $\alpha_i, \beta_i, \gamma_j, \delta_j \in (C_1 \cup C_2) \subset C$, $1 \leq i \leq m+1$, $1 \leq j \leq m$.
Case 1. $|u|$ is even. Note that any elements in $(C_1 \circ C_2)^*$ have even length. Therefore, $|v|$ is also even. Thus $u = x \circ y$, $v = x' \circ y'$ for some $x, y, x', y' \in A^+$ such that $|x| = |y| \geq 1$ and $|x'| = |y'| \geq 1$. It follows that

$$w = (\alpha_1\alpha_2 \cdots \alpha_{m+1}) \circ (\beta_1\beta_2 \cdots \beta_{m+1}) = (x\gamma_1\gamma_2 \cdots \gamma_m x') \circ (y\delta_1\delta_2 \cdots \delta_m y').$$

Therefore,

$$\alpha_1\alpha_2 \cdots \alpha_{m+1} = x\gamma_1\gamma_2 \cdots \gamma_m x' \in C^{m+1} \cap A^+C^mA^+.$$

This contradicts our assumption.
Case 2. $|u|$ is odd.
In this case $|v|$ is also odd. $u = x \circ y$, $v = x' \circ y'$ for some $x, y, x', y' \in A^+$ such that $|x| = |y| + 1$ and $|x'| = |y'| + 1$. It follows that

$$w = (\alpha_1\alpha_2 \cdots \alpha_{m+1}) \circ (\beta_1\beta_2 \cdots \beta_{m+1}) = (x\delta_1\delta_2 \cdots \delta_m y') \circ (y\gamma_1\gamma_2 \cdots \gamma_m x').$$

If $|u| = 1$, that is, if $u = x \in A$ and $y = 1$, then $y' \neq 1$ since $|u| + |v| = 2n \geq 4$. Thus $\alpha_1\alpha_2 \cdots \alpha_{m+1} = x\delta_1\delta_2 \cdots \delta_m y' \in C^{m+1} \cap A^+C^mA^+$. This is a contradiction.
If $|u| \neq 1$, then we have, $\beta_1\beta_2 \cdots \beta_{m+1} \in A^+C^mA^+$, a contradiction. Therefore, $(C_1 \circ C_2)^{m+1} \cap A^+(C_1 \circ C_2)^m A^+ = \emptyset$. Q.E.D.

Corollary 4. Let $C \subset A^n$, $n \geq 2$. If C is an intercode of index m, then $C \circ C$ is an intercode of index m.

Example 3. Let $C_1 = \{aba, ba^2\}$, $C_2 = \{ab^2\}$. Then C_1 is not circular. Thus C_1 is not an intercode. However, $(C_1 \circ C_2)^* = \{a^2b^2ab, ba^2bab\}$ is an intercode of index 1.

Proposition 8. Let $C \subset A^n$. If $C \circ C$ is an intercode of index m, then C is an intercode of index m.

Proof. Assume that $w \in C^{m+1} \cap A^+C^mA^+ \neq \emptyset$. Then $w = c_1c_2\cdots c_{m+1} = ud_1d_2\cdots d_mv$ for some $c_i, d_j \in C$, $1 \leq i \leq m+1, 1 \leq j \leq m$, $u, v \in A^+$.

$$\begin{aligned}(C \circ C)^{m+1} &\ni (c_1 \circ c_1)(c_2 \circ c_2)\cdots(c_{m+1} \circ c_{m+1}) \\ &= (ud_1d_2\cdots d_mv) \circ (ud_1d_2\cdots d_mv), \\ &= (u \circ u)((d_1d_2\cdots d_m) \circ (d_1d_2\cdots d_m))(v \circ v) \\ &\in (u \circ u)(C \circ C)^m(v \circ v).\end{aligned}$$

Therefore, $(C \circ C)^{m+1} \cap A^+(C \circ C)^m A^+ \neq \emptyset$. This contradicts our assumption. Thus C is an intercode of index m. Q.E.D.

A submonoid M of A^* is said to be *extractable* if for all $x, y, z \in A^*$,

$$z,\ xzy \in M \implies xy \in M.$$

It is obvious that M is biunitary. Therefore, the base of M is a bifix code. The base of an extractable submonoid is called an *extractable* code.

Example 4. (1) Let $\phi : A^* \to G$ be a morphism from a free monoid A^* onto a group G. Let H be a normal subgroup of G, and let C be the group code defined by $C^* = \phi^{-1}(H)$. Then C^* is an extractable submonoid of A^*.
(2) Let $C = \{ab, ba\}$. Then C^* is an extractable pure submonoid.
(3) Let $C = \{aba, b^2a\}$. Then C is (p, q)-limited for all $p, q \geq 0$ with $p + q = 3$, and C^* is extractable.
(4) Let $C = \{a^2, ab\}$. Then C^* is extractable, but it is not pure.
(5) Let $C = \{a^4, aba^2, a^2b^2\}$. Then C^* is not extractable.

Proposition 9. Let $C \subset A^n$. Then C is extractable if and only if $C \circ C$ is extractable.

Proof. Assume that C^* is extractable. Let $x,\ y \in A^*$, $z,\ xzy \in (C\circ C)^*$. We consider the following two cases.
Case 1. $|x|$ is even. Then, since $z \in (A^{2n})^*$, we have that y is even. Let $x = x_1 \circ x_2$, $|x_1| = |x_2|$, $y = y_1 \circ y_2$, $|y_1| = |y_2|$, and let $z = \alpha \circ \beta$, $|\alpha| = |\beta|$, $\alpha,\ \beta \in C^*$. Then

$$(x_1 \circ x_2)(\alpha \circ \beta)(y_1 \circ y_2) = (x_1 \alpha y_1) \circ (x_2 \beta y_2) \in (C \circ C)^*.$$

It follows that $\alpha,\ x_1\alpha y_1,\ \beta,\ x_2\beta y_2 \in C^*$. Hence $x_1y_1, x_2y_2 \in C^*$. Since $|x_1y_1| = |x_2y_2|$, we have

$$(x_1y_1) \circ (x_2y_2) = (x_1 \circ x_2)(y_1 \circ y_2) = xy \in (C \circ C)^*.$$

Thus $(C \circ C)^*$ is extractable.
Case 2. $|x|$ is odd. In this case, $|y|$ is also odd. Let $x = x_1 \circ x_2$, $|x_1| = |x_2|+1$, $y = y_1 \circ y_2$, $|y_1| = |y_2| + 1$, and let $z = \alpha \circ \beta$, $|\alpha| = |\beta|$, $\alpha,\ \beta \in C^*$. Then

$$(x_1 \circ x_2)(\alpha \circ \beta)(y_1 \circ y_2) = (x_1 \beta y_2) \circ (x_2 \alpha y_2) \in (C \circ C)^*.$$

It follows that $\beta,\ x_1\beta y_2,\ \alpha,\ x_2\alpha y_1 \in (C \circ C)^*$. Therefore, $x_1y_2, x_2y_1 \in C^*$. Since $|x_1y_2| = |x_2y_1|$, we have

$$(x_1y_2) \circ (x_2y_1) = (x_1 \circ x_2)(y_1 \circ y_2) = xy \in (C \circ C)^*.$$

Conversely, assume that $(C\circ C)^*$ is extractable. Let $x,\ y \in A^*$, and $z,\ xzy \in C^*$. If $z = 1$, then $xy \in C^*$. Thus, we consider the case $z \neq 1$. Then we have,

$$(xzy) \circ (xzy) = (x \circ x)(z \circ z)(y \circ y) \in (C \circ C)^* \ \text{ and } \ z \circ z \in (C \circ C)^*.$$

Since $(C \circ C)^*$ is extractable, we have $(x \circ x)(y \circ y) = (xy) \circ (xy) \in (C \circ C)^*$. Hence $xy \in C^*$. Q.E.D.

Let C be a code. If $uv \in C$ implies $vu \in C$, then C is called a *reflective code*.

Proposition 10. Let $C \subset A^n$. Then C is reflective if and only if $C \circ C$ is reflective.

Proof. Assume that C is reflective. Let $uv = x \circ y \in C \circ C$, $x = a_1a_2 \ldots a_n \in C$, $y = b_1b_2 \ldots b_n \in C\,(a_i, b_i \in A)$. Since C is reflective, $vu = (b_k \ldots b_nb_1 \ldots b_{k-1}) \circ (a_{k+1} \ldots a_na_1 \ldots a_k) \in C \circ C$ if the length $|u| =$

$2k-1(1 \le k \le n-1)$ of u, or $vu = (b_n b_1 \ldots b_{k-1}) \circ (a_1 \ldots a_n) \in C \circ C$ if $|u| = 2n-1$. Similarly we have $vu \in C \circ C$ if $|u| = 2k(0 \le k \le n)$.

Conversely, assume that $C \circ C$ is reflective. Let $uv = a_1 a_2 \ldots a_n \in C(a_i \in A)$ with $|u| = k$. We may suppose that $0 < k < n$. Since $(uv \circ uv) = {a_1}^2 {a_2}^2 \ldots {a_n}^2 \in C \circ C$ and $C \circ C$ is reflective, ${a_{k+1}}^2 \ldots {a_n}^2 {a_1}^2 \ldots {a_k}^2 \in C \circ C$. Moreover $C \subset A^n$ implies $vu = a_{k+1} \ldots a_n a_1 \ldots a_k \in C$. Q.E.D.

If a conjugacy class $C = Cl(w)$ of a word w is extractable, then C is an extractable reflective code. By Proposition 9 and Proposition 10, $C \circ C$ is an extractable reflective codes.

Corollary 11. Let $u \in A^+$ be a primitive word of length n. Let $w = u^m$, $m \ge 1$, and let

$$Cl(u) = \{u = u_0, u_1, \cdots, u_{n-1}\}, \qquad Cl(w) = \{w = w_0, w_1, \cdots, w_{n-1}\},$$

where $\Gamma(u_i) = u_{i+1}$ and $w_i = (u_i)^m$ for $0 \le i \le n-1$, and $w_{n+j} = w_j$. Furthermore, let

$$C = \begin{cases} Cl(w_0) \circ Cl(w_0) & \text{if } n = 1\,, \\ \bigcup_{j=1}^{p} Cl(w_0 \circ w_j) & \text{if } n = 2p \ \ , p \ge 1\,, \\ \bigcup_{j=1}^{p+1} Cl(w_0 \circ w_j) & \text{if } n = 2p+1,\ \ p \ge 1\,. \end{cases}$$

If $Cl(w)$ is extractable, then C is extractable.

Proof. $Cl(w) \circ Cl(w)$ is an extractable reflective code. We shall show that $C = Cl(w) \circ Cl(w)$. Note that $\Gamma(w_i \circ w_j) = w_j \circ w_{i+1}$, and

$$\Gamma^{2\alpha+1}(w_i \circ w_j) = w_{j+\alpha} \circ w_{i+\alpha+1}, \qquad \Gamma^{2\alpha}(w_i \circ w_j) = w_{i+\alpha} \circ w_{j+\alpha}.$$

If $n = 1$, then $u \in A$, and $C = Cl(w) \circ Cl(w) = \{u^{2m}\}$.

We consider the case $|u| = n = 2p$, $p \ge 1$. Let $1 \le j \le p$. Suppose that $\Gamma^{2\alpha+1}(w_0 \circ w_j) = w_0 \circ w_j$ for some $0 \le \alpha \le n-1$. Then $w_0 \circ w_j = w_{j+\alpha} \circ w_{\alpha+1}$. Therefore, $j + \alpha \equiv n (\operatorname{mod} n)$ and $\alpha + 1 \equiv j$. It follows that $n = 2\alpha - 1$. Since n is even, this cannot occur. Therefore $\Gamma^{2\alpha+1}(w_0 \circ w_j) \ne w_0 \circ w_j$ for all $0 \le \alpha \le n-1$. Suppose that $\Gamma^{2\alpha}(w_0 \circ w_j) = w_0 \circ w_j$ for some $1 \le \alpha \le n$. Then $w_\alpha \circ w_{j+\alpha} = w_0 \circ w_j$. It implies that $\alpha = n$ since $1 \le \alpha \le n$. Therefore, $\Gamma^{2\alpha}(w_0 \circ w_j) \ne w_0 \circ w_j$ for all $1 \le \alpha \le n-1$ and $\Gamma^{2n}(w_0 \circ w_j) = w_0 \circ w_j$. Thus $\Gamma^k(w_0 \circ w_j) \ne w_0 \circ w_j$ for all $1 \le k \le 2n-1$, Thus we have $|Cl(w_0 \circ w_j)| = 2n$ for $1 \le j \le p$.

Next we shall shows that $Cl(w_0 \circ w_j) \cap Cl(w_0 \circ w_k) = \emptyset$ for any two distinct j, k such that $1 \leq j \leq p$, $1 \leq k \leq p$. If $\Gamma^{2\alpha+1}(w_0 \circ w_j) = w_{j+\alpha} \circ w_{\alpha+1} = w_0 \circ w_k$, then $n - j + 1 = k$. Since $j \leq p$, we have

$$k = n - j + 1 \geq p + 1 > p.$$

This is a contradiction. Therefore, for $1 \leq j, k \leq p$ and $j \neq k$ we have $Cl(w_0 \circ w_j) \cap Cl(w_0 \circ w_k) = \emptyset$. Since $|\bigcup_{j=1}^{p} Cl(w_0 \circ w_j)| = 2np = n^2$, we have $Cl(w) \circ Cl(w) = \bigcup_{j=1}^{p} Cl(w_0 \circ w_j)$.

Now we consider the case $n = 2p+1$. We can show that $|Cl(w_0 \circ w_j)| = 2n$ for $1 \leq j \leq p$, and $Cl(w_0 \circ w_j) \cap Cl(w_0 \circ w_k) = \emptyset$ for any j and k such that $1 \leq j, k \leq p$ and $j \neq k$.
Let $0 \leq \alpha \leq p$. Suppose that

$$\Gamma^{2\alpha+1}(w_0 \circ w_{p+1}) = w_{\alpha+p+1} \circ w_{\alpha+1} = w_0 \circ w_{p+1}.$$

Then $\alpha = p$. Therefore, $\Gamma^{2\alpha+1}(w_0 \circ w_{p+1}) \neq w_0 \circ w_{p+1}$ for $1 \leq \alpha \leq p-1$ and $\Gamma^{2p+1}(w_0 \circ w_{p+1}) = w_0 \circ w_{p+1}$.
Let $1 \leq \alpha \leq p$. Suppose that $\Gamma^{2\alpha}(w_0 \circ w_{p+1}) = w_\alpha \circ w_{\alpha+p+1} = w_0 \circ w_{p+1}$. Then $\alpha = n > p$. This contradicts the condition $1 \leq \alpha \leq p$. Hence $\Gamma^{2\alpha}(w_0 \circ w_{p+1}) \neq w_0 \circ w_{p+1}$ for $1 \leq \alpha \leq p$. Thus $\Gamma^k(w_0 \circ w_{p+1}) \neq w_0 \circ w_{p+1}$ for all $1 \leq k \leq 2p$. Hence, $|Cl(w_0 \circ w_{p+1})| = n$. It follows that $|\bigcup_{j=1}^{p+1} Cl(w_0 \circ w_j)| = 2np + n = n^2$. Thus $Cl(w) \circ Cl(w) = \bigcup_{j=1}^{p+1} Cl(w_0 \circ w_j)$.
Q.E.D.

Remark. If $n = 2p+1$ and $u_0 = a_0 a_1 \cdots a_{2p}$, then $\Gamma^{p+1}(u_0) = a_{p+1} a_{p+2} \cdots a_p$. It follows that

$$u_0 \circ u_{p+1} = a_0 a_{p+1} \cdots a_r a_{r+p+1} \cdots a_p a_{2p+1} a_{p+1} a_{2p+2} \cdots a_{2p} a_{3p+1},$$

where $a_{n+s} = a_t$, $t \equiv n + s \bmod n$. Therefore

$$u_0 \circ u_{p+1} = (a_0 a_{p+1} a_1 a_{p+2} \cdots a_{p-1} a_{2p} a_p)^2.$$

References

[1] Berstel, J. and Perrin, D., 1985, *Theory of Codes*. Academic Press.

[2] Berard, B., 1987, Literal shuffle, *Theoret. Comput. Sci.* 51, pp. 281–299.

[3] Ito, M. and Tanaka, G., 1990, Dense property of initial literal shuffles, *Intern. J. Computer Math.*, Vol. 34, pp. 161–170.

[4] Ito, M., Thierrin, G. and Yu. S. S., 1996, Shuffle-closed languages, *Publ. Math. Debrecen*, 48/3-4, pp. 317–338.

[5] Lallement, G. 1979, *Semigroup and Combinatorial Applications*. Wiley.

[6] Restivo, A., 1974, On a Question of McNaughton and Papert, *Information and Control* 25, pp. 93–101.
[7] Shyr, H. J., 1991, *Free Monoids and Languages.* 2nd edition. Hon Min Book Company, Taichung, Taiwan.
[8] Tanaka, G., 1988, Alternating products of prefix codes, *Proc. 2nd Conference on Automata, Fromal Languages and Programming Systems*, Salgotarjan, Hungary. DM88-4, pp. 209–213.
[9] Yu, S. S., 1990, A characterization of intercodes, *Intern. J. Computer Math.*, Vol. 36, pp. 39–45.

Received: April 2, 2009

Revised: March 21, 2010